CHROMATOGRAPHIC
ENANTIOSEPARATION
Methods and Applications
Second Edition

ELLIS HORWOOD SERIES IN ANALYTICAL CHEMISTRY

Series Editors: Dr MARY MASSON, University of Aberdeen,
and Dr JULIAN F. TYSON, Amherst, USA
Consultant Editors: Prof. J. N. MILLER, Loughborough University of Technology, and
Dr R. A. CHALMERS, University of Aberdeen

CHROMATOGRAPHIC ENANTIOSEPARATION

Methods and Applications
Second Edition

STIG ALLENMARK B.Sc., Ph.D.
Laboratory of Microbiological Chemistry
University of Gothenburg, Sweden

ELLIS HORWOOD
NEW YORK   LONDON   TORONTO   SYDNEY   TOKYO   SINGAPORE

First published in 1991 by
ELLIS HORWOOD LIMITED
Market Cross House, Cooper Street,
Chichester, West Sussex, PO19 1EB, England

A division of
Simon & Schuster International Group
A Paramount Communications Company

Typeset in Times by Ellis Horwood Limited
Printed and bound in Great Britain
by Bookcraft Ltd, Midsomer Norton, Avon

British Library Cataloguing-in-Publication Data

Allenmark, Stig
Chromatographic enantioseparation: Methods and applications. — 2nd ed. —
(Ellis Horwood series in analytical chemistry)
I. Title. II. Series.
543.001
ISBN 0–13–132978–2

Library of Congress Cataloging-in-Publication Data

Allenmark, Stig G., 1936–
Chromatographic enantioseparation: methods and applications / Stig Allenmark. — 2nd ed.
p. cm. — (Ellis Horwood series in analytical chemistry)
Includes bibliographical references and index.
ISBN 0–13–132978–2
1. Chromatographic analysis. 2. Enantiomers — Separation.
I. Title. II. Series.
QD79.C4A48   1991
543′.089–dc20

91–31751
CIP

Table of contents

Preface

Chemists have been interested in optically active compounds ever since the recognition of Nature's remarkable ability to produce them. Likewise, optical resolutions of synthetic racemates have been a challenge and often considered to be small pieces of art, owing to the great difficulties of making any predictions concerning the possible success of various approaches. Today, we are still far from being able to consider a new optical resolution as an entirely straightforward task. During the last ten years, however, there has been a quite rapid development of chromatographic methods for optical resolution and thereby an accumulated knowledge of the prerequisites for a chiral recognition leading to enantioseparation. The purpose of this book is to provide the reader with a comprehensive treatment of chiral chromatography covering the basic theory as well as methods used, particularly the stationary phase design and various areas of application. Although a number of review articles dealing with the subject have appeared during recent years, a certain need for a monograph dealing entirely with the subject was felt at the start of my writing. Since no thorough discussion of the chiral recognition rationales proposed for enantioselective equilibria utilized in chiral chromatography can be made without fundamental knowledge of organic stereochemistry, the incorporation of the material found in the first three chapters seems to be justified. The treatment of this vast topic is, of course, by no means exhaustive, but aims only at giving the reader a suitable background to the following chapters.

The number of different methods in use or under investigation is far greater for optical resolution by liquid chromatography than by gas chromatography. Therefore, Chapter 7 is by far the most voluminous. Further, Chapter 9, in which preparative aspects are discussed, deals only with liquid chromatography.

Throughout the book the reader is referred to relevant publications in the chemical and chromatographic literature for possible supplementary reading.

It is my hope that this volume will bring together the accumulated knowledge in the field in a way which will be beneficial to the reader. New application areas of chiral chromatography are still developing, particularly within the life sciences, and it would be of great satisfaction if the present book could stimulate further progress.

I want to express my sincere thanks to my associates and colleagues whose active support and help has been of utmost value for the production of this volume. Particularly, I wish to acknowledge the significant contribution to the artwork made by Mrs. Shalini Andersson as well as the linguistic improvements of the text achieved thanks to Mr. Richard Thompson, who carefully read and commented on the manuscript. The invaluable help from the series editor, Dr. R. A. Chalmers, is also gratefully acknowledged. Finally, the contacts and discussions with many leading scientists in the field, especially Profs. G. Blaschke, A. Mannschreck and W. H. Pirkle, have further been highly beneficial and most stimulating.

August, 1987
Gothenburg, Sweden   STIG ALLENMARK

Preface to the second edition

Chiral chromatography has now developed into a well-established field within the separation sciences. Its progress has been rapid and fruitful, and particularly as a powerful analytical tool it has found many applications, not least for the evaluation of pharmacological differences between drug enantiomers. Cyclodextrin derivatives have been developed for use as gas chromatographic chiral stationary phases, with great success. Some new and promising protein phases for analytical chiral reversed-phase liquid chromatography have appeared. A number of new synthetic chiral stationary phases have been prepared, some of which are based on readily available bi- or trifunctional chiral synthons (diamines, tyrosine) from which a lot of structure variation can be obtained. Very high enantioselectivity has recently been obtained in optical resolutions with synthetic chiral stationary phases, designed by a kind of molecular modelling approach, in liquid chromatography. Chiral membranes for preparative optical enrichment by simple experimental techniques have been constructed. Achievements in the field of immobilization chemistry have led to valuable improvements in the attachment of chiral polymers to silica, of great importance for preparative-scale resolutions. Newly developed diode-laser-based micropolarimetric detectors for analytical use have appeared.

The rapid innovation taking place in the field has necessitated the publication of a second edition of the book rather soon after its first appearance. Although only some minor additions have been made to the first four chapters, a substantial expansion of the later chapters, in particular of Chapters 6–10, has been made. Also, the number of useful experimental procedures, described in Chapter 11, has been increased. The list of commercial suppliers of material for chiral GC and LC, given in the Appendix, has been completely updated. To strengthen the textbook role of the new edition, a number of new exercises have been added.

Finally, the importance of stereochemistry today is reflected in the increasing number of contributions in the field of asymmetric synthesis, chiral drug design, kinetic resolution via biocatalytic reactions, etc. In all these areas, chiral chromatographic techniques will be extremely useful for determination of stereochemical results on a very small scale. Another indication of the importance of these topics is the recent appearance of two new journals devoted entirely to stereochemical investigations, viz. *Chirality* and *Tetrahedron: Asymmetry*.

May, 1991
Gothenburg, Sweden   STIG ALLENMARK

List of symbols and abbreviations

A	absorbance
a.u.	absorbance unit
α	(1) measured optical rotation, (2) separation factor
$[\alpha]_\lambda^T$	specific rotation at T°C and wavelength λ nm
$[\alpha]_{max}$	maximum specific rotation (corresponding to 100% optical purity)
AGP	α_1-acid glycoprotein (orosomucoid)
BOC	*tert*-butyloxycarbonyl (protective group)
BSA	(1) bovine serum albumin, (2) *N,O*-bis(trimethylsilyl)acetamide
CBH	cellobiohydrolase
CD	(1) circular dichroism, (2) cyclodextrin
CLEC	chiral ligand exchange chromatography
CSP	chiral stationary phase
CT	charge-transfer
DANSYL	5-dimethylamino-1-naphthalenesulphonyl
DCC	dicyclohexylcarbodiimide
δ	NMR chemical shift
DIPTA	diisopropyl tartardiamide
DNB	3,5-dinitrobenzoyl
DOPA	3-(3,4-dihydroxyphenyl)alanine
DSC	differential scanning calorimetry
E	enantioselectivity
ECD	electron capture detector
e.e.	enantiomeric excess
EEDQ	*N*-ethoxycarbonyl-2-ethoxy-1,2-dihydroquinoline
ε	(1) absorption (extinction) coefficient, (2) dichroic absorption
FID	flame ionization detector
FLEC	1-(9-fluorenyl)ethyl chloroformate
FMOC	9-fluorenylmethoxycarbonyl
ΔG	free energy change
GC	gas chromatography
GLC	gas–liquid chromatography

H	plate height, i.e. height equivalent to one theoretical plate (HETP)
ΔH	enthalpy change
HPLC	high performance liquid chromatography
K	stationary/mobile phase equilibrium distribution constant
k'	capacity ratio (capacity factor)
L	column length
LC	liquid chromatography
MCTA	microcrystalline cellulose triacetate
MS	mass spectrometry
N	plate number, i.e. the number of theoretical plates in a column
N_{eff}	effective plate number
NMR	nuclear magnetic resonance spectroscopy
ORD	optical rotatory dispersion
OVM	ovomucoid
P	optical purity
PFP	pentafluoropropionyl
pI	isoelectric point
PTH	phenylthiohydantoin
q	phase ratio ($V_{\text{m}}/V_{\text{s}}$)
R	gas constant ($1.986\,\text{cal.K}^{-1}.\text{mole}^{-1}$; $8.314\,\text{J.K}^{-1}.\text{mole}^{-1}$)
RI	refractive index (detector)
R_{s}	resolution factor
ΔS	entropy change
SCOT	support-coated open tubular (capillary column)
SDS	sodium dodecyl sulphate
SFC	supercritical fluid chromatography
σ	standard deviation
T	absolute temperature (K)
TAPA	α-(2,4,5,7-tetranitrofluorenylideneaminooxy)propionic acid
TFA	trifluoroacetyl
THF	tetrahydrofuran
TID	thermionic detector
TLC	thin-layer chromatography
t_{R}	retention time
t_{W}	baseline peak width
θ	molecular ellipticity
V_{n}	net retention volume ($V_{\text{R}}-V_0$)
V_0 (or V_{m})	void volume (volume of the mobile phase in a column)
V_{R}	retention volume
V_{S}	volume of the stationary phase in a column
WCOT	wall-coated open tubular (capillary column)

1

Introduction

To fully comprehend the modern chromatographic methods for optical resolution available today, it is necessary to have a background knowledge of the most important advances in stereochemistry and separation methods that have so far been made. This is covered in the following three chapters. The next three chapters are devoted to the theory of chiral chromatography and the various principles used in gas and liquid chromatographic methods for the separation of optical antipodes. In the final chapters, analytical and preparative scale applications and the possible future role of the techniques are treated.

From Pasteur's very first optical resolution of a racemate to today's fast chromatographic techniques there has been a formidable accumulation of stereochemical knowledge (Fig. 1.1). Despite this, many of the most useful modern techniques are essentially based on empirical results.

Why is there such an interest in the separation of enantiomers? Part of the answer undoubtedly lies in pure scientific curiosity. It is a challenging problem, which can be dealt with from both theoretical and experimental points of view. The phenomena associated with the optical rotation characteristics of asymmetric molecules have long been studied by molecular spectroscopists. The importance of optically active compounds for the elucidation of reaction mechanisms and the dynamic behaviour of chiral molecules in organic chemistry has been enormous. Let us simply recall the fact that an elucidation of the mechanisms of nucleophilic substitution and elimination reactions (S_N1, S_N2, $E1$, $E2$, etc.) would hardly have been possible without studies of optically active compounds by means of polarimetry. The knowledge of mechanistic stereochemistry emerging from such studies has had a huge impact on advanced organic synthesis, leading to milestones in the art of total synthesis, such as reserpine (Fig. 1.2) or prostaglandin $F_{2\alpha}$ (Fig. 1.3), to mention just a few.

Nature's well-known ability to produce and convert chiral compounds with a remarkable stereospecificity has always been fascinating. The insight into enzyme structure and stereochemistry that mainly comes from kinetic and X-ray crystallographic studies has brought about a recognition of the importance of the spatial arrangement in multipoint active site–substrate interaction. Similarly, the inter-

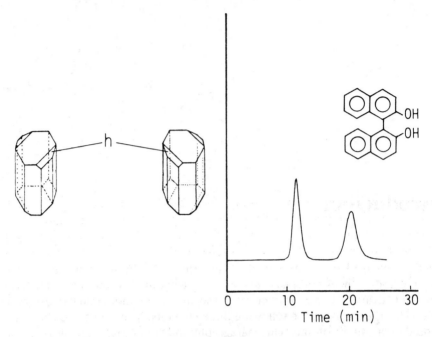

Fig. 1.1 — Left: the appearance of the enantiomorphous crystals (h=hemihedral facet) of racemic sodium ammonium tartrate separated by hand by Pasteur in 1848. From the exactly opposite optical rotations obtained from the solutions of the two fractions collected, non-identity of the molecules forming the respective fractions could be suspected. (From: L. Pasteur, *Ann. Chim. Phys.*, 1850, **28**, 56. Reproduced by courtesy of Masson, S. A., Paris). Right: liquid chromatographic optical resolution of racemic binaphthol by the use of a column containing right-handed helical (+)-poly(triphenylmethyl methacrylate) as a chiral stationary phase. (From Y. Okamoto and K. Hatada, *J. Liquid Chromatog.*, 1986, **9**, 369, reproduced by courtesy of Marcel Dekker Inc.).

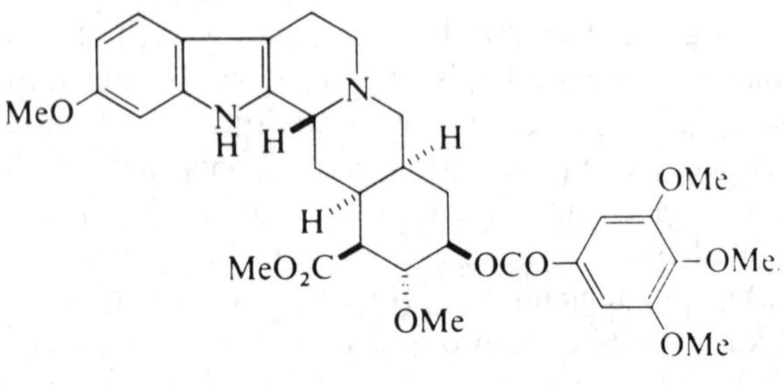

Fig. 1.2 — The structure of reserpine with its six chiral centres. The total synthesis of this alkaloid was achieved by Woodward *et al.* [1] and published in 1958.

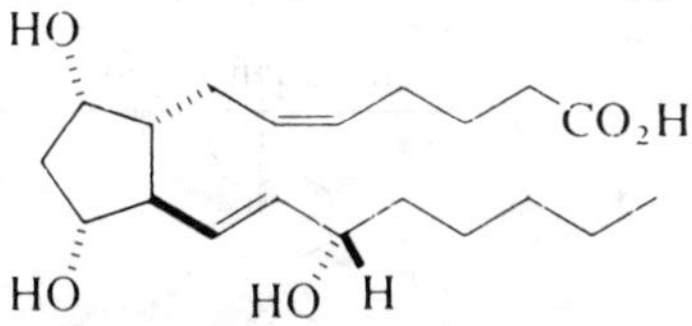

Fig. 1.3 — Five chiral centres are present in prostaglandin $F_{2\alpha}$. The prostaglandin hormones are medically highly important compounds. The $F_{2\alpha}$ variant was synthesized (together with the closely related E_2 species) by Corey and co-workers [2,3] in the late 1960s.

action between biologically active compounds and receptor proteins often shows a high or complete antipodal specificity. Old examples of the different physiological behaviour are found in the taste and smell of optical antipodes. It is perhaps not surprising that such data have been accumulated, in view of the common habit among chemists of the old school of tasting tiny amounts of the substances they isolated in their laboratory and carefully describing their findings. In Table 1.1 examples are given of the different tastes of the antipodes of some common amino-acids. The different odours of the enantiomers of particular terpenes such as carvone and limonene (Table 1.2) are also well documented in the literature [6,7].

Table 1.1 — Taste and absolute configuration of some amino-acids

Amino-acid	Taste/enantiomer	
	D-	L-
Asparagine	sweet	tasteless
Histidine	sweet	tasteless
Isoleucine	sweet	bitter
Leucine	sweet	bitter
Tryptophan	sweet	tasteless
Tyrosine	sweet	bitter
6-Chlorotryptophan (synthetic)	sweet*	tasteless

*The D-(+)-enantiomer has been reported to be ca. 10^3 times sweeter than sucrose [4, 5]

Even more striking examples of chiral discrimination in physiological reactions are found in the hormone field, to which we will return later in this text. Catechol-amines are human endogenous compounds, biosynthesized from tyrosine, which are extremely potent as regulatory hormones, affecting blood pressure and other important body functions. The immediate precursor of the catecholamines is L-dopa, i.e. 3-hydroxy-L-tyrosine, which yields dopamine after enzymatic decarboxylation (Fig. 1.4). A completely stereospecific enzymatic hydroxylation then converts dopamine into a chiral compound, norepinephrine (noradrenaline) which is finally changed into epinephrine (adrenaline) by N-methylation. Dopa-decarboxylase is totally stereospecific, acting only on the L-enantiomer. A commonly used treatment of Parkinson's disease is based on administration of the L-form to increase the blood

Table 1.2 — Odour and absolute configuration of two simple terpenes

Terpene	Absolute configuration	Odour
(R)-(−)-carvone		spearmint
(S)-(+)-carvone		caraway
(R)-(+)-limonene		orange
(S)-(−)-limonene		lemon

level of dopamine in the patients. The importance of the correct configuration at the chiral centre of epinephrine and its analogues is evident from the facts that the natural (R)-(−)-epinephrine is at least twenty times as active as its enantiomer and that (R)-(−)-isopropyl norepinephrine (isoprenaline) is ca. 800 times more effective as a bronchodilator than the isomer with the opposite configuration [8,9].

Numerous drugs are synthetic racemic compounds and used as such. Although this has often been quite adequate, there may exist the possibility that one of the two enantiomers is undesirable. A tragic example was the use of thalidomide (Fig. 1.5), a sedative and sleeping-drug used in the early 1960s, which caused serious malformations in newborns of women who had taken the drug during an early phase of pregnancy. In 1979 it was shown that it was only the (S)-(−)-enantiomer of thalidomide that possessed the teratogenic action [10,11]. Furthermore, this enantiomer was without importance for the desired sedative or sleep-inducing property. Therefore, if the (R)-(+)-enantiomer alone had been given, no teratogenic effects would have appeared and the drug might perhaps still have been used. For such reasons the pharmaceutical industry is becoming more and more interested in methods to resolve racemates into optical antipodes in order to be able to subject these individually to pharmacological testing. It is therefore also desirable to have reliable methods for a determination of the optical purity of the two forms. These topics will be dealt with in more detail later.

An elucidation of the structure of complex natural products will often need an application of chiral amino-acid analysis. Of the many remarkable compounds synthesized by lower, often procaryotic, organisms, we may choose the so-called transport antibiotics as an example. These compounds have quite unusual structures which contain D-amino-acid components. Two representatives are shown in Fig. 1.6. Valinomycin is cyclic, with a four-component thrice-repeating unit. Note the

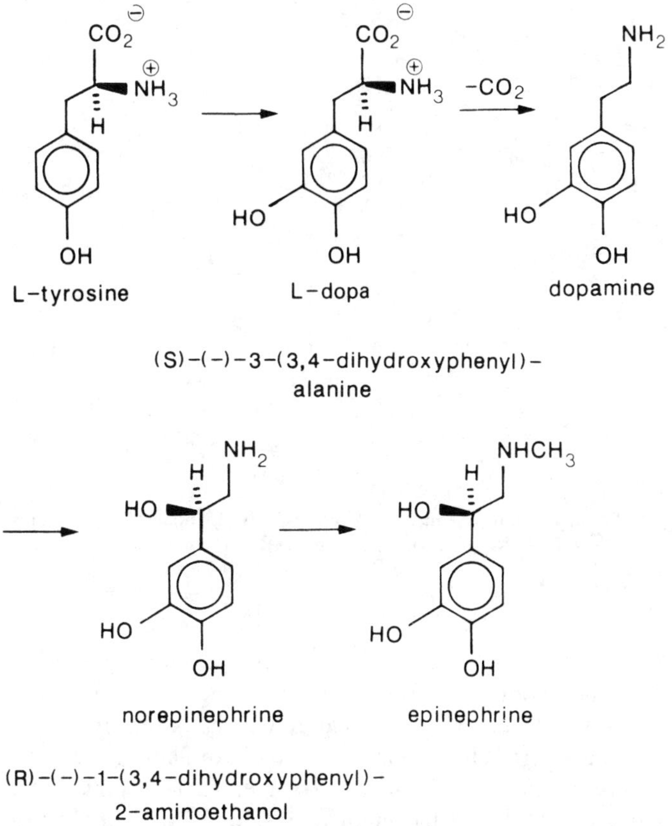

L-tyrosine   L-dopa   dopamine

(S)−(−)−3−(3,4−dihydroxyphenyl)−
alanine

norepinephrine   epinephrine

(R)−(−)−1−(3,4−dihydroxyphenyl)−
2−aminoethanol

Fig. 1.4 — Stereochemistry involved in the biosynthesis of the catecholamines.

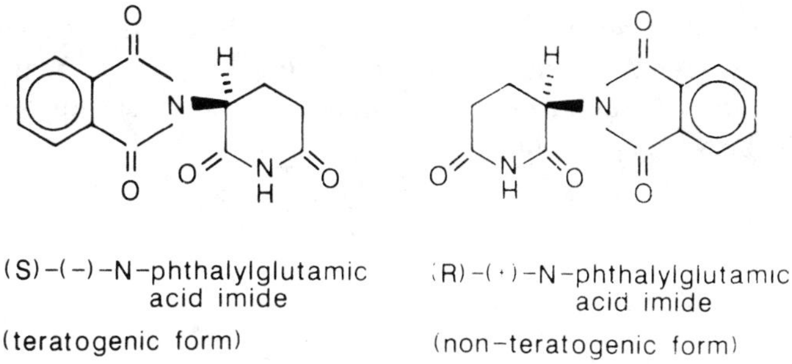

(S)−(−)−N−phthalylglutamic   (R)−(·)−N−phthalylglutamic
acid imide   acid imide

(teratogenic form)   (non−teratogenic form)

Fig. 1.5 — The two enantiomers of thalidomide.

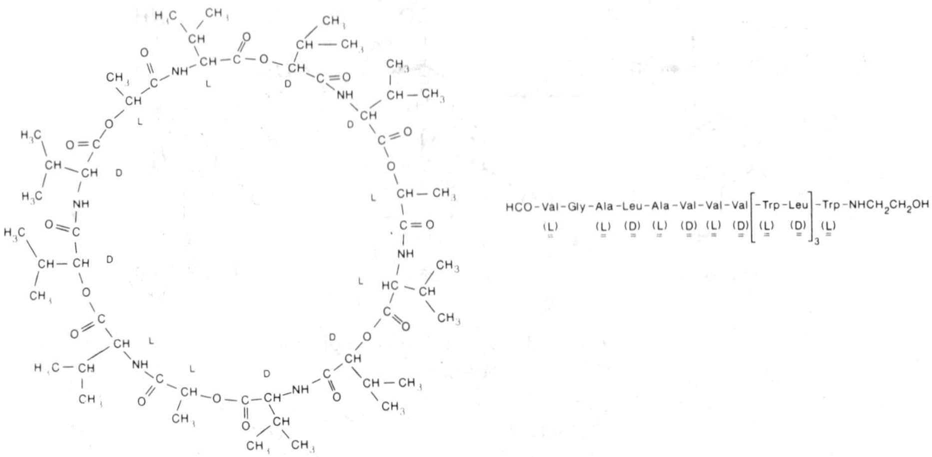

HCO - Val - Gly - Ala - Leu - Ala - Val - Val - Val ⎡- Trp - Leu⎤ - Trp - NHCH$_2$CH$_2$OH
(L)   (L) (D) (L) (D) (L) (D)⎣ (L) (D) ⎦₃ (L)

Fig. 1.6 — Valinomycin (left) is a macrocyclic compound with no less than 36 atoms in the ring. Gramicidin A (right) is composed of 15 amino-acids.

alternating amide and ester linkages, as well as the presence of both L- and D-valines at defined positions. Gramicidin A, on the other hand, is an open-chain polypeptide with an alternation of D- and L-amino-acid units. The *N*-terminus is formylated and the carboxyl end converted into an amide with ethanolamine.

Recently, a remarkable total synthesis of an extremely complex natural product has been achieved. After more than eight years of intensive work by in all 22 chemists at Harvard University, palytoxin (Fig. 1.7), a highly poisonous compound originally found in the Hawaiian coral *Palythoa toxica* [12], was synthesized [13]. Palytoxin is the most poisonous non-protein compound found to date. The molecule ($C_{129}H_{223}N_3O_{54}$, mol. wt. 2680.168) has in all 64 stereogenic centres. Even though many of these are located in ring systems, an almost astronomical number of stereoisomers can be generated. However, the naturally occurring compound consists of only one of all these stereoisomers. The stereochemistry of this is shown in Fig. 1.7.

With these final examples of the complexity of Nature's stereochemical pathways it should be evident from this short introductory chapter that chromatographic techniques for the separation and identification of enantiomers are of importance in many diverse areas of application.

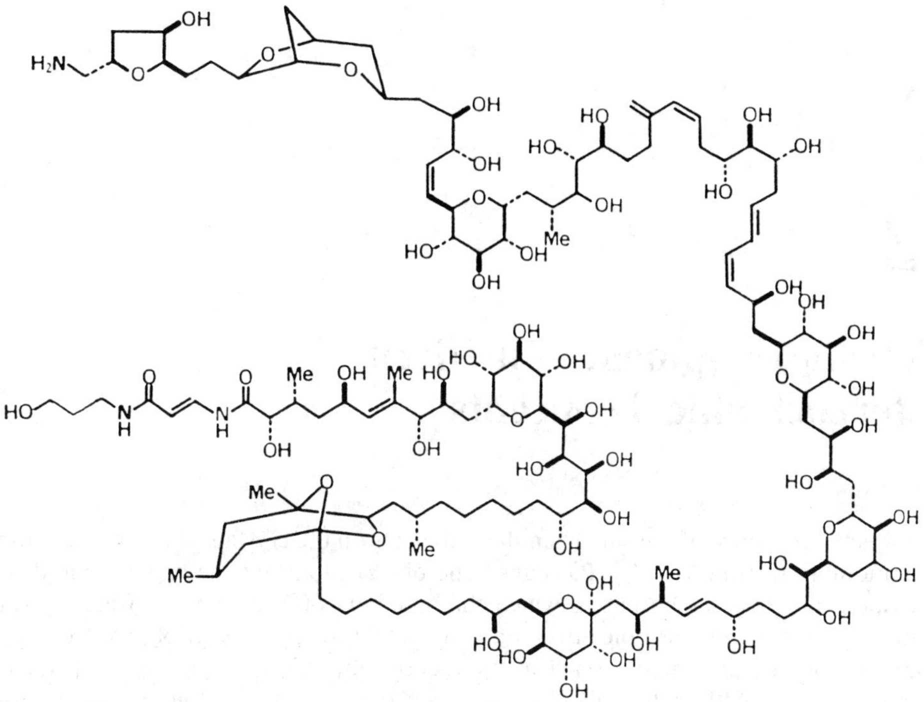

Fig. 1.7 — Molecular structure of palytoxin.

BIBLIOGRAPHY

R. Bentley, *Molecular Asymmetry in Biology*, Vols I, II, Academic Press, New York, 1969, 1970.
B. Vennesland, Stereospecificity in Biology, *Top. Current Chem.*, 1974, **48**, 39.
L. Stryer, *Biochemistry*, Freeman, San Francisco, 1975.
E. J. Ariëns, W. Soudijn and P. Timmermans, *Stereochemistry and Biological Activity of Drugs*, Blackwell, Oxford, 1983.

REFERENCES

[1] R. B. Woodward, F. E. Bader, H. Bickel, A. J. Frey and R. W. Kierstead, *Tetrahedron*, 1958, **2**, 1.
[2] E. J. Corey, N. H. Weinschenker, T. K. Schaaf and W. Huber, *J. Am. Chem. Soc.*, 1969, **91**, 5675.
[3] E. J. Corey, U. Koelliker and J. Neuffer, *J. Am. Chem. Soc.*, 1971, **93**, 1489.
[4] G. Chedd, *New Scientist*, 1974, **62**, 299.
[5] E. C. Kornfeld, J. M. Sheneman and T. Suarez, *Chem. Abstr.*, 1970, **72**, 30438c; 1971, **75**, 62341u.
[6] G. F. Russell and J. I. Hills, *Science*, 1971, **172**, 1043.
[7] L. Friedman and J. G. Miller, *Science*, 1971, **172**, 1044.
[8] A. Albert, *Selective Toxicity*, 5th Ed., Chapman & Hall, London, 1973.
[9] F. Luduera, L. von Euler, B. Tullar and A. Lauda, *Arch. Intern. Pharmacodyn.*, 1957, **11**, 392.
[10] G. Blaschke, H.-P. Kraft, K. Fickentscher and F. Köhler, *Arzneim.-Forsch.*, 1979, **29**, 1690.
[11] G. Blaschke, H.-P. Kraft and H. Markgraf, *Chem. Ber.*, 1980, **113**, 2318.
[12] R. E. Moore, *Prog. Chem. Org. Nat. Prod.*, 1985, **48**, 81.
[13] R. W. Armstrong, J.-M. Beau, S. H. Cheon, W. J. Christ, H. Fujioka, W.-H.Y. Ham, L. D. Hawkins, H. Jin, S. H. Kang, Y. Kishi, M. J. Martinelli, W. W. McWhorter, M. Mizuno, M. Nakata, A. E. Stutz, F. X. Talamas, M. Taniguchi, J. A. Tino, K. Ueda, J. Uenishi, J. B. White and A. Yonaga, *J. Am. Chem. Soc.*, 1989, **111**, 7525, 7531.

2

The development of modern stereochemical concepts

It was the presence of the small hemihedral facets on the crystals of racemic sodium ammonium tartrate that led Pasteur to the observation that the crystals could be divided into two categories, in which the members of one category were mirror images of the members of the other. Each crystal represents a chiral object (from the Greek word '*cheir*' = hand), which means that the object is not superimposable on its mirror image. Although van't Hoff has been regarded as the founder of stereo-chemistry, starting with his famous paper [1] of 1874, there is no doubt that the ideas of molecular asymmetry were already in Pasteur's mind, relating the asymmetric structure of the crystals to the molecules themselves. In modern terminology, molecules are chiral when they lack reflection symmetry. As will be shown later, however, it rather seldom happens that a racemate, comprised of mirror-image related chiral molecules, crystallizes with the formation of enantiomorphous crystals, i.e. chiral crystals composed of only one molecular form. More generally, the crystallization process leads to achiral crystals which are morphologically identical and built up from both molecular forms in equal amounts.

2.1 CHIRALITY AND MOLECULAR STRUCTURE

In the following section various relations between chirality and molecular structure will be discussed. Since molecules are often more or less flexible, it is also important to consider the stability of a chiral structure, i.e. the energy barriers that prevent interconversion between a given molecular structure and its mirror image.

2.1.1 Molecules with asymmetric atoms

Atoms which generate non-planar molecular structures by covalent bonds to other atoms may also create chirality. Thus, the tetrahedral structures formed by four different groups around atoms of elements such as carbon, silicon, nitrogen, phosphorus or sulphur are well-known, and numerous examples of optically active compounds of this type have been described in the literature.

Elements of groups V or VI in the periodic table can also form non-planar structures with three ligands, where the lone-pair electrons of the central atom can be regarded as the fourth ligand. In this case, a planar configuration is often possible, the formation of which will permit interconversion between the chiral pyramidal structures. Some common examples of central chirality are given in Fig. 2.1.

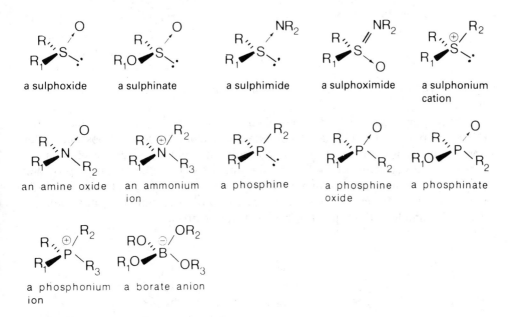

Fig. 2.1 — Examples of stable chiral molecular structures with central atoms other than carbon.

2.1.2 Other types of chiral molecular structures

One of the simplest cases of a chiral structure not involving an asymmetric central atom is derived from allene, $H_2C{=}C{=}CH_2$. It is readily seen that substitution of one of the hydrogen atoms at each carbon atom by a substituent R is sufficient to generate chirality, i.e. a non-superimposable mirror image will result from the particular geometry of this molecule:

A very similar type of optical isomerism is found in metal-ion–alkene co-ordination complexes and in metallocenes, as exemplified in Fig. 2.2.

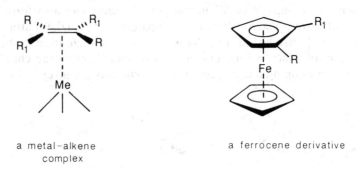

a metal–alkene   a ferrocene derivative
complex

Fig. 2.2 — Examples of the generation of chirality in organometallic complexes.

Of primary importance for the allenetype of isomerism, of course, is the molecular rigidity. Because no rotation around the double bonds is possible under normal conditions, any interconversion between the optical isomers is eliminated. Obviously, any structure of this type would generate stable optical isomers, provided only that interconversion through rotation is sufficiently restricted. This prerequisite is also fulfilled in certain substituted biaryls, o,o'-dinitrodiphenic acid being a classical example of this type of so-called atropisomerism:

In this case restricted rotation around the central bond is caused by the steric effect of the substituents. The groups are simply too large to pass each other, resulting in configurational stability.

A related case is found in the so-called 'push-pull' or polarized ethylenes, where electronic effects of the substituents cause a weakening of the π-bond, thus lowering the barrier to rotation. If the steric interactions between the two halves of the molecule are strong enough, it will adopt a twisted conformation in solution. This may often result in chiral conformations of sufficient stability to permit optical resolution of enantiomers even at room temperature [2]. The situation is shown in Fig. 2.3a.

A similar situation is present in other types of compounds with partial double-bond character. A good example is found in the extensively studied thioamides [3] (Fig. 2.3b).

The resemblance of atropisomerism to allene isomerism is evident if the pairs of identical substituents (A and B) are spatially represented:

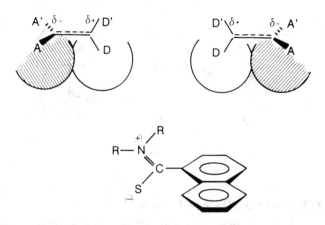

Fig. 2.3 — (a) Co-operation between electronic and steric effects in generating enantiometric stable conformations in a polarized ethylene. A and D denote acceptor and donor groups, respectively. (b) A chiral twisted conformation in a thioamide.

Molecular asymmetry in biaryls may also be caused by the bridging of two *ortho*-positions by derivative formation, as in binaphthylphosphoric acid (**1**).

Further, steric crowding in a molecule may give rise to molecular distortion and hence chirality. Very clear-cut and representative examples of this phenomenon are found in the condensed aromatic hydrocarbons called helicenes (**2**).

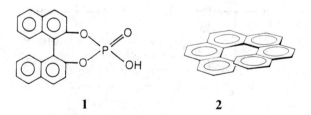

1   **2**

Steric constraints give rise to a helical form with an energy barrier to interconversion between the right- and left-handed enantiomers which is high enough to permit their resolution.

To sum up: chiral molecules may be of quite different nature and shape, but can, from considerations of symmetry elements, all be classified as belonging to one of three categories, viz. possessing central, axial or planar chirality. This means that three-dimensional space can be occupied asymmetrically about a chiral centre (**3**), a chiral axis (**4**) or a chiral plane (**5**).

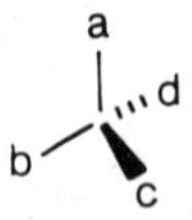

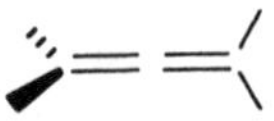

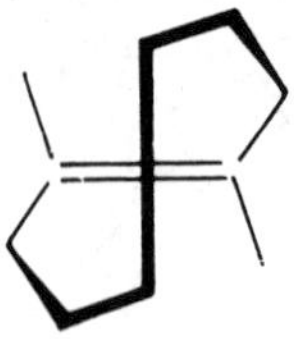

       3   **4**   **5**

2.2 DEFINITIONS AND NOMENCLATURE

At this point a very relevant question is how do we actually know the true arrangement in space of the atoms or groups in an optically active molecule? In fact, most of the fundamental research on optically active compounds and their chemical transformations has been made without this knowledge. An answer to the problem came in 1951 when Bijvoet *et al.*, making use of an anomalous dispersion effect, determined the absolute configuration of rubidium D-(+)-tartrate by X-ray crystallo-graphy [4]. It then turned out that the configuration of D-glucose (**6**), arbitrarily assigned by Fischer and configurationally related to D-glyceraldehyde (**7**), was by chance correct. The result has recently been verified on the basis of new stereo-chemical investigations [5].

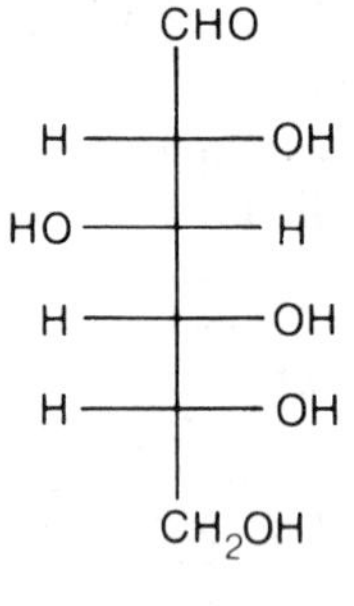

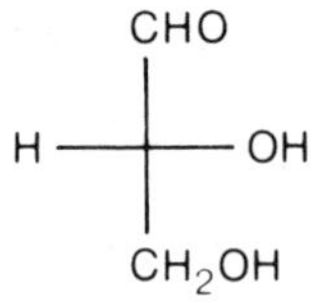

       6   **7**

Fischer projection formulae of D-glucose (aldehyde form) and D-glyceraldehyde. Vertical bonds are directed backwards from the plane of the paper, horizontal bonds outwards).

     This D,L-nomenclature system, which relates a variety of optically active com-pounds to each other (particularly carbohydrates, hydroxy-acids and amino-acids) is still very much used and has not been replaced in the case of carbohydrates (for obvious reasons shown later).

     However, the system is only applicable to compounds having asymmetric carbon atoms and therefore a more generally useful nomenclature system was needed. In

1956 Cahn, Ingold and Prelog [6–8] presented a new system, the (R,S)-nomencla-
ture, which is applicable to any chiral molecule and permits the absolute configuration to be determined directly from the (R)- or (S)-designation.

The system is briefly outlined below, although the reader is referred to the
original papers for further details.

Cahn–Ingold–Prelog system
The principle of this system lies in three steps, viz. (1) to arrange the ligands
associated with an element of chirality into a sequence, (2) to use this sequence to
trace a chiral path, and (3) to use the chiral sense of this path to classify the element of
chirality.

(1) The ligands around a centre of chirality are ordered in a sequence according to
    the following basic rules.
    (a) Higher atomic number is given priority.
    (b) Higher atomic mass is given priority.
    (c) *Cis* is prior to *trans*.
    (d) Like pairs [(R,R) or (S,S)] are prior to unlike pairs [(R,S) or (S,R)].
    (e) Lone-pair electrons are regarded as an atom with atomic number 0.
(2) By use of these rules, the ligands, ordered in a sequence A>B>C>D, are
    viewed in such a way that D (of lowest priority) is pointing backwards from the
    viewer:

(3) The remaining ligands are then counted, starting from the one of highest priority
     (i.e. A, B, C). If this operation is clockwise for the viewer the designation will be
     (R) (rectus), otherwise it will be (S) (sinister). Thus, the example shown above is
     an (R)-configuration.

The selection rule for axial chirality implies that the atoms closest to the axis are
considered in a priority sequence, e.g. the *ortho*-carbon atoms in a biaryl compound.

For a molecule showing planar chirality, a plane of chirality must first be selected.
The second step involves determination of a pilot atom P which should be bound
directly to an atom of the plane and located at the preferred ('nearer') side. P is
selected according to the sequence rules. The next step is to pass from P to the in-
plane atom to which it is directly bound (a). This atom is then the atom of highest
priority of the in-plane sequence. The second atom of this sequence is the in-plane

atom (b), bound directly to (a), which is most preferred by the standard subrules. After completion of the sequence the chirality rule can be applied. The paracyclophane below illustrates the principle used.

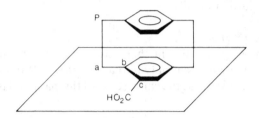

(R)-(−)-[2.2]paracyclophane-4-carboxylic acid

The helicenes can be treated as axially chiral molecules but are preferentially considered as secondary structures. Their chirality is best represented in terms of helicity. Thus, for hexahelicene as an example, the (−)-form represented below forms a left-handed helix [M (= minus) helicity] and is then denoted M-(−). The opposite enantiomer is called P (plus).

This M,P-nomenclature is also often used for chiral biaryls. First an axis is drawn through the single bond around which conformation is defined and the smallest torsion angle formed between the carbon atoms bearing the groups of highest priority is used to define the helix.

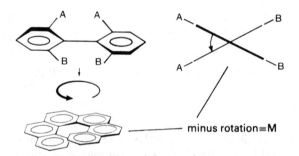

minus rotation≡M

There are certain disadvantages with the Cahn–Ingold–Prelog system, mainly associated with the priority rules. In the series of chiral epoxides (X = O) of the general formula and absolute configuration given below, the priority rules will cause the stereochemical designation according to the system to change from (R)- to (S)- when the Y-substituent is changed from CH_3 to Cl (cf. Section 8.4.1). The situation is even more unsatisfactory for a series of chiral thioepoxides (thiiranes; X = S), where the designation changes on going from Y = F to Y = Cl. However, it is not easy to find any alternative nomenclature systems where problems of a similar kind do not arise, although attempts to redesign the system have recently appeared in the literature [9].

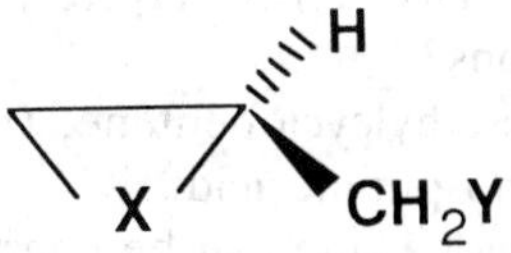

	X	Y	Absolute configuration
	O	CH_3	R
	O	F	S
	O	Cl	S
	O	OH	R
	S	CH_3	R
	S	F	R
	S	Cl	S
	S	OH	R

Some important relations and definitions are summarized below.

Stereochemical concepts used throughout this book
Stereoisomers: Molecules differing only in the spatial arrangement of the substituents.

(1) Enantiomers or optical antipodes: the molecules are chiral and related to each other by reflection symmetry, and have identical internal energies.
(2) Diastereomers: stereoisomers not related to each other by reflection symmetry; they have non-identical internal energies.

Chiral compounds: compounds composed of molecules which are either asymmetric or dissymmetric and have mirror images which are incongruent.

Asymmetric: a molecule which lacks any symmetry element.

Dissymmetric: a molecule which does not possess any S_n symmetry element. This means that there is no symmetry plane (S_1) or symmetry centre (S_2) present, but instead a C_n-axis, i.e. a simple axis of rotation ($n>1$).

Exercises
1. (−)-Benzylmethylsulphoxide can be represented by the projection formula:

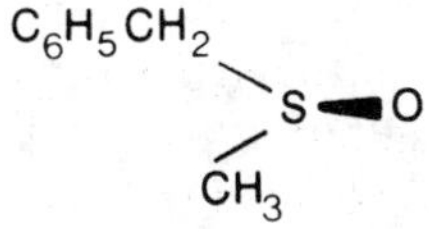

$$C_6H_5CH_2 \diagdown \atop {S \blacktriangleleft O} \atop {\diagup \atop CH_3}$$

Determine its designation according to the (R,S)-nomenclature.
2. By using the (R,S)-rules, draw the absolute configuration of the (S)-form of serine [$HOCH_2CH(NH_2)CO_2H$]. Convert it into a Fischer projection formula and determine whether it corresponds to D- or L-.

3. Draw stereoprojection formulae and name rationally (C–I–P-system) all possible isomers of 2,8-dimethyl-1,7-dioxaspiro[5.5]undecane, (a pheromone).
4. Deduce the absolute configurations of the (R)-forms of α-aminoethylphosphonic acid [$H_2NCH(CH_3)PO(OH)_2$] and alanine [$H_2NCH(CH_3)CO_2H$], respectively. What can be said about their relative configurations?
5. Draw a stereoprojection formula of (a) (S)-3-methylcyclopentene, (b) (S)-3-chlorocyclopentene and (c) (P)-1,1′-binaphthylphosphonic acid.
6. How many isomers of 1,2,3,4,5,6-hexachlorocyclohexane can be constructed? From symmetry considerations, can any chiral isomer(s) be found?

BIBLIOGRAPHY

K. Mislow, *Introduction to Stereochemistry*, Benjamin, Menlo Park, Cal. 1965.
G. Krow, The Determination of Absolute Configuration of Planar and Axially Dissymmetric Molecules, *Top. Stereochem.*, 1970, **5**, 31.
C. A. Mead, Symmetry and Chirality, *Top. Current Chem.*, 1974, **49**.
H. B. Kagan, *Organic Stereochemistry*, Arnold, London, 1979.
B. Testa, *Principles of Organic Stereochemistry*, Dekker, New York, 1979.
J. Jacques, A. Collet and S. H. Wilen, *Enantiomers, Racemates and Resolutions*, Wiley, London, 1981.
K. Schlögl, Planar Chiral Molecular Structures, *Top. Stereochem.*, 1984, **125**, 27.

REFERENCES

[1] O. T. Benfey (ed.), *Classics in the Theory of Chemical Combination*, Dover, New York, 1963, p. 151.
[2] J. Sandström, *Top. Stereochem.*, 1982, **14**, 83.
[3] A. Eiglsperger, F. Kastner and A. Mannschreck, *J. Mol. Structure*, 1985, **126**, 421.
[4] J. M. Bijvoet, A. F. Peerdeman and A. J. van Bommel, *Nature*, 1951, **168**, 271.
[5] H. Buding, B. Deppisch, H. Musso and G. Snatzke, *Angew. Chem.*, 1985, **97**, 503.
[6] R. S. Cahn, C. K. Ingold and V. Prelog, *Experientia*, 1956, **12**, 81.
[7] R. S. Cahn, C. K. Ingold and V. Prelog, *Angew. Chem. Int. Ed.*, 1966, **5**, 385, 511.
[8] V. Prelog and G. Helmchen, *Angew. Chem.*, 1982, **94**, 614; (*Int. Ed.* 1982, **21**, 567).
[9] H. Dodzink and M. Mirowitz, *Tetrahedron: Asymmetry*, 1990, **1**, 171.

3

Techniques used for studies of optically active compounds

3.1 DETERMINATION OF OPTICAL OR ENANTIOMERIC PURITY

One of the most important questions in connection with studies of optically active compounds is how to determine their purity. There are two aspects: the purity of the compound as determined by the common analytical procedures used in organic chemistry, and the enantiomeric purity, i.e. the purity with respect to the content of only one enantiomer. Only the second problem, and the main methods that are available, will be dealt with here. These methods can be divided into two fundamentally different categories depending upon whether they involve separation of the enantiomers or not.

3.1.1 Methods not involving separation

The principal methods used for determination of enantiomeric composition which do not require separation of the enantiomers are polarimetry, nuclear magnetic resonance (NMR), isotopic dilution, calorimetry, and enzyme techniques. Of these the first mentioned is the oldest and most widely applied. With the exception of NMR, all these methods rely upon measurement of a net effect, and therefore data from an optically pure enantiomer are needed for comparison.

3.1.1.1 Polarimetry

This method makes use of the unique property of a chiral compound to rotate the plane of polarization of plane-polarized light. Historically, Biot was the first to demonstrate that certain naturally occurring organic compounds possess this property [1]. Very briefly, the phenomenon can be made understandable by regarding the polarized light, i.e. light with its vibrational plane limited to one direction, as composed of two vectors which are right- and left-handedly circularly polarized, respectively (Fig. 3.1). Because these vectors represent chiral objects which are mutual antipodes, they do not behave identically on interaction with chiral

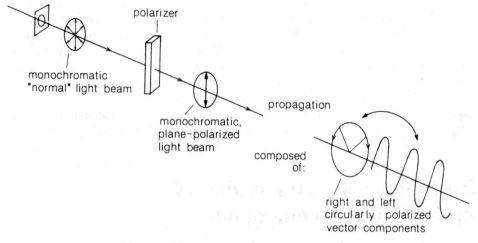

Fig. 3.1 — Dissection of a plane-polarized light-wave.

molecules. On passage through a medium containing chiral molecules they are therefore propagated at different velocities. As a result of vector addition, the plane of polarization will be rotated.

The rotation of the plane, or the **optical rotation**, α, is dependent on a variety of factors such as solute concentration, optical path-length, solvent, temperature, wavelength used, etc. In order to obtain a quantity suitable for the characterization of an optically active compound, the conditions used have to be specified. The **specific rotation**, $[\alpha]$, is defined according to Eq. (3.1):

$$[\alpha]_\lambda^T = \frac{100\alpha}{lc} \tag{3.1}$$

where α = the measured optical rotation, T = the temperature (°C), l = the cell path-length (dm), c = the concentration of the compound (g/100 ml).

The specific rotation is often highly solvent-dependent and may also be dependent upon the concentration used. These conditions should therefore always be given. A statement such as $[\alpha]_D^{20} = +192$ (c. 2, dioxan) [which is the specific rotation of the (optically pure) hormone progesterone], contains all necessary information. It is then easy for anyone to analyse a preparation of the same compound.

If the specific rotation of an optically pure compound is denoted by $[\alpha]_{max}$, it follows that the optical purity, P (%), of a given sample of specific rotation $[\alpha]$ can be obtained from

$$P = \frac{100[\alpha]}{[\alpha]_{max}} \tag{3.2}$$

Equation (3.2) defines the optical purity. Because it is based upon experimentally determined chiroptical properties it may be associated with systematic errors and may not always correspond to the actual enantiomeric composition or enantiomeric purity. Strictly speaking, optical purity is linearly related to enantiomeric purity (representing the *actual* composition) only when there is no molecular association between the enantiomers in solution [2]. Therefore, methods which permit a distinction between the enantiomeric forms are generally more reliable. It is also obvious that the accuracy with which the optical purity can be determined is not as high as with techniques involving separation of the enantiomers.

Further, the use of polarimetry for the purpose of optical purity determination requires knowledge of the specific rotation of the compound in optically pure form. If no such data are available and no optically pure sample is at hand, $[\alpha]_{max}$ has to be determined by an indirect method, viz. the isotope dilution technique, described later in this section.

3.1.1.2 Nuclear magnetic resonance

An NMR spectrum does not differentiate between enantiomers unless these have been transformed into some sort of diastereomeric states, which of necessity will require an optically active partner. Apart from the technique of converting the enantiomers into diastereomers by reaction with a suitable chiral reagent (which is the oldest one, developed by Mislow and Mosher [3]), two 'direct' methods have been used. The first of these, developed by Pirkle and co-workers [4–8], relies on the use of a chiral solvent. A particularly useful solvent is (R)-(−)- 2,2,2-trifluoro-1-phenylethanol. The solvent induces a chemical-shift difference between the enantiomers (of otherwise identical nuclei) and accordingly the enantiomer ratio can be directly obtained by integration.

It is important to realize that the optical purity of the solvent does not influence the integration result, i.e. the peak *ratio*, but only the peak *separation*. The principle of the technique is illustrated in Fig. 3.2, which shows a practical example.

The peak-splitting resulting from the solvent-induced chemical-shift difference is, in fact, a consequence of the preferential interaction of one of the enantiomers with the chiral solvent. This is, as we will see later, the same principle that is utilized in chiral liquid chromatography, where preferential interaction between one enantiomer and the chiral stationary phase (CSP) is of fundamental importance. The NMR studies actually led Pirkle and his group to the design of suitable CSPs.

The chemical shift difference and its use can be deduced as follows.

Let the compound be called C (enantiomers C_S and C_R) and the solvent S (enantiomers S_S and S_R). The interaction equilibria give:

Species	Molar concentration	
C_S	p	
C_R	p'	
C_S–S_S	q	$p + q + r = [C_S]_{tot}$
C_R–S_R	q'	$p' + q' + r' = [C_R]_{tot}$
C_S–S_R	r	
C_R–S_S	r'	

The chemical shifts, δ, of a certain nucleus in C_S and C_R, respectively, are denoted by δ_{C_S} and δ_{C_R}. Then:

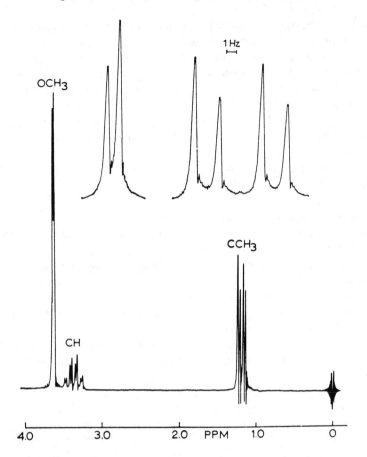

Fig. 3.2 — Determination of the enantiomeric composition of optically enriched methyl alaninate by NMR in optically active 2,2,2-trifluoro-1-phenylethanol. (Reproduced from W. H. Pirkle and S. D. Beare, *J. Am. Chem. Soc.*, 1969, **91**, 5150, with permission. Copyright 1969, American Chemical Society.)

$$\delta_{C_S}^{obs} = (p\delta_{C_S} + q\delta_{C_S-S_S} + r\delta_{C_S-S_R})/[C_S]_{tot} \qquad (3.3)$$

$$\delta_{C_R}^{obs} = (p'\delta_{C_R} + q'\delta_{C_R-S_R} + r'\delta_{C_R-S_S})/[C_R]_{tot}$$

We know that $\delta_{C_S} = \delta_{C_R}$, $\delta_{C_S-S_S} = \delta_{C_R-S_R}$ and $\delta_{C_S-S_R} = \delta_{C_R-S_S}$, which means that for a racemic solvent $[S_S]_{tot} = [S_R]_{tot}$) $q = q'$ and $r = r'$, leading to $\delta_{C_S}^{obs} = \delta_{C_R}^{obs}$. If the solvent is not racemic, i.e. $[S_S]_{tot} \neq [S_R]_{tot}$, then $q \neq q'$ and $r \neq r'$ and therefore $\delta_{C_S}^{obs} \neq \delta_{C_R}^{obs}$, which means splitting of the resonance line. Therefore, if either $[S_S]_{tot}$ or $[S_R]_{tot}$ is zero (use of an optically pure solvent), the peak separation will be maximal.

It should be remembered, however, that the peak separation resulting from the interaction with the chiral solvent is often rather small and not always of practical use. Furthermore, the various nuclei in the diastereomeric solvates formed are affected quite differently and usually only those close to the chiral centres are of interest (cf. Fig. 3.2).

An even more powerful NMR method is the use of optically active lanthanide-shift reagents; a technique which combines the high resolution obtained through pseudocontact downfield shifts [9] with splitting of resonance lines by enantioselective interaction with the chiral lanthanide complex [10]. The principle is outlined in Fig. 3.3. Generally, β-diketones (in enol form) form stable complexes with metal

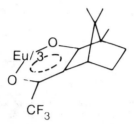

Fig. 3.3 — Structure of a chiral shift reagent, [tris(3-trifluoromethylhydroxy-methylene)-(+)-camphorato]-europium(III) (Eu[(+)facam]₃), used to create a downfield shift and peak splitting in the NMR of compounds able to interact with the metal atom.

ions of the lanthanide series, such as Eu^{3+} and Pr^{3+}. Such paramagnetic complexes can combine or interact with compounds containing electron-donating groups, such as amines, amino-acids, alcohols, ketones and esters, resulting in a considerable downfield shift for nuclei which are not too far from the site of interaction. However, since chiral β-diketones are readily available, optically active lanthanide shift reagents can be made. The example given in Fig 3.3 is based on (+)-camphor. The reagents are readily soluble in typical NMR solvents such as carbon tetrachloride or deuteriochloroform and the spectral shifts produced are often studied as a function of the amount added. A substantial increase in spectral resolution may be achieved when chiral lanthanide-shift reagents are used and consequently resonance lines corresponding to two enantiomers are better separated and the enantiomer composition can be accurately determined by integration. An application of the technique is shown in Fig. 3.4.

Results obtained from NMR methods by peak integration give the concentration ratio (r) of the enantiomers, and the enantiomeric purity or enantiomeric excess (e.e.) is calculated as:

$$\text{e.e.} = \frac{(1-r)}{(1+r)} \times 100\% \qquad (3.4)$$

3.1.1.3 *Isotope dilution*
This method requires the determination of two variables, viz. the specific rotation and the isotope content. The first is determined by polarimetry, and the second can

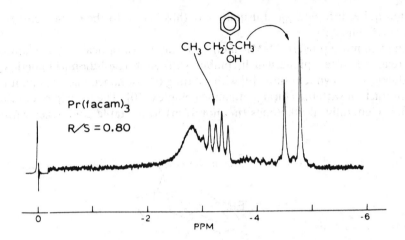

Fig. 3.4 — Direct observation of the enantiomer ratio of partially resolved 2-phenyl-2-butanol by NMR in carbon tetrachloride in the presence of $0.3M$ Pr[(+)facam]$_3$. (Reproduced from H. L. Goering, J. N. Eikenberry, G. S. Koermer and C. J. Lattimer, *J. Am. Chem. Soc.*, 1974, **96**, 1493, with permission. Copyright 1974, American Chemical Society.)

be obtained by mass spectrometry or by liquid scintillation counting for β-emitting radioisotopes. The most general principle is to mix the sample of unknown enantiomeric purity with an isotopically labelled racemate of the same compound. Let us assume that a is the weight of the sample (of enantiomeric purity P and specific rotation $[\alpha]$, the (R)-form being assumed to predominate), mixed with a weight b of the racemic compound having an isotope content I_0. The isotope contents of the enantiomers in the mixture $a + b$ will then be given by

$$I(R) = \frac{bI_0}{b + a(1 + P)} \quad \text{and} \quad I(S) = \frac{bI_0}{b + a(1 - P)} \tag{3.5}$$

As these isotope contents are not changed by any further processing of the mixture, Eq. (3.5) can be used in the further calculations.

The mixture is subjected to recrystallization to yield a new sample with another optical purity P' and specific rotation $[\alpha']$. If the isotope content I' of this sample is determined, it is then possible to calculate the maximum specific rotation $[\alpha]_{max}$ of the compound and hence the optical purity of the original sample. It is easily deduced that I' will be given by

$$I' = I(R)\frac{(1 + P')}{2} + I(S)\frac{(1 - P')}{2} \tag{3.6}$$

A combination of Eqs. (3.5) and (3.6) with elimination of $I(R)$ and $I(S)$ will give

$$I' = I_0 \frac{b^2 + ab - abPP'}{b^2 + 2ab + a^2(1 - P^2)} \tag{3.7}$$

But $P = [\alpha]/[\alpha]_{max}$ and $P' = [\alpha']/[\alpha]_{max}$ [cf. Eq. (2)]. Substitution and rearrangement will then give

$$[\alpha]_{max} = \left(\frac{I'a^2[\alpha]^2 - I_0 ab[\alpha][\alpha']}{I'(a+b)^2 - I_0 b(a+b)} \right)^{\frac{1}{2}} \tag{3.8}$$

The value of $[\alpha]_{max}$ thus calculated is then inserted into Eq. (3.2), which gives the optical purity.

3.1.1.4 Calorimetry

As a detailed account of this method is outside the scope of this book, only a brief outline will be given. The most important variant in this category is called differential scanning calorimetry (DSC). In principle, the energy absorbed or evolved by a sample is determined as a function of the temperature. The DSC instrument contains two cells (sample and reference) and the energy input difference required to maintain the two cells at the same temperature is scanned during a linear temperature change with time. Such a microcalorimetric device gives the temperatures as well as the enthalpies of phase transitions (such as melting). For racemic compounds, DSC traces for the sample of unknown optical purity as well as for the racemate are recorded. This gives the necessary data (melting point and enthalpy of fusion of the racemate as well as the termination-of-fusion temperature, T_f, of the sample of interest) for a calculation of the enantiomeric composition from the Prigogine–Defay equation [11].

The method is, of course, only applicable to solid compounds and is, very broadly interpreted, a simplified way of constructing a melting point diagram. The relation between a DSC trace and the racemic-compound branch of a melting point diagram is shown in Fig. 3.5.

3.1.1.5 Enzyme techniques

Many enzymatic reactions are highly stereoselective and, particularly in the case of amino-acids, enzymes may discriminate completely between a pair of enantiomers [12]. Therefore, techniques based on enzyme catalysis of amino-acid transformations are particularly useful for an exact determination of high enantiomeric purities, based on study of the reaction of the contaminating enantiomer. It is then possible to detect as little as 0.1% of this enantiomer in the presence of 99.9% of its optical antipode.

Two types of reaction have been favoured.

(1) Oxidation:

$$H_2N—CHR—CO_2H \xrightarrow{\frac{1}{2}O_2, \text{AA-oxidase}} R—CO—CO_2H + NH_3$$

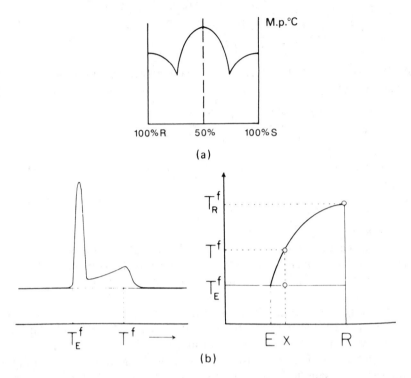

Fig. 3.5 — (a) Typical melting point diagram of a racemic compound. (b) Principle of determination of enantiomer composition by DSC. (Reprinted, with permission, from J. Jacques, A. Collet and S. H. Wilen, *Enantiomers, Racemates and Resolutions*, Wiley, New York, 1981, p. 152. Copyright 1981, John Wiley & Sons Ltd.)

(2) Decarboxylation:

$$H_2N\text{---}CHR\text{---}CO_2H \xrightarrow{\text{AA-decarboxylase}} R\text{---}CH_2\text{---}NH_2 + CO_2$$

For the first reaction there are L- and D-amino-acid oxidases commercially available. Decarboxylation of only L-amino-acids can be enzymically catalysed by the second reaction, which means that only the optical purity of D-amino-acids can be determined.

3.1.2 Methods based on separation
The only practically useful methods in this category are those based on chromatographic separation. Since these are treated in the following chapters, only the principal aspects are given here.

The basic forms of chromatography differ with respect to the aggregation state of the mobile phase and sample. In gas chromatography (GC) the sample has to be volatile enough to be transported by the mobile gas phase, which generally requires an elevated column temperature. It is therefore often necessary to increase the volatility of polar compounds by suitable derivatization. In the liquid chromatography (LC) methods there is usually no need for derivatization, except sometimes in order to increase the sensitivity of detection.

Until quite recently almost all stationary phases used in GC and LC were achiral, i.e. they were not based on any optically active material. This meant that separation of enantiomers could *not* be directly performed chromatographically. Therefore, chromatographic separations aiming at the determination of enantiomeric composition were limited to diastereomeric derivatives formed by a reaction with an optically pure reagent (Scheme 3.1).

Enantiomers	Optically pure reagent	Products
(R)-A—X	(R)-B—Y	(R)-A∗(R)-B
	$\longrightarrow$	
(S)-A—X	– XY	(S)-A∗ (R)-B

Scheme 3.1 — The principle of chiral derivatization of enantiomers to produce diastereomers which are chromatographically separable on non-chiral stationary phases.

Because a variety of derivatization techniques have been developed since the introduction of GC, it is not surprising to find that many of these reactions, but utilizing an optically active reagent, have been applied to chiral derivatization of enantiomers. A number of such derivatization reactions are also available for LC separations.

The success of such *indirect* methods for determination of enantiomeric composition is dependent on a number of factors which will be treated in the next chapter.

A fundamentally different approach is found in the *direct* chromatographic methods, where enantiomers are separated by **chiral chromatography**. In these cases the stationary phase is usually an optically active compound, causing a difference in retention between the solute enantiomers. This is always the case in GC. In LC enantiomers may also be directly separated by the use of a chiral mobile phase or of chiral constituents added to the mobile phase. This will be treated in more detail in Chapter 7. In Table 3.1 the different modes of chromatography for determination of optical purity are summarized.

Table 3.1 — Different methods used for chromatographic determination of optical purity

Principle	Method	Technique
Chiral derivatization	Indirect	Use of conventional (achiral) columns in GC and LC
No chiral derivatization (conventional derivatization sometimes used to enhance selectivity and/or sensitivity)	Direct	GC: use of column containing a chiral stationary phase LC: either use of a column containing a chiral stationary phase or of an achiral column together with a chiral mobile phase system

3.2 DETERMINATION OF ABSOLUTE CONFIGURATION

The *absolute* configuration of a chiral molecule tells us the true orientation in space of the various atoms or groups of atoms. By *relative* configuration we mean that a molecule A can be stereochemically related to another molecule B without the necessity for knowledge of the absolute configuration. This is conceptually simpler, because we only need to know the stereochemistry involved in the reaction steps which transform A into B or vice versa. Such configurational relationships have been obtained in very elegant ways, to correlate compounds of unknown chirality to those with previously known absolute configuration. An example is given in Fig. 3.6. Here

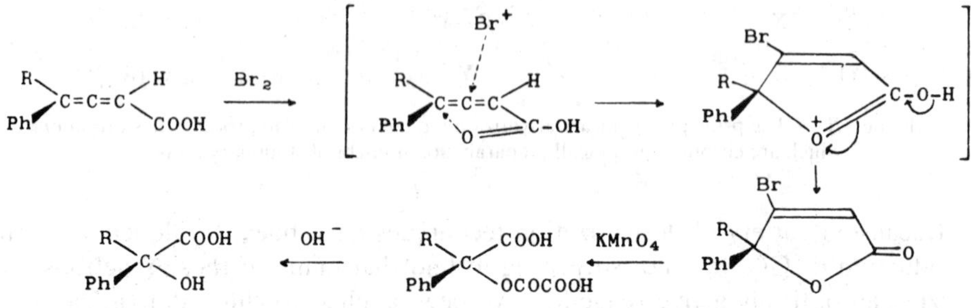

Fig. 3.6 — Configurational relationship established between phenylallenecarboxylic acids and 2-substituted mandelic acids. (Reprinted, with permission, from K. Shingu, S. Hagishita and M. Nakagawa, *Tetrahedron Lett.*, 1967, 4371. Copyright 1967, Pergamon Journals Ltd.)

a series of (+)-phenylallenecarboxylic acids was shown to be converted by permanganate oxidation of the bromolactone intermediate into (+)-alkylmandelic acids. As the absolute configuration of the latter is known and the reaction stereochemistry is unequivocal, the stereochemical correlation obtained will also give us the absolute configuration of the (+)-allenes.

As already mentioned (cf. Section 2.2), the problem of determining absolute configuration, i.e. how to obtain directly information that tells us how the atoms in the molecules of an optically pure compound are spatially oriented, remained unsolved for many years. The solution came from the observation of a special effect in X-ray diffraction patterns.

3.2.1 X-Ray crystallography with anomalous scattering
This technique, which has been in use since its discovery by Bijvoet in 1949, relies on the interpretation of an anomalous dispersion effect caused by heavy atoms present in the crystal lattice.

Although this effect had long been known and used as early as 1928 to determine the absolute polarity of a polar zinc sulphide crystal [13], it was regarded as fundamentally impossible to determine the absolute configuration of a chiral molecule by X-ray crystallography until Bijvoet showed how the effect could be used to establish an absolute polarity in a crystal structure composed of chiral molecules [14]. In the first example reported, a rubidium salt of (+)-tartaric acid was used [15]. Since there are no simple means by which stereochemical relationships between compounds belonging to different types of chirality can be established, applications of the Bijvoet technique to compounds of axial and planar chirality as well, have been of the greatest importance.

The Bijvoet technique has now been highly refined by the use of modern computerized diffractometers and it is no longer necessary to introduce heavy atoms as anomalous scatterers in the crystal lattice. The accuracy of intensity measurements obtained with present-day technology has made it possible to use atoms as light as oxygen as anomalous X-ray scatterers. Consequently, the absolute configuration of (+)-tartaric acid has been redetermined with tartaric acid itself [16].

Some of the key determinations are shown in Table 3.2.

An example of the use of modern computerized X-ray diffractometry for determining absolute configuration in a homochiral compound is given in Fig. 3.7.

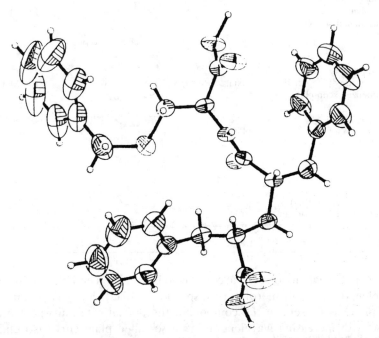

Fig. 3.7 — Stereoprojection of *N*-[(S,S)-2,4-dibenzyl-4-carboxybutyryl]-(S)-*S*-benzylcysteine (a 2,4-dibenzylglutaric acid amide acting as an enkephalinase inhibitor) as obtained from X-ray crystallography of a single crystal of the enantiomerically pure compound. (Reprinted, with permission, from G. M. Ksander, C. G. Diefenbacher, A. M. Yuan, F. Clark, Y. Sakane and R. D. Ghai, *J. Med. Chem.*, 1989, **32**, 2519. Copyright 1989, American Chemical Society.)

This stereoprojection also shows the thermal motion of the atoms in the structure.

3.2.2 Spectroscopic (ORD, CD) and chromatographic methods based on comparison

These methods, like the chemical correlations, are all indirect and will require a reference compound of known absolute configuration and should really be called correlation methods.

Table 3.2 — Absolute configurations determined by the anomalous X-ray dispersion technique with reference compounds belonging to different chirality categories

Compound	Chirality type	Structure determined	Absolute configuration	Reference
(+)-tartaric acid (as Rb-salt)	central	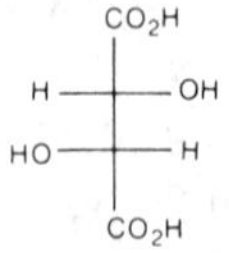	l.- = (R,R)	[15]
(+)-2,2'-diamino-6,6'-dimethylbiphenyl (as Co(salicylaldehyde)$_2$ complex)	axial	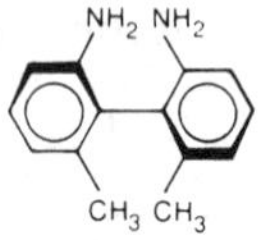	R- (M-)	[17]
(+)-2,2'-dihydroxy-3,3'-dimethoxycarbonylbinaphthyl (as bromobenzene complex)	axial	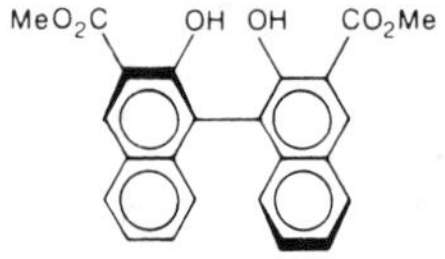	R- (M-)	[18]
(−)-[2.2]paracyclophane-4-carboxylic acid	planar	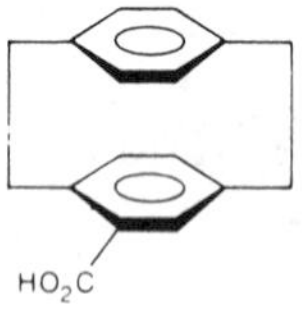	R-	[19]
(+)-[2](1,3)benzeno[0](3,4)-thiopheno[0](1,3)benzenophan	planar	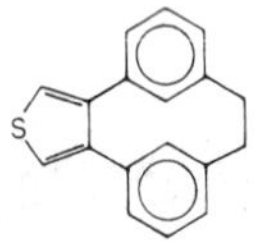	M-	[20]

If the optical activity of a chiral compound is measured and plotted as a function of wavelength, an optical rotatory dispersion (ORD) curve is obtained. If no chromophore is present in the compound, the optical rotation will continuously decrease with increasing wavelength and a so-called plain curve is obtained. If, however, the spectral range investigated covers an absorption band of the compound, that band will give rise to a Cotton effect [21], i.e. the curve will show one or more peaks or troughs (extrema). The *sign* and *intensity* of the Cotton effect, the molecular amplitude (a), is defined according to Eq. (3.9). Here $[\Phi]_1$ and $[\Phi]_2$ denote the molecular rotation at the extremum of the longer (subscript 1) and shorter (subscript 2) wavelengths, respectively. The molecular rotation, in turn, is defined according to Eq. (3.10), where M is the molecular weight of the compound.

$$a = \frac{[\Phi]_1 - [\Phi]_2}{100} \qquad (3.9)$$

$$[\Phi] = \frac{[\alpha]M}{100} \qquad (3.10)$$

By a related technique, taking advantage of the circular dichroism (CD) effect, i.e. the fact that a right circularly polarized beam is differently absorbed from its left circularly polarized counterpart, the differential dichroic absorption ($\Delta\varepsilon = \varepsilon_L - \varepsilon_R$), and the molecular ellipticity ($[\theta] = 3300\Delta\varepsilon$) can be measured as a function of wavelength. Often the Cotton effect is better evaluated from such a CD curve.

The appearance of ORD and CD curves, their relation to an ultraviolet-absorption curve, and how the sign and magnitude of the Cotton effect are evaluated, are illustrated in Fig. 3.7. The compound studied contains two chromophores with absorption maxima at 217 and 293 nm, respectively. If we consider only the latter band, it is readily seen that this shows a positive Cotton effect ($a = +214$) and that the λ_0 value (290 nm) corresponds well to the CD-maximum and UV-absorption maximum. If the optical antipode of this compound had been investigated, the ORD and CD curves would have been completely inverted in sign along the x-axis.

It has been customary to distinguish between chromophores which are chiral, or inherently dissymmetric, and those which are inherently symmetric but asymmetrically perturbed, i.e. where an asymmetric environment makes the transition optically active. A wide variety of compounds containing groups of either kind have been studied by both theoretical and experimental methods. From considerations of the molecular geometry around the chromophore giving rise to the Cotton effect, certain rules for the prediction of positive or negative contributions to the sign of the Cotton effect have been established. It would be beyond the scope of this chapter to develop this field further and therefore the interested reader is referred to the treatments given by Djerassi [22], Crabbé [23] and other authorities on the subject [24, 25]. However, it is important to realize that while the optical rotation as usually measured at a long wavelength [589 nm (Na D-line) or 546 nm (Hg-line)] is of very little value for the determination of absolute configuration, the sign of the Cotton effect can be used either to correlate a compound of unknown stereochemistry with a structurally related one of known absolute configuration or sometimes also to directly determine the absolute configuration by making use of the rules (octant, sector, quadrant) established for the particular chromophore investigated.

The use of chromatographic methods for stereochemical correlation has only recently begun but shows great promise. As a more detailed treatment will be given in later chapters only the principle is outlined here.

For a limited number of cases the preferential adsorption of one of the two enantiomers on a chiral chromatographic stationary phase can be well rationalized at the molecular level and a 'chiral recognition' mechanism of some generality may be established. Therefore, for a series of structurally similar compounds, a correlation between elution order and configuration can be found. As long as the differences in retention are substantial, this method can be used with confidence and has the advantage of requiring only very small amounts of the compound. The technique resembles the correlation methods based on kinetic resolution developed by Horeau [26], also based on energy differences between diastereomeric states, which are reasonably well understood. It may be expected that with the aid of molecular modelling and calculation methods these techniques will gain even more importance in the future.

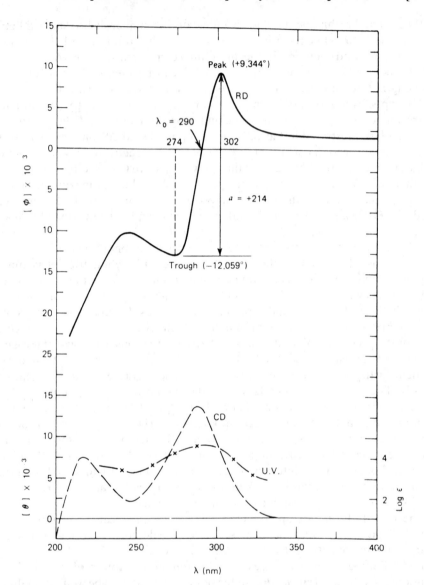

Fig. 3.8 — A comparison between the ORD, CD and UV spectra of an optically active compound. (Reprinted, with permission, from P. Crabbé and A. C. Parker, *Optical Rotatory Dispersion and Circular Dichroism*, in *Physical Methods of Chemistry*, A. Weissberger and B. W. Rossiter (eds.), Part IIIC, p. 183, Wiley–Interscience, New York, 1972. Copyright 1972, John Wiley & Sons Ltd.)

Exercises

1. Deduce the stereochemistry in the steps outlined below, where (2R,3R)-dimethyl tartrate is converted into trans-(2S,3S)-2,3-dimethyloxirane. What does the Cahn–Ingold–Prelog nomenclature tell you about the configurations in the

starting material as compared with the final product? Which are the stereochemically important steps? (Draw stereoprojection formulae.)

2. Reduction of (S)-2-chloropropanoic acid with lithium aluminium hydride gives the corresponding 2-chloropropanol, which in alkaline solution transforms into optically active methyloxirane. Which configuration can be expected in the product? Draw stereoprojection formulae and show which step is the stereochemically important one.

BIBLIOGRAPHY

P. Crabbé, Recent Applications of Optical Rotatory Dispersion and Optical Circular Dichroism in Organic Chemistry, *Top. Stereochem.*, 1967, **1**, 93.

P. Crabbé and A. C. Parker, Optical Rotatory Dispersion and Circular Dichroism, in *Techniques of Chemistry*, Vol. I, *Physical Methods of Chemistry*, A. Weissberger and B. W. Rossiter (eds.), Part IIIC, Wiley-Interscience, New York, 1972, p. 183.

Y. Izumi and A. Tai, *Stereodifferentiating Reactions*, Kodansha, Tokyo; Academic Press, New York, 1977.

H. B. Kagan (ed.), *Stereochemistry, Fundamentals and Methods*, Vols. 1–3, Thieme Verlag, Stuttgart, 1977.

W. Klyne and J. Buckingham, *Atlas of Stereochemistry*, 2nd Ed., Oxford University Press, Oxford, 1978.

W. H. Pirkle and D. J. Hoover, NMR Chiral Solvating Agents, *Top. Stereochem.*, 1982, **13**, 263.

J. D. Morrison (ed.), *Asymmetric Synthesis*, Vol. 1, *Analytical Methods*, Academic Press, New York, 1983.

REFERENCES

[1] J. B. Biot, *Mem. Acad. Sci. Inst. Fr.*, 1838, **15**, 93.

[2] G. Consiglio, P. Pino, L. I. Flowers and C. U. Pittman, Jr., *J. Chem. Soc., Chem. Commun.*, **1983**, 612.

[3] M. Raban and K. Mislow, *Top. Stereochem.*, 1967, **2**, 216.

[4] W. H. Pirkle, *J. Am. Chem. Soc.*, 1966, **88**, 1837.

[5] T. G. Burlingame and W. H. Pirkle, *J. Am. Chem. Soc.*, 1966, **88**, 4294.

[6] W. H. Pirkle and S. D. Beare, *J. Am. Chem. Soc.*, 1967, **89**, 5485.

[7] W. H. Pirkle and S. D. Beare, *J. Am. Chem. Soc.*, 1968, **90**, 6250.

[8] W. H. Pirkle and S. D. Beare, *J. Am. Chem. Soc.*, 1969, **91**, 5150.

[9] C. C. Hinckley, *J. Am. Chem. Soc.*, 1969, **91**, 5160.

[10] G. M. Whitesides and D. W. Lewis, *J. Am. Chem. Soc.*, 1970, **92**, 6979.

[11] I. Prigogine and R. Defay, *Chemical Thermodynamics*, 4th Ed., Longmans, London, 1967.

[12] J. P. Greenstein and M. Winitz, *The Chemistry of Amino Acids*, Vol. 2, Wiley, New York, 1961, p. 1738.

[13] S. Nishikawa and R. Matsukawa, *Proc. Imp. Acad. Japan*, 1928, **4**, 96.

[14] J. M. Bijvoet, *Proc. Koninkl. Ned. Wetenschap.*, 1949, **B52**, 313.

[15] J. M. Bijvoet, A. F. Peerdeman and A. J. van Bommel, *Nature*, 1951, **168**, 271.

[16] H. Hope and U. de la Camp, *Acta Crystallog.*, 1972, **A28**, 201.
[17] L. H. Pignolet, R. P. Taylor and W. DeW. Horrocks Jr., *Chem. Commun.*, **1968**, 1443.
[18] H. Akimoto, T. Shiori, Y. Iitaka and S. Yamada, *Tetrahedron Lett.*, **1968**, 3967.
[19] J. Tribout, R. H. Martin, M. Doyle and H. Wynberg, *Tetrahedron Lett.*, **1972**, 2839 (footnote).
[20] F. Vögtle, M. Palmer, E. Fritz, U. Lehmann, K. Meurer, A. Mannschreck, F. Kastner, H. Irngartinger, U. Huber-Patz, H. Puff and E. Friedrichs, *Chem. Ber.*, 1983, **116**, 3112.
[21] A. M. Cotton, *Ann. Chim. Phys.*, 1896, **8**, 347.
[22] C. Djerassi, *Optical Rotatory Dispersion: Applications to Organic Chemistry*, McGraw-Hill, New York, 1960.
[23] P. Crabbé, *Optical Rotatory Dispersion and Circular Dichroism in Organic Chemistry*, Holden-Day, San Francisco, 1965.
[24] L. Velluz, M. Legrand and M. Grosjean, *Optical Circular Dichroism: Principles, Measurements and Applications*, Verlag Chemie, Weinheim, 1965.
[25] G. Snatzke (ed.), *Optical Rotatory Dispersion and Circular Dichroism in Organic Chemistry*, Heyden, London, 1967.
[26] A. Horeau, *Tetrahedron*, 1975, **31**, 1307.

4

Modern chromatographic separation methods

4.1   A REVIEW OF BASIC CHROMATOGRAPHIC THEORY

Before entering into discussions concerning chiral chromatography it will be worthwhile to consider some of the basic concepts of chromatographic theory. The origin of chromatography dates back to 1900 when the Russian chemist Mikhail Tswett discovered that coloured bands were formed when plant extracts were passed through glass columns packed with calcium carbonate. However, it was not until the 1930s that research within the carotenoid field revealed the importance of these results [1]. Since then there has been an extensive development, and today chromatography is the most powerful and diversified separation technique in chemistry.

The concept of chromatography relies basically on the distribution of a compound between two phases, one of which (the mobile phase) is moving with respect to the other (the stationary phase). The various modes of chromatography depend on the respective nature of the two phases. The classification is shown in Fig. 4.1.

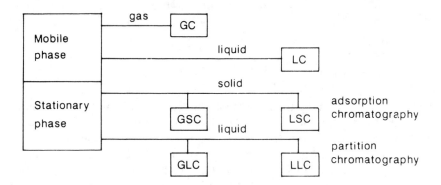

Fig. 4.1 — Origin of the different modes of chromatography.

The essential part of a chromatographic system is the column which contains the stationary phase over which the mobile phase flows and where the separation of the mixture into the individual components takes place. The sample is introduced onto the column by an injection device and the separated components are monitored by a suitable detection system.

Ideally, the profile of a chromatographic band, as registered by the detector, should have a Gaussian distribution, resulting in a completely symmetrical peak. Theoretically, this is the case when the sorption isotherm is linear, i.e. the phase distribution ratio is independent of concentration. In practice this is seldom the case, for a variety of reasons which will not be treated here.

As a chromatographic band moves along a column it is broadened by a number of dispersion effects. A detailed account of these phenomena is outside the scope of this book; readers who wish to enter more deeply into this area are referred to the textbooks listed at the end of this chapter. However, the fundamental equations used in chromatography, which are needed for an understanding of the relative merits of different modes of enantioseparation, will be reviewed.

A chromatographic column may be characterized by its efficiency, which is a measure of its ability to transport a compound with little peak broadening. Column efficiency is expressed as height equivalent to one theoretical plate, or plate height (H). H is readily calculated from a chromatogram (Fig. 4.2) by using Eq. (4.1).

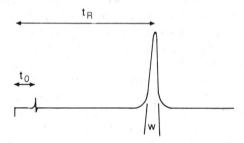

Fig. 4.2 — The retention and peak width parameters used for evaluation of column efficiency and plate height.

$$N=16(t_R/w)^2, \quad H=L/N \tag{4.1}$$

where t_R is the retention time, w the baseline peak width, and L the column length.

The column efficiency is consequently inversely proportional to the magnitude of H. Very low H-values are achieved in modern open tubular capillary columns for GC as well as in packed analytical LC-columns (typical values are ca. 0.025 mm). The GC capillaries are usually of such a length ($L=20$–200 m) that the plate number of the column may be extremely high ($N=2\times10^5$ or more is not uncommon). Such columns are also capable of accommodating a large number of chromatographic bands, i.e.

their peak capacity is high. This is not possible to achieve with packed analytical LC-columns since their lengths are usually less than 0.3 m; the particle size (3–10 μm diameter) of the packing materials results in back-pressures which preclude the use of longer columns. The situation has been improved, however, by the use of long open capillary columns in LC [2], but they have the disadvantage of very low flow-rates (often <1 μl/min) and consequently extremely long retention times, usually several hours.

The retention of a compound on a column can be expressed by its retention time (t_R), retention volume ($V_R=t_R F$ where F is the flow-rate) or capacity ratio (k'), which is directly related to its equilibrium distribution constant (K) in the stationary–mobile phase system. The capacity ratio is defined by:

$$k'=A_s/A_m \tag{4.2}$$

where A_s and A_m denote the amount of the compound in the stationary and mobile phase, respectively. Let V_s and V_m be the volumes of the respective phases (in adsorption chromatography V_s is often approximated by the surface area); then

$$k'=C_s V_s/C_m V_m = K V_s/V_m \tag{4.3}$$

V_m is commonly written V_0 and represents the dead or void volume in the column, which does not contribute to the separation. Consequently, the net retention volume, V_n, can be written as $V_n=V_R-V_0$. Since $K=V_n/V_s$, combination with Eq. (4.3) gives:

$$k'=(V_R-V_0)/V_0 \tag{4.4}$$

This expression permits determination of the capacity ratio directly from the chromatogram.

Let us now consider the chromatographic separation of two components (1 and 2) as illustrated by Fig. 4.3. It can be seen that $k_2'/k_1'=K_2/K_1=\alpha$, where α is the

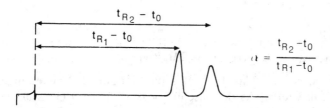

Fig. 4.3 — The separation factor (which directly relates to the different interaction of a pair of compounds with the stationary phase) and its evaluation from a chromatogram.

separation factor. Further, since $\Delta G = -RT \ln K$, an expression $\Delta\Delta G = RT \ln \alpha$ can be obtained. The separation factor, which is readily determined from the chromatogram, is consequently equivalent to an energy quantity which may be expressed in kJ/mole. We will return to this later.

From Eq. (4.4), α can be formulated as:

$$\alpha = (V_{R2} - V_0)/(V_{R1} - V_0) \tag{4.5}$$

Thus, the separation factor is simply the ratio of the net retention volumes of the two components. If V_0, the void volume, has been determined, the separation factor is easily calculated from Eq. (4.5).

It is important to recognize that the separation factor, α, is a measure of relative peak separation and is constant under given analytical conditions (stationary and mobile phase, temperature etc.). It is independent of factors that do not affect the equilibrium constants of the system, such as flow-rate, column dimensions, particle size etc. Such factors, on the other hand, are of fundamental importance for the column efficiency, N, defined in Eq. (4.1).

The resolution (R_s) of two peaks in a chromatogram is therefore dependent upon α as well as on N and can be easily evaluated from a chromatogram (Fig. 4.4)

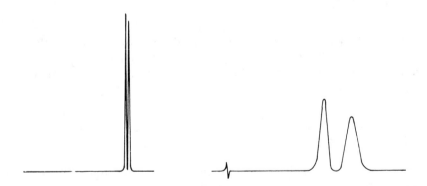

Fig. 4.4 — Resolution is a function of both column efficiency and the separation factor given by the stationary phase as illustrated by two different cases giving the same resolution. Left: A high column efficiency coupled with a small separation factor. Right: a low column efficiency coupled with a large separation factor.

according to:

$$R_s = 2(t_{R2} - t_{R1})/(w_1 + w_2) = 2\Delta t_R/(w_1 + w_2) \tag{4.6}$$

As shown, R_s is defined as the peak separation, divided by the mean value of the baseline peak widths. As the baseline peak width of a Gaussian band equals 4σ (where σ is the standard deviation), it follows that a Δt_R equal to 4σ corresponds to $R_s = 1.0$, equivalent to only 2% peak overlap. A '6σ resolution', i.e. $R_s = 1.5$, gives complete baseline separation, whereas $R_s < 0.8$ is generally considered insufficient.

Equation (4.6) can be transformed into an expression where R_s is a function of the parameters defined earlier:

$$R_s = \left(\frac{\alpha - 1}{4\alpha}\right)\left(\frac{k'_2}{1 + k'_2}\right)\sqrt{N} \tag{4.7}$$

It should be noted that for α-values close to 1, small changes in α will result in large changes in R_s. Accordingly, a change in α from 1.1 to 1.2 means roughly a doubling of R_s, owing to the influence on the $(\alpha - 1)/\alpha$ factor. Equation (4.7) is often simplified by the introduction of the concept N_{eff}, the effective number of theoretical plates. Because of the definition $N_{eff} = 16[(t_R - t_0)/w]^2$, cf. Eq. (4.1), it can be shown that Eq. (4.7) can be rearranged into the simpler form

$$R_s = \frac{\alpha - 1}{4\alpha}\sqrt{N_{eff}} \tag{4.8}$$

where the factor $k'_2/(1 + k'_2)$ is incorporated into N_{eff}: $N_{eff} = N[k'/(1 + k')]^2$. From graphical evaluation of Eq. (4.8) at given R_s values, the effective plate number required to obtain a certain resolution as a function of α can be found. An illustrative example is given in Fig. 4.5. Thus, to obtain a 6σ resolution ($R_s = 1.5$) an α-value of 1.05 will require 15700 effective theoretical plates, whereas the corresponding number needed for $\alpha = 1.15$ will be only 2110.

4.2 INSTRUMENTATION

The basic instrumental requirements for GC as well as LC are rather simple: a system for mobile phase delivery to the column, an injection device for application of the sample, a separation column and a system for detection of the separated components. In GC it is of course also essential to use a thermostatically controlled oven to contain the column, and separate temperature controls for the injection and detection units. The extraordinarily rapid development of GC and LC instruments, however, has almost made instrument technology a science of its own. It is not the purpose of this chapter to give any detailed account of modern instrumentation, but merely to concentrate upon what is essential in this context with respect to separation of enantiomers. For further details on instrumentation the reader is referred to the exhaustive treatment by Poole and Schuette (see Bibliography).

4.2.1 Gas chromatographic instrumentation

The principal components in a gas chromatographic system are outlined in Fig. 4.6. The mobile phase or carrier gas should be inert under the conditions used and its selection will be dependent mainly on the type of detector used. Nitrogen, hydrogen, helium or argon is preferred. Hydrogen has the lowest viscosity of all, which means that it is used with advantage in long capillary columns where relatively high flow-rates are required. This is sometimes of practical importance, e.g. when the

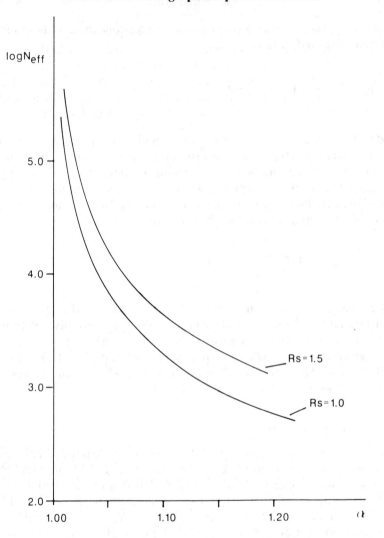

Fig. 4.5 — Plots of plate numbers *vs.* α-values at different constant resolution values. Note the
drastic increase in plate numbers needed to maintain resolution at very low α-values.

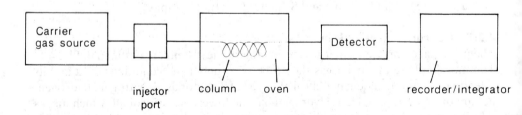

Fig. 4.6 — Essential features of a gas chromatograph.

temperature stability of the stationary phase is limited, since at a low column temperature retention times would otherwise be too long. The van Deemter curve, which relates the column plate height to the linear flow-rate of the mobile phase and from which the column efficiency can be optimized, is very different for hydrogen and nitrogen. As shown in Fig. 4.7 there is a much flatter minimum for hydrogen, which leads to considerably higher column efficiency at high flow-rates.

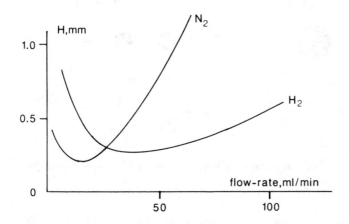

Fig. 4.7 — Difference in diffusion properties of two carrier gases as shown by the van Deemter curves.

The carrier gas enters the column through the injector block and leaves through the detector system. There is a wide variety of injection modes, particularly for capillary columns. Normally, however, the injector block is kept at a temperature ca. 50°C higher than the column to ensure a fast vaporization of the solute and immediate transfer to the column. The sample is usually introduced as a solution by means of a microsyringe, through a septum in the injector port.

Whereas sample injection onto packed columns usually presents no particular problems, the much smaller amounts that are necessary for capillary columns give rise to certain difficulties. One way to introduce a small, defined, amount onto a capillary column is to use a 'split' system, i.e. only a certain amount of the injected sample will reach the column. This amount is determined by the split ratio, which normally is in the range 1/20–1/200. Since split injection causes discrimination against higher boiling components, it is often not completely suitable for quantitative analysis. It should be remembered, however, that for a determination of enantiomer composition by the use of a *chiral* stationary phase, any type of sample introduction may be used. For obvious reasons the ratio of peak areas of the enantiomers will not be affected by the experimental conditions.

Other principal sample introduction techniques for capillaries are splitless injection, which is particularly useful for very dilute samples since it concentrates the sample at the column inlet, and on-column injection. In the latter technique the sample is introduced directly, by a special syringe needle, onto the column, without prior heating or admixture with carrier gas.

As already discussed (Section 4.1), capillary columns have considerable advantages over conventional packed columns in column efficiency (and plate height). To recall the main column modifications used so far in GC, consider Fig. 4.8, where

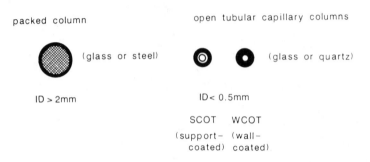

Fig. 4.8 — Different column modifications used in gas chromatography.

different columns are shown in cross-section. The packed column always contains the liquid stationary phase on an inert support, usually a diatomaceous earth, and is completely filled. Open columns, on the other hand, are characterized by an open space in the centre. They may contain support material (SCOT-columns, which are Support-Coated Open Tubular columns) or be directly wall-coated with the stationary phase (WCOT-columns). Most GC work today is performed with the latter type of column.

Column tubing is usually made from glass, metal or fused silica, the last of which was introduced in 1979 for capillaries. The nature of the inner wall surface of the tubing is critical for maintenance of the stationary phase. The inner wall surface can be coated with the stationary phase liquid by various techniques to give a desired film thickness which then determines the capacity and retention ability of the column. Normally, the film thickness is $0.1-0.3 \mu$m. When fused silica in particular is used, it is essential to 'immobilize' the stationary phase in some way, otherwise column bleeding will become a problem at higher temperatures. The various techniques used for this purpose are aimed at creating cross-links in the stationary phase polymer. Although the mechanisms of these reactions are not known in detail, it is likely that covalent bonding to the wall surface will take place to some extent. The main result is a column with very low bleeding at elevated temperatures, and high solvent resistance. Further, immobilization techniques make it possible to increase the film layer thickness considerably.

Since resolution is of primary importance for enantiomer separation by GC, WCOT-columns are by far of greatest interest. Consequently, details of the wall coating and temperature stability of the stationary phase are important and will be considered further in later chapters.

A variety of detectors have been developed for the monitoring of separated components in a GC column effluent. The most widely used belong to the ionization detector category. The principle applied is the measurement of change in electric conductivity caused by changes in ion currents generated in the detector. The flame ionization detector (FID) is the most universally used. It meets all requirements of a good GC detector, such as high sensitivity, very good stability, fast response (1 msec), low dead volume (1 μl) and a broad linear response range. It operates by means of combustion of the effluent components in a hydrogen flame, thereby generating positive ions which cause an increase in the conductivity of the applied electric circuit.

An even more sensitive and more selective ionization detector is the electron capture detector (ECD). As expected from the name, it operates through capture of electrons by the analytes, which means that certain demands are placed on the organic structure of these compounds. In ECD, the carrier gas molecules are ionized by a β-emitting radiation source. This ionization produces thermal electrons which generate a stable background current when subjected to a potential difference in the ECD-cell. When an electron-capturing compound is eluted from the column, it will diminish the background current and produce a signal on the recorder. The ECD can be said to operate in reverse mode to the FID.

The ECD, which was originally applied to high-sensitivity detection of halogenated hydrocarbons, has proved excellent for use in combination with derivatization of amines, amino-acids, hydroxy-acids and similar compounds. Halogenated, notably perfluorinated, acylating reagents are used to introduce EC-sensitive groups into amino- and hydroxy-compounds by the formation of volatile amides and esters. The sensitivity of the EC-detector is highly dependent on the analyte structure. A basic requirement is the ability of the compound to accommodate the negative charge produced by the electron capture. Accordingly, halogenated compounds, nitroaromatics, polycyclic aromatic hydrocarbons and conjugated carbonyl compounds are among the structures giving a high detector response.

4.2.2 Liquid chromatographic instrumentation

Since a liquid mobile phase is used, the solvent delivery system forms an important part of an LC instrument. As high-efficiency columns usually produce a significant back-pressure, a high-pressure pump must be used to force the mobile phase through the column at a controlled flow-rate. The common instrumental set-up is shown in Fig. 4.9. The sample is introduced as a solution by a syringe through an injection device — usually a loop system. Normally, packed columns of 4–5 mm internal diameter and 15–25 cm length are used for analytical work. Many different types of column packing materials have been used, but the majority today are based on a silica support. If the LC modes based on molecular sieving effects are excluded, LC separations are completely dominated by the use of bonded phase materials, i.e. those in which the stationary phase is integrated with the (silica) support by covalent bonding. The chemistry of ligand coupling to silica particle surfaces has developed rapidly and a multiplicity of bonded phases for use in ion-exchange, reversed-phase and affinity chromatography is now available. The immobilization chemistry will be dealt with in further detail in connection with the presentation of chiral stationary

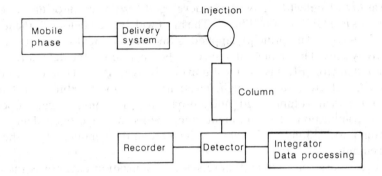

Fig. 4.9 — Main components of a liquid chromatograph.

phases for LC. In Table 4.1 some common structures of silica-based column packing materials are given.

For historical reasons LC has been characterized as performed in normal-phase or reversed-phase modes, the former being the original technique, where an organic mobile phase of lower polarity than the stationary phase is used (Section 4.1). With respect to the nature of the solute–stationary phase interaction, normal-phase LC has been further characterized as liquid–solid, i.e. adsorption chromatography, or liquid–liquid, i.e. partition chromatography, depending on whether the solid surface or adsorbed water (or some other polar liquid) should be regarded as the stationary phase. Such distinctions are often hard to make, however, and borderline cases are quite common.

Underivatized silica and alumina are the usual adsorbents in normal-phase LC. The operation of polar adsorption sites in the retention process is indicated by the often relatively large separation factor shown for geometric isomers.

In reversed-phase LC the stationary phase is by definition less polar than the mobile phase, which is usually an aqueous solution. This has the advantage that quite a number of mobile phase parameters can be varied in order to influence and regulate retention. Thus, in aqueous mobile phase systems variations can be made in pH, buffer types and strength, added salt, organic modifier, etc. This gives many interesting possibilities in connection with chiral separations, as we shall see later. An attempt to summarize the principal modes of LC is given in Scheme 4.1.

The species eluted can be detected by a variety of techniques, but four types of detector are the most common. The refractive index (RI) detector is universal as it measures a bulk property, but has rather low sensitivity. The UV-detector is the standard detector in LC, equivalent to the FID in GC. It can be used to monitor most LC separations, but the sensitivity obtained is highly dependent on the structure of the solute. Particularly with the variable wavelength version, covering the range 190–350 nm, there are generally excellent possibilities for obtaining good chromatograms. Recently, so-called diode-array detectors have been introduced, by means of which it is possible to obtain a complete UV-spectrum from any part in the chromatogram. The technique is especially valuable as a means of establishing the identity and also the homogeneity of peaks, when peak overlap is suspected.

Table 4.1 — Various types of functionalized silica used as sorbents in liquid chromatography

Terminal (adsorbing) function	Type of chromatography
Alkyl Phenyl	Reversed-phase
Nitrile Ether Nitro Diol	Normal-phase or reversed-phase
Amino Dimethylamino	Normal- or reversed-phase, anion-exchange
Quaternary ammonium	Anion-exchange
Sulphonic acid	Cation-exchange

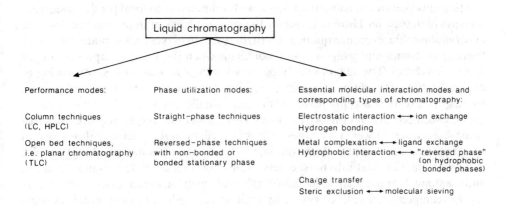

Scheme 4.1 — Basic modes of liquid chromatography.

Fluorescence detectors are highly selective and often extremely sensitive. Besides detection of naturally fluorescent compounds, use can be made of derivatives containing a fluorophore, prepared by either pre-column or post-column reactions.

Further, electrochemical detectors have become increasingly important as a highly selective and sensitive means of monitoring compounds which are susceptible to electro-oxidation or -reduction. They can only be used with conductive aqueous mobile phases and are therefore best suited to reversed-phase or ion-exchange chromatography.

Finally, a word about planar chromatography, another important mode of liquid chromatography (see Scheme 4.1). Today thin-layer chromatography (TLC) is the most widely used method of this kind, having developed from the earlier often practised paper chromatography. Since TLC is an open-bed liquid chromatography technique, it has obvious advantages related to its simplicity of performance. One of these is the widely used possibility of running several samples simultaneously. This is very convenient, not only from the point of view of increased sample throughput, but also since the positions of spots can be easily compared with those found from reference samples run in parallel. Further, since detection of the separated components is not done during the chromatographic process, the separation requires only quite simple equipment. Chromatographic development is, unlike that in column chromatography, a relatively fast process, since it is independent of the migration velocity of the components. Thus, large t_R-values are converted into small migration distances on the plate. Accordingly, the high sample throughput rate makes TLC an ideal technique for routine screening work. It is also excellent for rapidly checking the purity of a compound or for following the progress of a synthetic reaction. Retention is expressed as R_f-values, defined as the migration distance of the compound divided by the migration distance of the solvent front. Thus, $R_f = 1$ and 0 correspond to $k' = 0$ and infinity, respectively.

Detection in TLC offers a multitude of various possibilities. There exist numerous reagents that can react with the separated components to form products suitable for optical detection. However, one of the simplest ways of detection involves the observation of fluorescence quenching. In this case, the layer on the plate contains a fluorescent compound which emits visible light when the plate is exposed to light from a UV lamp. The separated compounds are then seen as dark spots, owing to their reduction of the fluorescence intensity. The compounds must, however, possess some UV-absorbance at the wavelength used (usually 254 nm).

Quantitative evaluation of thin-layer chromatograms requires more sophisticated techniques, however, and these are not as precise and sensitive as those used in LC. This results, of course, from the fact that the analyte is adsorbed on the solid matrix on the TLC plate. In most cases, a reflectance mode of detection is required and scanning over the plate then yields a relatively high background noise level. This can be compensated for, however, by dual wavelength operation which permits recording of the *difference* between the signals at two wavelengths (one of these is chosen outside the absorption band of the analyte and will give only the scattering from the background). Subtraction of the recorded signals therefore gives a significantly lower background. In Fig. 4.10 the optical arrangements of two common types of scanning densitometers are shown.

4.3 SEPARATION OF ENANTIOMERS BY MEANS OF COVALENT DIASTEREOMERS — A SURVEY

In Section 3.1.2 it was noted that enantiomers can be separated as the diastereomeric derivatives produced by reaction with an optically active reagent. Because of their different physical and chemical properties, such derivatives can be separated by all commonly used chromatographic techniques. Often such methods are simple to apply, particularly in GC if derivatization is necessary anyway. There are, however, some drawbacks which necessitate very careful interpretations of the results.

First, it is of utmost importance to know the optical purity of the derivatization reagent. Only if this is 100% will the analytical results be directly representative of

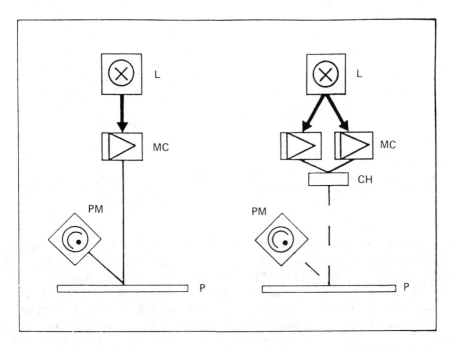

Fig. 4.10 — Examples of arrangements of optical scanning densitometers for TLC plates. (*a*)
Single wavelength (*b*) dual wavelength configuration. (Reproduced from D. C. Fenimore and
C. M. Davis, *Anal. Chem.*, 1981, **53**, 252A, with permission. Copyright 1981, American
Chemical Society.)

the enantiomer composition. The reason for this is shown in Scheme 4.2. If the
enantiomers of A are to be separated and determined as their diastereomeric
derivatives (I and II), obtained by reaction with (+)-B, any contamination with (−)-
B will be deleterious. The products (III and IV) produced with (−)-B form
enantiomeric pairs with the two main products (IV with I and III with II) and are
therefore added to the corresponding peaks. The possible effect of this is best shown
by an example. Let us assume that A consists of the (+)-form in 99% optical purity.
The reagent (+)-B is assumed to be 97% optically pure. By definition (see Section
3.1.1), there will be 99.5% of (+)-A, 0.5% of (−)-A, 98.5% of (+)-B and 1.5% of
(−)-B present in the reaction mixture. Assuming complete reaction, the proportions
of I–IV will then be: 98.0075%, 0.4925%, 1.4925% and 0.0075% respectively. Since
I and IV are superimposed in the chromatogram, as well as II and III, the two peaks
will show the proportions 98.015% and 1.985%, respectively. Therefore, if the
optical purity of the derivatization reagent is not taken into consideration, A will be
found to contain 98% of (+)-form, which corresponds to an optical purity of 96%, a
considerable error relative to the true value.

Secondly, the quantitative analysis relies upon the assumption that the reactions
are complete. If this is not the case, differences in product yields may result in large
errors. It is also very important to be sure that the reactions are not associated with
racemization or epimerization. Compounds containing asymmetric carbon centres

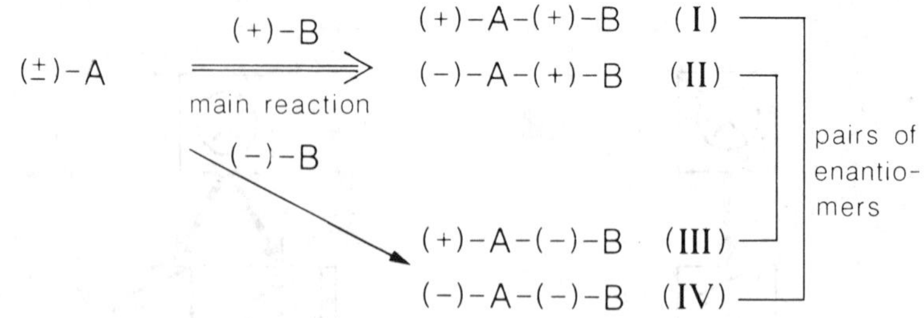

$$(\pm)-A \xrightarrow[\text{main reaction}]{\substack{(+)-B \\[1em] (-)-B}} \begin{array}{l} (+)-A-(+)-B \quad (\text{I}) \\[1em] (-)-A-(+)-B \quad (\text{II}) \\[3em] (+)-A-(-)-B \quad (\text{III}) \\[1em] (-)-A-(-)-B \quad (\text{IV}) \end{array} \left. \begin{array}{l} \\ \\ \\ \\ \end{array} \right\} \begin{array}{l} \text{pairs of} \\ \text{enantio-} \\ \text{mers} \end{array}$$

Scheme 4.2—Illustration of the effect of the use of a chiral derivatization reagent with less than 100% optical purity. Because enantiomers cannot be separated on a non-chiral stationary phase, products I and IV give only one peak, as do II and III.

which could change configuration through carbanion or carbonium ion intermediates have to be treated particularly carefully as they may not possess the necessary configurational stability under conditions that are too basic or too acidic.

Another factor of primary importance in chiral derivatization for the separation of diastereomers is the distance between the two chiral centres in the derivatives. As a general rule the centres should be as close as possible to each other in order to maximize the difference in chromatographic properties. Generally, three bonds separate the two centres and derivatives having distances exceeding four bonds are not often useful.

For a detailed and comprehensive treatment of the chromatographic separation of diastereomers, the reader is referred to the recent book by Souter (see Bibliography).

4.3.1 Gas chromatography

It was shown as early as 1960 that it was possible to separate the L-alanyl derivatives of D,L-phenylalanine by GC as their N-trifluoroacetylated methyl esters [3]. This led to the use of amino-acid derivatives as chiral N-derivatizing agents for the GC resolution of amino-acids as diastereomeric dipeptide derivatives. A commonly used reagent is N-trifluoroacetyl-L-prolyl chloride, which may be prepared in high optical purity from L-proline, thanks to the resistance of this cyclic amino-acid to racemization. Other chiral reagents for the derivatization of the amino group in amino-acids are various 2-chloroacyl chlorides. These compounds can also be prepared from their corresponding amino-acids.

For chiral derivatization of the carboxyl group, various optically active secondary alcohols have been used, such as 2-butanol or higher homologues. (−)-Menthol has also been used for the preparation of diastereomeric esters prior to GC [4].

Hydroxy functions, as in alcohols or hydroxy-acids, are readily derivatized with (−)-menthyl chloroformate [5] to give diastereomeric esters or with (+)- or (−)-phenylethyl isothiocyanate [6] to yield carbamates.

A number of other possibilities for chiral derivatization exist which have been found useful. A summary is given in Table 4.2, together with references to relevant original articles.

Table 4.2 — Chiral derivatization reactions used in gas chromatography

Functional group transformation	Chiral derivatization reagent	Product structure	Type of analyte	References
$-NH_2 \longrightarrow \overset{\text{O}}{\overset{\|}{-N}}=\overset{*}{C}-R$ (with H on N)	2-Chloroisovaleryl chloride	$-N-\overset{O}{\overset{\|}{C}}-\overset{Cl}{\overset{\|}{CH}}CHMe_2$ (H on N)	Amino-acids Amines	[7–9] [10]
	Drimanoyl chloride	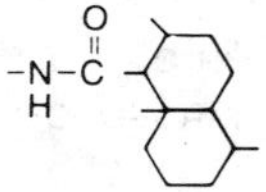	Amino-acids Amines	[11] [11]
	Chrysanthemoyl chloride	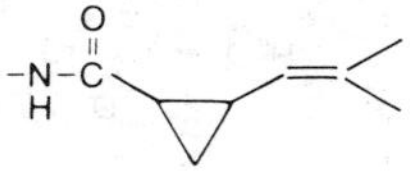	Amino-acids Amines	
	N-TFA-Prolyl chloride	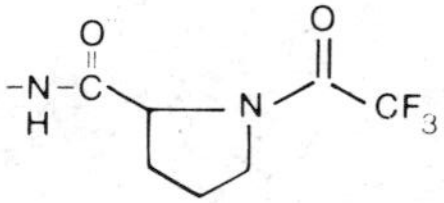	Amino-acids Amines	[10, 12, 13] [14–16]
	N-(Pentafluorobenzoyl)prolyl chloride	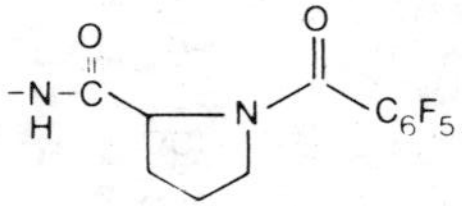	Amines	[17, 18]
	N-TFA-alanyl chloride and homologues (R_1=CH$_3$, CHMe$_2$, CH$_2$CHMe$_2$, C$_6$H$_5$)	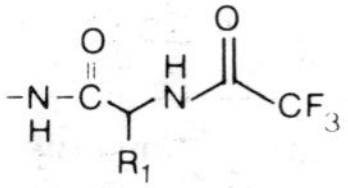	Amines	[16]

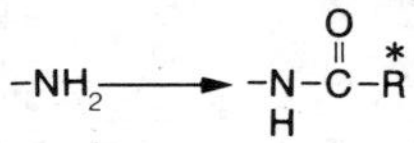

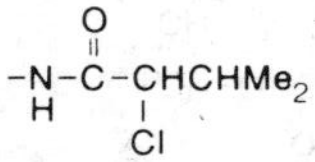

Table 4.2 (*contd.*)

Functional group transformation	Chiral derivatization reagent	Product structure	Type of analyte	References
$-NH_2 \longrightarrow \overset{\overset{O}{\parallel}}{-N}-\overset{*}{C}-OR$ (H)	(S)-2-Methoxy-2-trifluoromethylphenylacetyl chloride (MTPA-Cl)	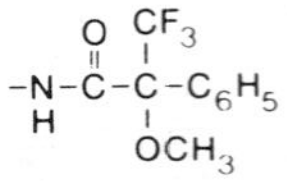 $\overset{\;CF_3}{\underset{\;OCH_3}{-N-C-C-C_6H_5}}$ (O, H)	Amines	[19–21]
	(−)-Menthyl chloroformate	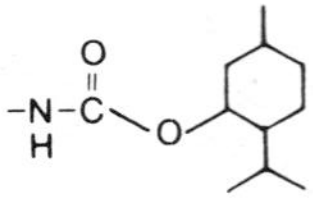 (menthyl chloroformate structure) $O=C-O$, $N-H$	Amino-acids Amines	[22] [22]
$-OH \longrightarrow \overset{\overset{O}{\parallel}}{-O}-\overset{*}{C}-R$	2-Phenylpropionyl chloride	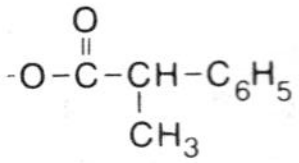 $O=C-CH-C_6H_5$, CH_3	Hydroxy-acids Alcohols	[23] [23]
	2-Phenylbutyryl chloride	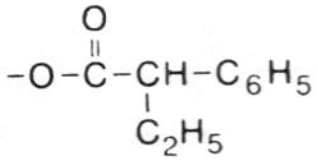 $-O-C-CH-C_6H_5$, O, C_2H_5	Alcohols	[24–26]
	O-Acetyllactic acid chloride	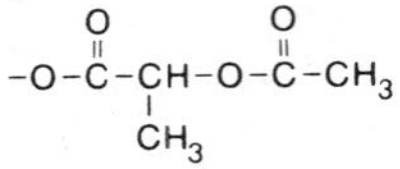 $-O-C-CH-O-C-CH_3$, O, CH_3, O	Alcohols	[27–29]
	Drimanoyl chloride	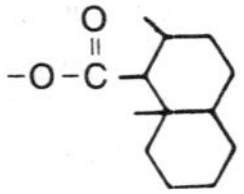 (drimanoyl chloride structure) $O=C$, $-O-C$	Hydroxy-acids	[11]

Table 4.2 (*contd.*)

Functional group transformation	Chiral derivatization reagent	Product structure	Type of analyte	References
$-OH \longrightarrow -O-\overset{\displaystyle O}{\overset{\|}{C}}-\overset{*}{O}R$	Chrysanthemoyl chloride	$-O-\overset{O}{\overset{\|}{C}}$ ...	Hydroxy-acids Alcohols	[11] [11]
	MTPA-Cl	$-O-\overset{O}{\overset{\|}{C}}-\overset{CF_3}{\underset{OCH_3}{C}}-C_6H_5$	Amphetamines (OH)	[19–21]
$-OH \longrightarrow -O-\overset{\displaystyle O}{\overset{\|}{C}}-\overset{*}{\underset{H}{N}}-R$	(−)-Menthyl chloroformate	$-O-\overset{O}{\overset{\|}{C}}-O$ ...	Hydroxy-acids Alcohols	[22] [22, 30–32]
	1-Phenylethyl isocyanate	$-O-\overset{H}{\overset{\|}{C}}-\underset{O}{\overset{\|}{N}}-CH-C_6H_5$, CH_3	Hydroxy-acids	[6, 33]
$-CO_2H \longrightarrow -\overset{*}{C}O_2R$	2-Alkanols ($R_1=$n-C_2–C_6, Me_2CH, Me_3C)	$-\overset{O}{\overset{\|}{C}}-O-\overset{R_1}{\underset{}{CH}}-CH_3$	Amino-acids α-Hydroxy-acids Keto-acids	[27, 28, 34–37] [29, 38–40] [41]

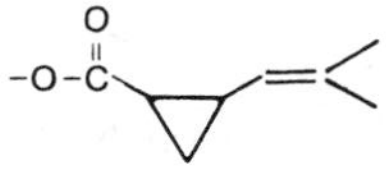

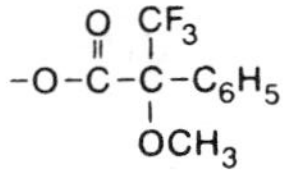

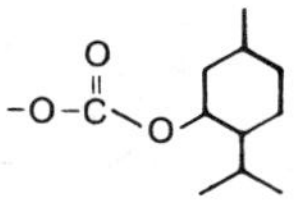

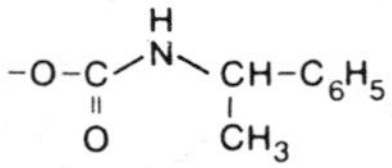

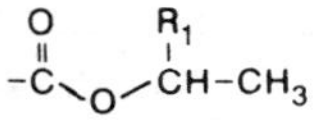

Table 4.2 (*contd.*)

Functional group transformation	Chiral derivatization reagent	Product structure	Type of analyte	References
$-CO_2H \longrightarrow$ $\begin{array}{c} O \\ \parallel \\ =C-N-R \\ \quad\; \mid \\ \quad\; H \end{array}$ (*)	(−)-Menthol	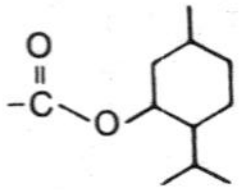 (menthyl ester) $\begin{array}{c} O \\ \parallel \\ -C-O- \end{array}$	Amino-acids α-Hydroxy-acids Other acids	[4, 42–44] [45] [46–51]
	(+)-4-Methyl-2-pentylamine	$\begin{array}{c} CH_3 \\ \mid \\ O=C-N-CH-CH_2-CHMe_2 \\ \quad\;\; \mid \\ \quad\;\; H \end{array}$	Amino-acids	[52]
	Amino-acid methyl esters	$\begin{array}{c} R \\ \mid \\ O=C-N-CH-CO_2\,Me \\ \quad\;\; \mid \\ \quad\;\; H \end{array}$	Amino-acids	[3, 53]
$>C=O \longrightarrow\; >C=N-N-R$ $\qquad\qquad\qquad\quad\; \mid$ $\qquad\qquad\qquad\quad\; H$ (*)	(+)-2,2,2-Trifluoro-1-phenylethylhydrazine	$\begin{array}{c} CF_3 \\ \mid \\ >C=N-N-CH-C_6H_5 \\ \qquad\;\; \mid \\ \qquad\;\; H \end{array}$	Ketones	[54]
$>C=O \longrightarrow\; >C=N-O-R$ (*)	O-(−)-Menthylhydroxylamine	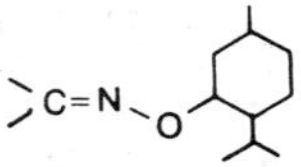 $>C=N-O-$ (menthyl)	Carbohydrates	[55]

4.3.2 Liquid chromatography

Much of the early work in this area was, as in the case of GC, centred on amino-acids, particularly stereochemical analysis in peptide synthesis. A very comprehensive investigation of the separation of diastereomeric dipeptides on cation-exchangers was performed by Manning and Moore [56, 57]. *N*-Carboxy-anhydrides of various amino-acids were used as chiral derivatization agents. The technique was later modified by the use of BOC-L-amino acid *N*-hydroxysuccinimide esters, followed by removal of the BOC group with trifluoroacetic acid [58]. It was found that on reversed-phase LC of such dipeptides, the D,D- or L,L- forms were eluted before their D,L- or L,D- diastereomers. NMR investigations showed that in their most stable conformation only the latter have the hydrophobic side-chains *cis*-oriented. It was assumed that such an orientation would yield a more effective interaction with the hydrophobic stationary phase [59,60]. Other very useful chiral derivatization re-agents for amino-acids and RP-LC include the isothiocyanates of tetra-acetylglucose (GITC) and of triacetylarabinose (AITC) [61, 62]. The latter are also suitable for amines [63] and amino-alcohols [64]. Optically active 1-phenylethyl isocyanate [65] has also been found useful for RP-LC of chiral amines. It has also been used, as has the 1-naphthyl analogue, for derivatization of alcohols to diastereomeric carbamates which may be separated by normal-phase LC [66, 67].

A variety of diastereomeric amide and ester derivatives of chiral carboxylic acids have been used and methyl esters of L-amino-acids, such as L-phenylalanine [68] or L-phenylglycine [69], are readily available chiral reagents.

A useful strategy for design of chiral derivatizing agents for liquid chromato-graphy has been to try to combine excellent detection properties with high reactivity and a stereogenic centre as close as possible to the reaction site. For this purpose, it is often favourable to start from a simple, achiral, bifunctional reagent that can easily be transformed into a monosubstituted, chiral reagent. Two reagents, both particu-larly useful for chiral derivatization of amino acids and amines, may exemplify this strategy.

By reacting the symmetric 5-chloro-2,4-dinitrochlorobenzene with L- (or D-) alanine amide, Marfey [70] prepared optically active *N*-(5-chloro-2,4-dinitropheny-l)alanine amide (Marfey's reagent) which still has a reactive site for nucleophilic substitution (the 5-position) in the aromatic ring. Since the dinitro-substituents create favourable UV-absorption properties and the leucine amide chirality causes significant differences in the chromatographic properties of the two diastereomers formed from a pair of analyte enantiomers, the desired properties of the reagent are achieved. This reagent can be regarded as an analogue to Sanger's classic amino acid reagent, but having the additional bonus of also permitting separation of the amino acid enantiomers.

A nice example of diastereomer separation after use of this reagent is shown in Fig. 4.11. Note the large separation factor obtained, despite the long distance between the two stereogenic centres in the analyte. The rigid structural part interconnecting the centres is likely to be an important factor here.

A second variation on the theme is the FLEC reagent [(+)- or (−)-1-(9-fluorenyl)ethyl chloroformate] which was launched in 1987 [71,72]. This reagent is a chiral analogue of the fluorogenic amino acid reagent FMOC-Cl (9-fluorenylmethyl

Table 4.3 — Chiral derivatization reactions used in liquid chromatography

Functional group transformation	Chiral derivatization reagent	Product structure	Type of analyte	References
$-NH_2 \longrightarrow -\overset{\overset{\displaystyle O}{\|\|}}{N}-\overset{*}{C}-R$	N-Carboxy-anhydrides or BOC-derivatives of L-amino acids	$-\underset{H}{N}-\overset{\overset{\displaystyle O}{\|\|}}{C}-\underset{R}{CH}-NH_2$	Amino-acids	[56–60]
	N-TFA-prolyl chloride	$-\underset{H}{N}-\overset{\overset{\displaystyle O}{\|\|}}{C}$ (pyrrolidine ring) $N-\overset{\overset{\displaystyle O}{\|\|}}{C}-CF_3$	Amino-alcohols	[73, 74]
	(S)-O-Methylmandelic acid chloride	$-\underset{H}{N}-\overset{\overset{\displaystyle O}{\|\|}}{C}-\underset{OCH_3}{CH}-C_6H_5$		[75]
	(−)-1-Methoxy-1-methyl-1-(1-naphthyl)acetic acid	$-\underset{H}{N}-\overset{\overset{\displaystyle O}{\|\|}}{C}-\overset{\overset{\displaystyle CH_3}{\|}}{\underset{OCH_3}{C}}$ (naphthyl)		[76, 77]
	(−)-1-Methoxy-1-methyl-1-(2-naphthyl)acetic acid	$-\underset{H}{N}-\overset{\overset{\displaystyle O}{\|\|}}{C}-\overset{\overset{\displaystyle CH_3}{\|}}{\underset{OCH_3}{C}}$ (naphthyl)	Amino-acid esters	
$-NH_2 \longrightarrow -\overset{\overset{\displaystyle S}{\|\|}}{\underset{H}{N}}-\overset{*}{\underset{H}{N}}-R$	2,3,4,6-Tetra-O-acetyl-β-D-glucopyranosyl isothiocyanate (GITC)		Amino-acids Amino-alcohols	[61, 62] [63, 64]

Table 4.3 (*contd.*)

Functional group transformation	Chiral derivatization reagent	Product structure	Type of analyte	References
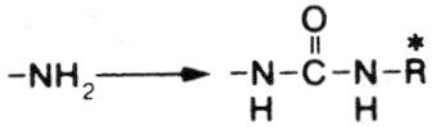 $-NH_2 \longrightarrow -N-\underset{H}{\overset{O=C-N-R}{\overset{\ast}{}}} $	2,3,4-Tri-O-acetyl-β-D-arabino-furanosyl isothiocyanate (AITC)		Ephedrines Amphetamines	[78] [79]
	1-Phenylethyl isocyanate	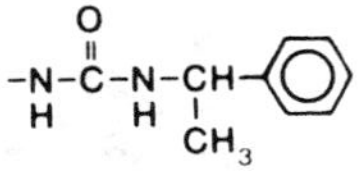 $O=C-N-\overset{\ast}{C}H-\underset{CH_3}{\overset{}{}}$	Amines	[66]
	1-(1-Naphthyl)ethyl isocyanate	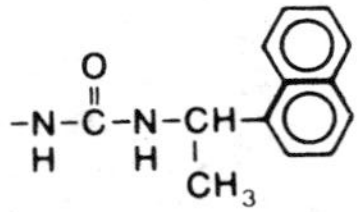 $O=C-N-\overset{\ast}{C}H-\underset{CH_3}{\overset{}{}}$		
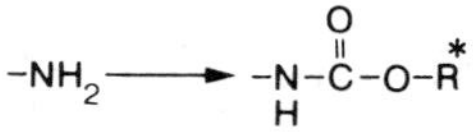 $-NH_2 \longrightarrow -N-\underset{H}{\overset{O=C-O-\overset{\ast}{R}}{}}$	1-(9-Fluorenyl)ethyl chloroformate (FLEC)	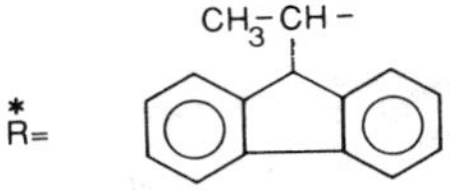 $\ast R = CH_3-\overset{\ast}{C}H-$ (fluorenyl)	Amines, amino acids, alcohols	[71,72,75]
 $-NH_2 \longrightarrow -N-\overset{\ast}{Ar}$ (H)	N-(5-Chloro-2,4-dinitrophenyl)-L-alanine amide	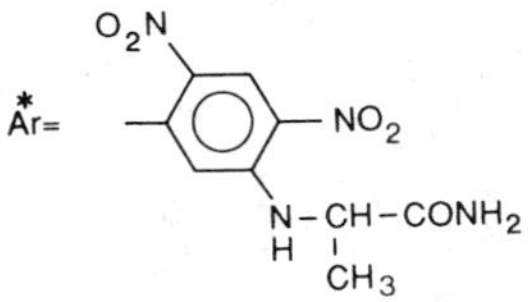 $\ast Ar = $ (2,4-dinitrophenyl)–N-CH-CONH₂, CH₃	Amines, amino acids	[70]
 $-OH \longrightarrow -O-\overset{\ast}{C}-\overset{\ast}{R}$ (O=C)	MTPA-Cl	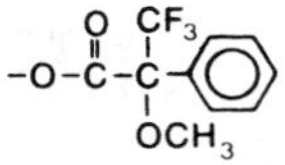 $-O-C(=O)-\overset{CF_3}{\underset{OCH_3}{C}}-$	Terpenoid alcohols	[80–82]

Table 4.3 (*contd.*)

Functional group transformation	Chiral derivatization reagent	Product structure	Type of analyte	References
$-OH \longrightarrow -O-\overset{O}{\overset{\|}{C}}-\overset{H}{\overset{\|}{N}}-\overset{*}{R}$	1-Phenylethyl isocyanate	$-O-\overset{O}{\overset{\|}{C}}-\overset{H}{\overset{\|}{N}}-\overset{CH-}{\underset{CH_3}{\|}}$ (phenyl)		[66, 83]
	1-(1-Naphthyl)ethyl isocyanate	$-O-\overset{O}{\overset{\|}{C}}-\overset{H}{\overset{\|}{N}}-\overset{CH-}{\underset{CH_3}{\|}}$ (naphthyl)		[83]
$-CO_2H \longrightarrow -\overset{O}{\overset{\|}{C}}-\overset{H}{\overset{\|}{N}}-\overset{*}{R}$	1-Phenylethylamine	$-\overset{O}{\overset{\|}{C}}-\overset{H}{\overset{\|}{N}}-\overset{CH-}{\underset{CH_3}{\|}}$ (phenyl)	Carboxylic acids	[84]
	1-(1-Naphthyl)ethylamine	$-\overset{O}{\overset{\|}{C}}-\overset{H}{\overset{\|}{N}}-\overset{CH-}{\underset{CH_3}{\|}}$ (naphthyl)	Carboxylic acids	[85]
	1-(4-Nitrophenyl)ethylamine	$-\overset{O}{\overset{\|}{C}}-\overset{H}{\overset{\|}{N}}-\overset{CH-}{\underset{CH_3}{\|}}$ (4-nitrophenyl, NO_2)	Carboxylic acids	[86]
	Methyl phenylalaninate	$-\overset{O}{\overset{\|}{C}}-\overset{H}{\overset{\|}{N}}-\overset{CH-CO_2CH_3}{\underset{CH_2C_6H_5}{\|}}$	Carboxylic acids	[87]

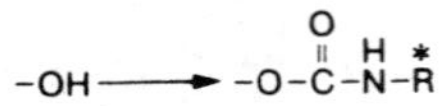

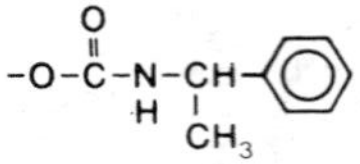

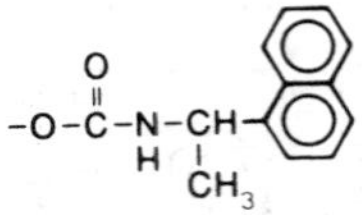

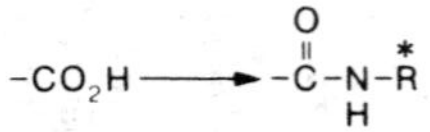

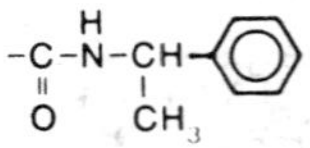

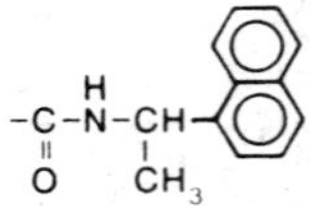

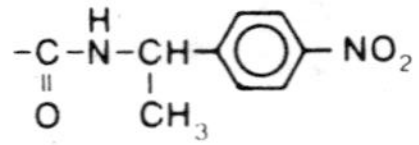

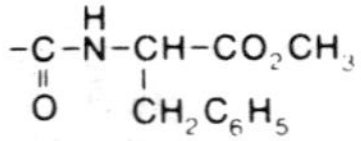

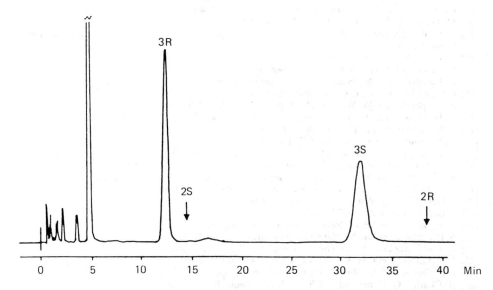

Fig. 4.11 — Separation of (R)- and (S)-β-leucine, after derivatization with Marfey's reagent, by liquid chromatography on an octadecylsilica column. UV detection at 340 nm. Peak positions for the corresponding derivatives of α-leucine are also indicated. (Reprinted, with permission, from D. J. Aberhart, J.-A. Cotting and H.-J. Lin, *Anal. Biochem.*, 1985, **151**, 88. Copyright 1985, American Chemical Society.)

chloroformate). Like the latter, it is obtained by reaction of phosgene with an aromatic alcohol to substitute only one of the chlorine atoms. FLEC reacts rapidly at room temperature with amines and amino acids and can also be used for alcohols and phenols.

A summary of the most useful chiral derivatization techniques in liquid chromatography is given in Table 4.3.

BIBLIOGRAPHY

E. Gil-Av and D. Nurok, Resolution of Optical Isomers by Gas Chromatography of Diastereoisomers, *Adv. Chromatog.*, 1972, **10**, 99.
L. R. Snyder and J. J. Kirkland, *Introduction to Modern Liquid Chromatography*, Wiley, New York, 1974.
K. Blau and G. S. King, *Handbook of Derivatives for Chromatography*, Heyden, London, 1978.
C. F. Poole and S. A. Schuette, *Contemporary Practice of Chromatography*, Elsevier, Amsterdam, 1984.
R. W. Souter, *Chromatographic Separation of Stereoisomers*, CRC Press, Boca Raton, 1985.
M. Ahnoff and S. Einarsson, Chiral Derivatization, in *Chiral Liquid Chromatography*, W. J. Lough (ed.), Blackie, Glasgow, 1989, p. 37.

REFERENCES

[1] L. S. Ettre and A. Zlatkis (eds.), *75 Years of Chromatography. A Historical Dialogue*, Elsevier, Amsterdam, 1979.
[2] M. Novotny, *Anal. Chem.*, 1981, **53**, 1294A.
[3] F. Weygand, B. Kolb, A. Prox, M. Tilak and I. Tomida, *Z. Physiol. Chem.*, 1960, **322**, 38.
[4] B. Halpern and J. W. Westley, *Chem. Commun.*, 1965, 421.

[5] R. G. Annett and P. K. Stumpf, *Anal. Biochem.*, 1972, **43**, 515.
[6] M. Hamberg, *Chem. Phys. Lipids*, 1971, **6**, 152.
[7] J. W. Westley and B. Halpern, in *Gas Chromatography*, C. L. A. Harbourn (ed.), Inst. of Petrol., London 1968, p. 119.
[8] B. Halpern and J. W. Westley, *Chem. Commun.*, 1965, 246.
[9] S. Lande and R. A. Landowne, *Tetrahedron*, 1966, 3085.
[10] B. Halpern and J. W. Westley, *Chem. Commun.*, 1966, 34.
[11] C. J. W. Brooks, M. T. Gilbert and J. Gilbert, *Anal. Chem.* 1973, **45**, 896.
[12] B. Halpern and J. W. Westley, *Biochem. Biophys. Res. Commun.*, 1965, **19**, 361.
[13] B. Halpern and J. W. Westley, *Tetrahedron Lett.*, 1966, 2283.
[14] J. W. Westley and B. Halpern, *Anal. Chem.*, 1968, **40**, 2046.
[15] R. W. Souter, *J. Chromatog.*, 1975, **108**, 265.
[16] B. L. Karger, R. L. Stern and W. Keane, *Anal. Chem.*, 1967, **39**, 228.
[17] S. B. Matin, M. Rowland and N. Castagnoli, Jr., *J. Pharm. Sci.*, 1973, **62**, 821.
[18] S. B. Matin, S. H. Wan and J. B. Knight, *Biomed. Mass. Spectrom.*, 1973, **4**, 118.
[19] J. A. Dale, D. L. Dull and H. S. Mosher, *J. Org. Chem.*, 1969, **34**, 2543.
[20] J. Gal and M. M. Ames, *Anal. Biochem.*, 1977, **83**, 266.
[21] J. Gal, *J. Pharm. Sci.*, 1977, **66**, 169.
[22] J. W. Westley and B. Halpern, *J. Org. Chem.*, 1968, **33**, 3978.
[23] S. Hammarström and M. Hamberg, *Anal. Biochem.*, 1973, **52**, 169.
[24] C. J. W. Brooks and J. D. Gilbert, *Chem. Commun.*, 1973, 194.
[25] J. D. Gilbert and C. J. W. Brooks, *Anal. Lett.*, 1973, **6**, 639.
[26] J. P. Guette and A. Horeau, *Tetrahedron Lett.*, 1965, 3049.
[27] R. Charles, G. Fisher and E. Gil-Av, *Isr. J. Chem.*, 1963, **1**, 234.
[28] E. Gil-Av, R. Charles and G. Fisher, *J. Chromatog.*, 1965, **17**, 408.
[29] S. Julia and J. M. Sans, *J. Chromatog. Sci.*, 1979, **17**, 651.
[30] S. Hammarström, *FEBS Lett.*, 1969, **5**, 192.
[31] M. Hamberg, *Anal. Biochem.*, 1971, **43**, 515.
[32] R. G. Annett and P. K. Stumpf, *Anal. Biochem.*, 1972, **47**, 638.
[33] W. Pereira, *Anal. Lett.*, 1970, **3**, 23.
[34] M. E. Bailey and H. B. Hass, *J. Am. Chem. Soc.*, 1941, **63**, 1969.
[35] G. E. Pollock and V. I. Oyama, *J. Gas Chromatog.*, 1966, **4**, 126.
[36] G. E. Pollock and A. H. Kawauchi, *Anal. Chem.*, 1968, **40**, 1356.
[37] F. Raulin and B. N. Khare, *J. Chromatog.*, 1973, **75**, 13.
[38] E. Gil-Av and D. Nurok, *Proc. Chem. Soc.*, 1962, 146.
[39] B. L. Karger, R. L. Stearn, H. C. Rose and W. Keane, in *Gas Chromatography*, A. B. Littlewood (ed.), Inst. of Petrol., London, 1967, p. 240.
[40] J. P. Kamerling, M. Duran, G. J. Gerwig, D. Ketting, L. Bruinvis, J. F. G. Vliegenhart and S. K. Wadman, *J. Chromatog.*, 1981, **222**, 276.
[41] H. Schweer, *J.Chromatog.*, 1982, **243**, 149.
[42] S. V. Vitt, M. B. Saporovskaya, J. P. Gudkova and V. M. Belikov, *Tetrahedron Lett.*, 1965, 2575.
[43] M. Hasegawa and I. Matsubara, *Anal. Biochem.*, 1975, **63**, 308.
[44] R. Klein, *J.Chromatog.*, 1979, **170**, 468.
[45] J. P. Kamerling, G. J. Gerwig, J. F. G. Vliegenhart, M. Duran, D. Ketting and S. K. Wadman, *J. Chromatog.*, 1977, **143**, 117.
[46] I. Maclean, G. Eglinton, K. Douraghi-Zadeh, R. G. Ackman and S. N. Hooper, *Nature*, 1968, **218**, 1019.
[47] R. E. Cox, J. R. Maxwell, G. Eglinton and C. T. Pillinger, *Chem. Commun.*, 1970, 1639.
[48] R. G. Ackman, R. E. Cox, G. Eglinton, S. N. Hooper and J. R. Maxwell, *J. Chromatog. Sci.*, 1972, **10**, 392.
[49] A. Murano, *Agr. Biol. Chem.*, 1972, **36**, 2203.
[50] M. Miyakado, N. Onno, Y. Okuno, M. Hirano, K. Fujimoto and H. Yoshioka, *Agr. Biol. Chem.*, 1975, **39**, 267.
[51] M. Horiba, H. Kitaharo, H. Takahashi, S. Yamamota, A. Murano and N. Oi, *Agr. Biol. Chem.*, 1979, **43**, 2311.
[52] B. Halpern, L. F. Chew and J. W. Westley, *Anal. Chem.*, 1967, **39**, 399.
[53] F. Weygand, A. Prox, L. Schmidhammer and W. König, *Angew. Chem.*, 1963, **75**, 282.
[54] W. E. Pereira, M. Solomon and B. Halpern, *Aust. J. Chem.*, 1971, **24**, 1103.
[55] H. Schweer, *J. Chromatog.*, 1982, **243**, 149.
[56] J. M. Manning and S. Moore, *J. Biol. Chem.*, 1968, **243**, 5591.
[57] J. M. Manning, *J. Biol. Chem.*, 1971, **246**, 2926.

[58] A. R. Mitchell, S. B. H. Kent, I. C. Chu and R. B. Merrifield, *Anal. Chem.*, 1978, **50,** 637.
[59] E. P. Kroeff and D. J. Pietrzyk, *Anal. Chem.*, 1978, **50,** 1053.
[60] E. Lundanes and T. Geibrokk, *J. Chromatog.*, 1978, **149,** 214.
[61] N. Nimura, H. Ogura and T. Kinoshita, *J. Chromatog.*, 1980, **202,** 375.
[62] T. Kinoshita, Y. Kasahara and N. Nimura, *J. Chromatog.*, 1981, **210,** 77.
[63] A. J. Sedman and J. Gal, *J.Chromatog.*, 1983, **278,** 199.
[64] N. Nimura, Y. Kasahara and T. Kinoshita, *J. Chromatog.*, 1981, **213,** 327.
[65] J. A. Thompson, J. L. Holtzman, M. Tsuru and J. L. Holtzman, *J. Chromatog.*, 1982, **238,** 470.
[66] W. H. Pirkle and J. R. Hauske, *J. Org. Chem.*, 1977, **42,** 1839.
[67] W. H. Pirkle and K. A. Simmons and C. W. Boeder, *J. Org. Chem.*, 1979, **44,** 4891.
[68] F. P. Schmidtchen, P. Rauschenbach and H. Simon, *Z. Naturforsch.*, 1977, **32b,** 98.
[69] G. Helmchen, H. Völter and W. Schule, *Tetrahedron Lett.*, 1977, 1417.
[70] P. Marfey, *Carlsberg Res. Commun.*, 1984, **49,** 591.
[71] S. Einarsson, B. Josefsson, P. Möller and D. Sanchez, *Anal. Chem.*, 1987, **59,** 1191.
[72] D. Sanchez, P. Möller, S. Einarsson and B. Josefsson, *Janssen Chimica Acta*, 1988, **6,** no. 1.
[73] J. Hermansson and C. von Bahr, *J. Chromatog.*, 1980, **221,** 109.
[74] B. Silber and S. Riegelman, *J. Pharm. Exp. Ther.*, 1980, **215,** 643.
[75] G. Helmchen and W. Strubert, *Chromatographia*, 1974, **7,** 713.
[76] J. Goto, M. Hasegawa, S. Nakamura, K. Shimada and T. Nambara, *Chem. Pharm. Bull.*, 1977, **25,** 847.
[77] J. Goto, M. Hasegawa, S. Nakamura, K. Shimada and T. Nambara, *J. Chromatog.*, 1978, **152,** 413.
[78] J. Gal, *J. Chromatog.*, 1984, **307,** 220.
[79] K. J. Miller, J. Gal and M. M. Ames, *J. Chromatog.*, 1984, **307,** 335.
[80] P. Loew and W. S. Johnson, *J. Am. Chem. Soc.*, 1971, **93,** 3765.
[81] M. Koreeda, G. Weiss and S. Nakanishi, *J. Am. Chem. Soc.*, 1973, **95,** 239.
[82] K. Imai and S. Marumo, *Tetrahedron Lett.*, 1976, 1211.
[83] A. Rüttiman, K. Schiedt and M. Vecci, *J. High Resol. Chromatog.*, *Chromatog. Commun.*, 1983, **612,** 612.
[84] S. Tamura, S. Kuzuna, K. Kawai and S. Kishimato, *J. Pharm. Pharmacol.*, 1981, **33,** 701.
[85] M. Vecci and R. K. Müller, *J. High Resol. Chromatog.*, *Chromatog. Comm.*, 1979, **2,** 195.
[86] C. G. Scott, M. J. Petrin and T. McCorcle, *J. Chromatog.*, 1976, **125,** 157.
[87] B. J. Bergot, R. J. Anderson, D. A. Schooley and C. A. Henrick, *J. Chromatog.*, 1978, **155,** 97.

5

Theory of chiral chromatography for direct optical resolution

In the previous chapter it was noted that separation of enantiomers, by the formation of diastereomeric derivatives, is associated with a number of disadvantages which make its use relatively unattractive. Therefore, *direct* separations, i.e. by the use of a chiral stationary phase, are far more interesting from an analytical, as well as a preparative point of view. Particularly in LC, a technique that is very suitable for separation without prior derivatization, development of new chiral stationary phases has been a rapidly expanding field during the last few years. It is the purpose of this chapter to give a general treatment of the theories underlying the various efforts and achievements made in this area.

5.1 THE PREREQUISITE FOR ENANTIOSELECTIVE INTERACTION WITH THE CHIRAL STATIONARY PHASE

In all forms of chromatography, separation is based on differences in retention of some kind by the stationary phase. We have already seen how the chromatographic process can be regarded as a series of equilibria and how the equilibrium constant describing the distribution of a compound between the stationary and mobile phases is related to the chromatographic capacity ratio. It has been customary to distinguish between partition and adsorption chromatography, depending upon whether the stationary phase is a liquid or a solid. With the introduction of bonded organic phases in LC and immobilized coatings in fused silica capillary GC, this distinction is no longer obvious. Nevertheless the consideration of primary importance should be the types of molecular interactions with the stationary phase that cause retention.

It should be helpful for later discussions, however, to consider the fact that there are cases in which retention is not due to bonding interactions with a stationary phase but rather to differences in the distance the components under separation have to travel through the column. In principle, molecular sieve chromatography, of particular importance for protein separation, operates essentially by means of steric exclusion, i.e. the larger molecules travel a shorter distance because they are unable to diffuse into the finer pores of the matrix and are therefore eluted faster than the smaller molecules.

The theory of chiral chromatography (i.e. separation in which a chiral stationary phase selectively retains more of one enantiomer than the other) is still rather rudimentary. A number of chiral recognition models have been proposed to account for optical resolutions by GC and LC, which are often based on the 'three-point interaction' theory advanced by Dalgliesh [1] in 1952. According to this postulate, three simultaneously operating interactions between an enantiomer and the stationary phase are needed for chiral discrimination. The enantioselective situation is visualized in Fig. 5.1. It is obvious that this is a sufficient condition for enantioselection to occur. Now it may be asked whether it is always necessary. As we shall see later, there are many instances where this is probably not the case.

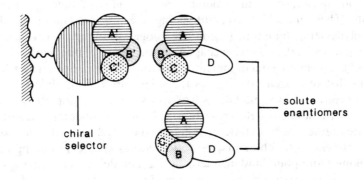

Fig. 5.1 — The three-point interaction model advanced by Dalgliesh.

Dalgliesh arrived at his conclusions from studies of certain aromatic amino-acids by paper chromatography. He assumed that the hydroxyl groups of the cellulose were hydrogen-bonded to the amino and carboxyl groups of the amino-acid. A third interaction was caused, according to these views, by the aromatic ring substituents.

The 'three-points rule' has often been used in a very uncritical way to rationalize experimental results. It is therefore important to try to analyse the situation in more detail. It is readily understood that in order to determine the configuration of a chiral object by matching with some probe, a minimum of three simultaneous, spatially significant contacts or interactions must be present. This minimum number of contact points, however, does not necessarily mean points of attachment when it comes to molecular interactions. In principle, a situation where only steric interactions cause a molecular steric discrimination is quite possible. In adsorption chromatography, though, there must always exist some kind of bonding interaction with the sorbent. This may arise through any non-covalent attachment possible under the prevailing conditions. Thus, hydrogen bonding as well as ionic or dipole attraction is enhanced by non-polar solvents, whereas hydrophobic interactions may be important in aqueous media, etc.

5.2   SOME GENERAL ASPECTS REGARDING CHIRAL RECOGNITION MODELS AND CHROMATOGRAPHIC ENANTIOSELECTIVITY

Optical resolution by chromatography is possible through reversible diastereomeric association between a chiral environment, introduced into a column, and solute enantiomers. The multiplicity of experimental conditions under which direct chromatographic optical resolutions have been achieved also tells us that the difference in association which is necessary can be obtained by means of many types of molecular interactions. The association, which may be expressed quantitatively as an equilibrium constant, will be a function of the magnitudes of the binding as well as of the repulsive interactions involved. The latter are usually steric, although dipole–dipole repulsions may also occur, whereas various kinds of binding interactions may operate. These include hydrogen bonding, electrostatic and dipole–dipole attractions, charge-transfer interaction and hydrophobic interaction (in aqueous systems). As we shall see in the following chapters, a single *type* of bonding interaction may be sufficient to promote enantiomer differentiation. For example, it appears to be quite evident that hydrogen bonding, as the sole source of attraction, is sufficient for optical resolution in some GC as well as LC modes of separation. The fact that enantiomeric solutes, bearing only one hydrogen bonding substituent, can be separated under such conditions, points to the conclusion that only *one* attractive force is necessary for chiral discrimination in this type of chromatography.

Taking a one-point binding interaction as a model, we may envisage a difference in the equilibrium constant of the two enantiomers at the chiral binding site as due to effects from the site forcing one of the enantiomers to take an unfavourable conformation. This resembles a situation often assumed to be present in enzyme–substrate interactions to account for substrate specificity.

Let us follow the reasoning a bit further. Is it possible to base enantiomer differentiation entirely on steric fit? In other words, can chiral cavities be constructed for the preferential inclusion of only one enantiomer? Although no chiral stationary phase (CSP) has yet been prepared that is based entirely on steric exclusion from chiral cavities (cf. Section 7.1.3), some recent work with the use of 'molecular imprinting' techniques is very interesting in this respect [2]. The idea is to create rigid chiral cavities in a polymer network in such a way that only one of two enantiomers will find the environment acceptable. Other types of CSPs, where steric fit is of primary importance, include those based on inclusion phenomena, such as cyclodextrin and crown-ether phases. These are described in Chapter 7.

In the following sections some theoretical aspects of important binding types present in enantioselective sorption processes are given.

5.2.1   Co-ordination to transition metals

The transition metals are characterized by having unfilled inner-shell *d*-orbitals. A transition metal complex is formed by ligands which may donate electrons to these unfilled orbitals. Such co-ordination complexes possess a very well-defined geometry, such that the ligands can only occupy certain given positions in space. The donor ligand atoms in the complex are thus held at strictly fixed distances from the metal atom and in defined orientations. This so-called co-ordination sphere is therefore densely packed with the ligands and with the solvent molecules. The latter also form a second (outer) highly organized sphere. This, in turn, means that the stability of the complex should be highly stereochemically dependent since if two or more chiral ligands are present in the co-ordination sphere, their mutual interaction, either directly or through the solvent molecules of the first or second sphere, will probably give rise to differences in complex stability and thereby cause enantioselectivity.

This principle has been experimentally utilized in both gas and liquid chromatography. Apart from the necessary difference in diastereomeric complex stability, another prerequisite for its successful use is that the complexes formed should be sufficiently kinetically labile, i.e. their formation and dissociation should be fast on the chromatographic time-scale. Although to a certain extent this condition can be achieved by temperature adjustment, this is not always possible. The complex stability is also highly dependent on the transition metal used and generally Cu(II) complexes are the most stable and preferred in LC, whereas the less stable complexes formed with Ni(II), Co(II) or Zn(II) have been explored in GC complexation techniques.

In both cases the basic idea is to use a chiral stationary phase composed of an immobilized chiral ligand forming a co-ordination complex with the transition metal ion. During passage of a suitable racemic mixture through the column, diastereomeric mixed-ligand sorption complexes will form by a displacement or exchange mechanism. This ligand-exchange process is schematically visualized in Fig. 5.2.

5.2.2   Charge-transfer interaction

This particular type of interaction requires π-electron systems. Stable CT-complexes are often formed between aromatic rings acting as donor and acceptor components. Such an aromatic π–π interaction, together with additional polar interactions (hydrogen bonding, dipole interactions), forms the basis of very efficient chiral selectors used in LC.

Nitroaromatics are usually quite good π-acceptors because a negative charge is effectively delocalized by participation of the substituents in resonance stabilization. Good π-donors are aromatics carrying electron-releasing substituents such as amino or alkoxy groups.

Although π–π interaction as the sole source of retention has been sufficient for the optical resolution of condensed aromatic hydrocarbons of planar chirality (helicenes; cf. Section 7.2.3), further bonding interactions increase the enantioselectivity exerted by the selector. A very successful selector principle, operating in organic solvents (hexane/2-propanol mixtures) often with large α-values, is shown in Fig. 5.3. The matching of the groups around the stereogenic centres in the two partners corresponds to simultaneous interactions at three points, of which at least two are bonding (the situation shown is that for the most strongly retained

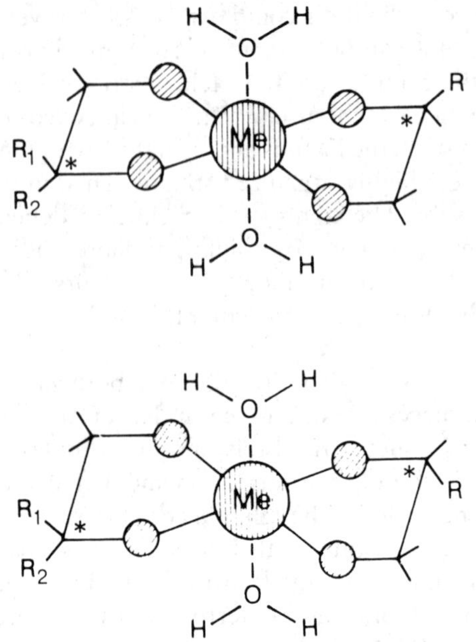

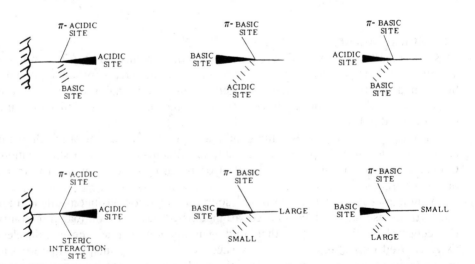

Fig. 5.2 — Illustration of the principle of chiral ligand exchange used for chromatographic optical resolution. Reversible formation of diastereomeric metal complexes is obtained in a reaction between a chiral selector (left-hand side) and the respective enantiomers (right-hand side).

Fig. 5.3 — Model used to interpret the stereoselectivity obtained with charge-transfer based chiral selectors. (From W. H. Pirkle, M. H. Huyn and B. Bank, *J. Chromatog.*, 1984, **316**, 585. Copyright 1984, Elsevier Science Publishers B.V.).

enantiomer). Evidence for this multiple interaction has been obtained from inter-molecular nuclear Overhauser effects in NMR [3], as well as from molecular mechanics calculations [4].

Further support for the postulated mechanism has come from X-ray crystal structure analyses of analogous chiral charge-transfer complexes. With the use of (R)- or (S)-N-3,5-dinitrobenzoylleucine methylamide as a soluble model of a π-accepting CSP and (S)-N-acetylleucine-2-naphthylamide as the π-donating partner, crystals of the 1:1 complexes (R,S and S,S) were prepared and investigated. From chromatographic data the (S,S)-complex should be expected to be the more stable. X-ray data of this complex showed the crystal to contain two π-donor (D1, D2) and two π-acceptor (A1, A2) molecules in each asymmetric unit. When the environments of the two donors presented in Fig. 5.4 are looked at, it is clearly seen that in each case the donor–acceptor interactions (D1–A1 and D2–A2) involve three simulta-neous interactions [two hydrogen bonds (indicated by dotted lines) and a nearly parallel orientation of the aromatic ring systems]. In the structure of the (R,S)-complex, however, only pairs of donor–acceptor molecules with two mutual inter-actions were found.

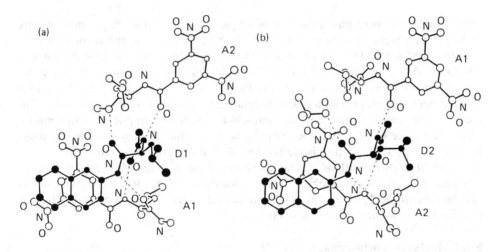

Fig. 5.4 — X-ray crystallographic structure of the S,S-diastereomeric complex formed from (S)-N-acetylleucine-2-naphthylamide and (S)-N-3,5-dinitrobenzoylleucinemethylamide. (Repro-duced from R. Däppen, G. Rihs and C. W. Mayer, *Chirality*, 1990, **2**, 185, with permission. Copyright 1990, Wiley-Liss, Inc.)

An important feature of these systems is their reciprocity, i.e. the enantioselecti-vity shown should be independent of which partner is the one immobilized. Use of this reciprocity principle will lead to new selectors of different applicability.

5.2.3 Hydrogen bonding
Intermolecular association via bidentate hydrogen bonding, as shown in Fig. 5.5, will yield diastereomeric complexes differing in free energy. Since bifunctional amides of this type are readily available from common amino acids as starting materials, it is perhaps not surprising that most of the first successful chiral separations with gas

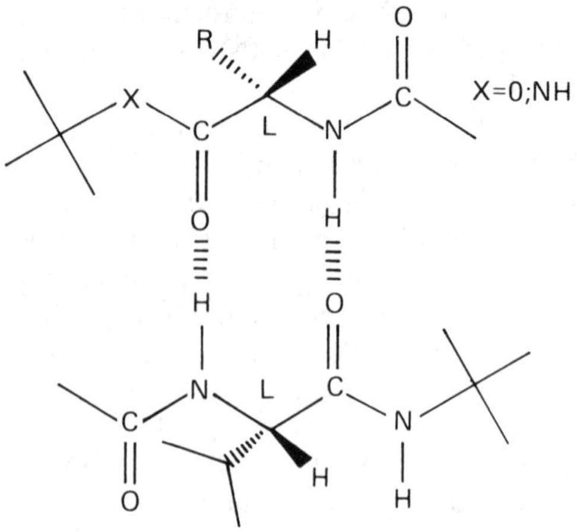

Fig. 5.5 — Association model of an L-L-diastereomeric complex formed by bidentate hydrogen bonding.

chromatography were based on this principle [5]. Two equivalent simultaneous hydrogen bonds are present in many dimeric structures, like in the carboxylic acid dimers, for example. In this case, it is also well known that they can exist even in the gas phase. A low polarity of the surrounding medium will, of course, improve the stability of such associates and therefore the principle can also be utilized in liquid chromatography [6], provided that only relatively non-polar solvents are used. The chiral selector can be used here either immobilized onto the support or in the mobile phase. It is quite obvious that the diastereomeric complexes shown in Fig. 5.5 must behave like cyclic *cis–trans* isomers, and that the steric differences will cause different interactions with an achiral stationary phase in a chromatographic system. The dual hydrogen bond association and its importance for chiral recognition in gas and liquid chromatography has also been demonstrated by X-ray crystal-structure determination on diastereomeric complexes [7].

5.2.4   Inclusion phenomena
The ability of certain compounds to use their particular structures to include suitable guest molecules has long been known. Classical examples are the host properties of urea and starch. Crystal analyses have shown that urea molecules form complexes with a channel-like interior into which unbranched alkanes fit nicely. Such n-alkane –urea complexes therefore form spontaneously. Branched alkanes do not fit into such interiors and consequently this phenomenon can be used to separate n-alkanes from mixtures of isomers. Starch is well known for its inclusion of iodine. The cyclodextrins (Schardinger's dextrins) are crystalline degradation products of starch, which are obtained through the action of micro-organisms (see Section 7.1.1.1.). Whereas the α-form, composed of a ring with six glucose units, is of the correct size to include iodine or benzene, it is too small to include bromobenzene. The β-form (composed of seven units), on the other hand, is (contrary to the α-form) precipitated by bromobenzene as a consequence of inclusion complex formation.

The strict steric requirements for the formation of such 'host–guest' complexes imply that these phenomena should be highly stereoselective. Thus, by the use of a chiral host, enantiomeric guest molecules might be separated. These principles are utilized fully or partially in some of the liquid chromatographic techniques described in Chapter 7. A short general treatment of the inclusion phenomena present, or thought to be present, in these enantiomer-differentiating phases is given below.

Let us consider two different types of host molecules, one with a hydrophilic interior and hydrophobic exterior and the other with the opposite polarity configuration (Fig. 5.6).

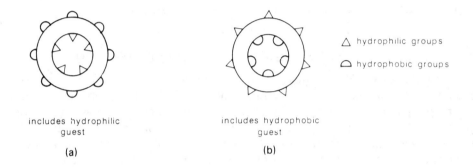

△ hydrophilic groups

◠ hydrophobic groups

includes hydrophilic
guest

includes hydrophobic
guest

(a)

(b)

Fig. 5.6 — Simplified models to represent different types of host–guest inclusion complex formation.

The hydrophilic interior in (**a**) means that the cavity contains hetero-atoms such as oxygen, where lone-pair electrons are able to participate in bonding to electron acceptors such as metal or organic cations. The hydrophobic exterior makes the host–guest complex as a whole soluble in organic media, a phenomenon which has been exploited in so-called phase-transfer catalysis (cf. Section 7.2.1). One type of these host compounds is found in the naturally occurring macrocyclic polyethers, which are known to bind alkali-metal cations. Synthetic chiral analogues of these compounds, chiral crown ethers, have been found to exert remarkable enantioselectivity towards organic ammonium ions. In these cases the ammonium ion is held within the cavity by hydrogen bonds to the ether oxygen atoms. Thus, in this case, the structural and steric requirements of the guest are high.

The hydrophobic interior of (**b**), on the other hand, means that a cavity suitable for inclusion of hydrocarbon-rich parts of a molecule is present. No bonds are involved in the complex formation, as the inclusion is merely a result of the hydrophobic effect and the structural demands are not as pronounced as in the previous case. This type of host is found in the cyclodextrins, which will be treated in Sections 7.1.1.1 and 7.3.2. Under reversed-phase conditions (aqueous media) the combination of hydrophobic interaction, which generates inclusion, with steric effects from substituents present in a chiral structure on the cavity entrance, is assumed to be the cause of enantioselection.

A rather special case of inclusion effects is found in chiral matrices composed of swollen, microcrystalline cellulose derivatives. The triacetate, prepared by heterogeneous acetylation in order to preserve microcrystallinity, has been shown to act partially by steric exclusion effects. Thus, of a series of aromatic hydrocarbons (with essentially no bonding properties), benzene is highly retained, mesitylene (1,3,5-trimethylbenzene) is much less retained and 1,3,5-tri-*tert*-butylbenzene is totally excluded (no retention). This phenomenon can be explained by considering the lamellar arrangement of the polysaccharide chains. These yield a kind of two-dimensional molecular sieve, permitting inclusion of flat aromatic compounds in particular, but excluding sterically more demanding structures. The higher retention of benzene (compared to that of toluene) has further led to the suggestion of a secondary effect, viz. an action by pockets in the chain structure.

5.3   COMPUTATIONAL AND MOLECULAR MODELLING STUDIES

In recent years, with the progress of advanced computer technology, an increased interest in the application of computational methods to the problem of chiral discrimination has developed. These methods are based on evaluation of the potential energy contours obtained in molecular docking processes. Thus, by computerized matching of the contours found from docking of two enantiomeric molecules with the same chiral entity, an energy difference between the two complexes (at their potential energy minima) can be calculated. This offers, at least in principle, a means of predicting the elution order in a chromatographic system. However, solvent effects cannot be taken specifically into account by such methods, and so far the predictive value is rather limited. Furthermore, even in very well defined chromatographic systems containing a simple chiral selector, different retention mechanisms may compete [8], making the situation more complex than assumed in the theoretical approach. Nevertheless, data from computational work on chiral discrimination will continue to add to our understanding of the molecular interactions behind the mechanism of differential adsorption and will certainly gain in importance in the future. The reader who wishes further insight into this field is referred to the recent work by Lipkowitz [9–12], Topiol [13–15] and Norinder [16].

5.4   SOME THERMODYNAMIC AND KINETIC CONSIDERATIONS

5.4.1   Temperature effects on α

Recalling our treatment in the previous chapter and the definition of α, we may readily arrive at Eq. (5.1) by considering the equilibrium of each enantiomer between the mobile and the stationary phase. If the equilibrium constants are denoted by K_R and K_S, respectively, the expression for the change in free energy, $\Delta G° = -RT\ln K$, will give the free energy difference: $\Delta\Delta G° = \Delta G°_R - \Delta G°_S$, as $-\Delta\Delta G° = RT\ln K_R/K_S$, where $K = C_s/C_m$ and $K_R > K_S$ (arbitrary assumption). From our definition of k' we know that $k' = KV_s/V_m$ which then gives Eq. (5.1):

$$-\Delta\Delta G° = RT\ln k'_R/k'_S = RT\ln\alpha \qquad (5.1)$$

Thus, the free energy difference associated with a given α-value is easily computed from chromatographic data. The figures given in Table 5.1 are quite illustrative as they show the very small energy difference that is needed for complete optical resolution, provided a reasonable column efficiency is available.

Table 5.1 — Free energy differences necessary to produce separation factors >1

α	$\Delta\Delta G$, cal/mole (J/mole)
1.05	29 (121)
1.10	56 (236)
1.50	240 (1005)
2.00	410 (1717)
10.0	1364 (5705)

In the previous chapter, we have also seen that modern column technology can provide us with extremely efficient capillary columns, which in good analytical GC instruments give baseline resolution of peaks even at α-values < 1.05. In columns of effective plate numbers around 2×10^5 an α-value of 1.01 means only 2% peak overlap ($R_s = 1.11$), which, in turn, corresponds to an energy difference of only 5.9 cal/mole (24.7 J/mole). Such minute energy values are at least an order of magnitude lower than those normally associated with conformational changes in a molecule. It is obvious therefore, that binding of two enantiomers to a given chiral site may involve different amounts of energy simply because one of the enantiomers, for steric reasons, might be forced to adopt an energetically less favourable conformation.

The much lower column efficiency in LC is often more than compensated for by the considerably larger α-values that can be obtained. In certain cases α-values of 30 or more have been found, which then correspond to $\Delta\Delta G°$ values in the range of 2 kcal/mole (8.4 kJ/mole). Generally, such values are obtained owing to very low retention of the first enantiomer eluted. This means that a very enantioselective adsorption process is operating in the column, i.e. one of the enantiomers is virtually unbound by the CSP for steric reasons. Such phenomena are not easily explained by the three-point interaction model, but rather indicate the operation of a sort of 'chiral steric exclusion' mechanism, more in line with a 'steric fit' concept involving only one binding interaction.

An expansion of Eq. (5.1) to involve the enthalpy and entropy terms (by an application of the Gibbs–Helmholtz equation: $G = H - TS$) yields:

$$\ln\alpha = -\frac{\Delta\Delta H°}{RT} + \frac{\Delta\Delta S°}{R} \tag{5.2}$$

Thus, from a study of the dependence of α on temperature, lnα may be plotted as a function of $1/T$. The slope of the line will then be proportional to the enthalpy difference and lnα = 0 will give the temperature at which the enthalpy and entropy contributions cancel each other. Such studies are easily performed by GC [17]. Figure 5.7 shows the general appearance of such a plot.

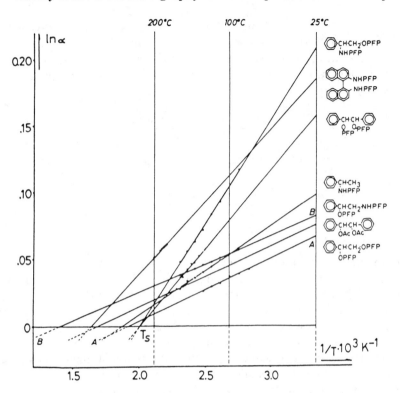

Fig. 5.7 — Illustration of the decrease in the enantiomeric separation factor with increasing temperature for chiral GC separations. The plots permit evaluation of the enthalpy and entropy contributions, respectively, to the separation. Different compounds were studied, with the same column. (Reprinted, with permission, from B. Koppenhöfer and E. Bayer, *Chromatographia*, 1984, **19**, 123. Copyright 1984, Fr. Vieweg & Sohn Verlagsgesellschaft mbH).

5.4.2 Peak coalescence due to enantiomerization phenomena

There is always a possibility that the enantiomers of a compound may be interconverted by some mechanism which generates an achiral intermediate. In solution such processes are often acid- or base-catalysed and thus related to the stereochemical fate of positively or negatively charged transient intermediates. Common examples are enantiomerization reactions through carbonium ion or carbanion formation (Scheme 5.1).

In these and similar cases, the free energy barrier to enantiomer interconversion will be dependent on substituent effects in the transition state leading to formation of the achiral intermediate.

Even simpler mechanisms of enantiomerization are found in those cases where chirality is caused predominantly by steric hindrance, as in compounds with axial or planar chirality. Thus, if an atropisomeric compound is taken as an example (Scheme 5.2), the free energy barrier to internal rotation around the central bond may be high enough to prevent any significant racemization of an optically active form, in solution at room temperature, and thereby permit its isolation. In many cases, however, the

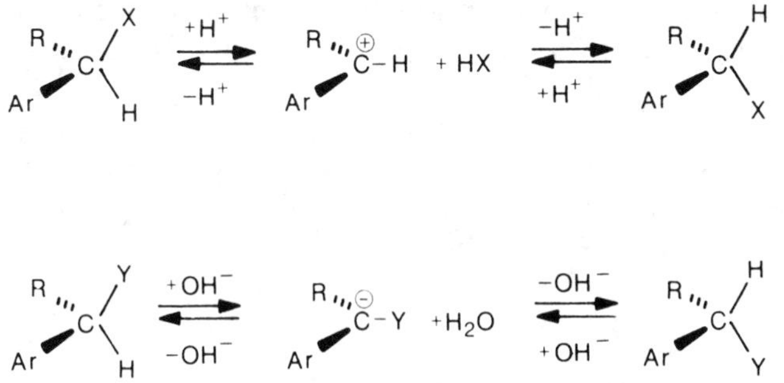

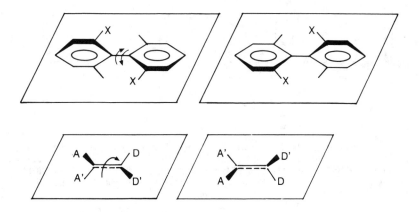

Scheme 5.1 — Examples of (a) an acid-catalysed and (b) a base-catalysed reaction yielding enantiomerization due to formation of achiral charged intermediates.

Scheme 5.2 — Thermal enantiomerization reactions by changes in molecular conformation by internal rotation, (a) in a biaryl system, and (b) in a polarized alkene system.

temperature has to be lowered considerably in order to decrease the rate of internal rotation leading to enantiomer interconversion.

Taking such enantiomerization reactions into consideration in relation to chromatographic methods for direct optical resolution, it is obvious that there are many cases where the chromatographic conditions used play an important role. In general terms, if the chromatographic conditions are such that the rate of enantiomerization is significant on the chromatographic time-scale, the elution pattern will deviate from the normal one. During passage of the first-eluted enantiomer through the column, it will be partially transformed into the last-eluted enantiomer. This process will result in tailing of the leading peak. Conversely, the enantiomer eluted last will also be transformed into the first-eluted enantiomer at the same rate. This will give rise to a 'fronting' of the second peak. The net result will be an extended peak overlap, such that the baseline will not be reached between the two peaks. If the enantiomerization reaction is fast enough in comparison with the chromatographic process, complete coalescence of the peaks will take place. The situation is visualized in Fig. 5.8.

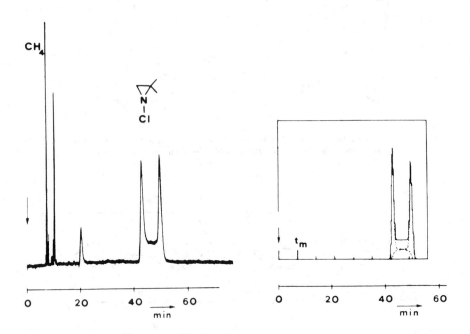

Fig. 5.8 — Chromatographic pattern arising from on-column enantiomerization. Left: Experimentally obtained chromatogram of 1-chloro-2,2-dimethylaziridine. Right: Computer simulation of the chromatographic process. (Reprinted, with permission, from W. Bürkle, H. Karfunkel and V. Schurig, *J. Chromatog.*, 1984, **288**, 1. Copyright 1984, Elsevier Science Publishers, B.V.).

Consequently, enantiomerization phenomena are readily detected by chromatography on chiral phases. Conversely, enantiomerization rates may be calculated from chromatographic peak coalescence data by use of suitable computer programs. However, very little work has yet been done in this field. It should also be kept in mind that the rate observed may not always be representative of that in bulk solution, because the reaction process may be catalysed by the surfaces with which the solute comes into contact during passage through the column [18].

We will return to the chiral chromatography of enantiomerization-labile compounds in Section 8.5.

BIBLIOGRAPHY

J. Porath, Explorations into the Field of Charge-Transfer Adsorption, *J. Chromatog.*, 1978, **159**, 13 (*Chromatog. Rev.*, 1978, **22**, 13).
V. A. Davankov, Resolution of Racemates by Ligand-Exchange Chromatography, *Adv. Chromatog.*, 1980, **18**, 139.

J. L. Atwood, J. E. D. Davies and D. D. McNicol (eds.), *Inclusion Compounds*, Academic Press, London, 1984.
D. Worsch and F. Vögtle, Separation of Enantiomers by Clathrate Formation, *Top. Curr. Chem.*, 1987, **140**, 21.

REFERENCES

[1] C. Dalgliesh, *J. Chem. Soc.*, 1952, 137.
[2] G. Wulff, in *Polymeric Reagents and Catalysts*, W. T. Ford (ed.), ACS, Washington DC, 1986, p. 186.
[3] W. H. Pirkle and T. C. Pochapsky, *J. Am. Chem. Soc.*, 1986, **108**, 5627.
[4] K. B. Lipkowitz, D. A. Demeter, C. A. Parish and T. Darden, *Anal. Chem.*, 1987, **59**, 1731.
[5] E. Gil-Av and B. Feibush, *Tetrahedron Lett.*, 1967, 3345.
[6] A. Dobashi and S. Hara, *Tetrahedron Lett.*, 1983, **24**, 1509.
[7] Y. Dobashi, S. Hara and Y. Iitaka, *J. Org. Chem.*, 1988, **53**, 3894.
[8] W. H. Pirkle and R. Däppen, *J. Chromatog.*, 1987, **404**, 107.
[9] K. B. Lipkowitz, D. A. Demeter, C. A. Parish, J. M. Landwer and T. Darden, *J. Comput. Chem.*, 1987, **8**, 753.
[10] K. B. Lipkowitz, D. A. Demeter, R. Zegarra, R. Larter and T. Darden, *J. Am. Chem. Soc.*, 1988, **110**, 3446.
[11] K. B. Lipkowitz, B. Baker and R. Zegarra, *J. Comput. Chem.*, 1989, **10**, 718.
[12] K. B. Lipkowitz and B. Baker, *Anal. Chem.*, 1990, **62**, 774.
[13] S. Topiol, *Chirality*, 1989, **1**, 69.
[14] S. Topiol, M. Sabio, W. B. Caldwell and J. Moroz, *J. Am. Chem. Soc.*, 1988, **110**, 8367.
[15] S. Topiol and M. Sabio, *J. Chromatog.*, 1989, **461**, 129.
[16] U. Norinder and E. G. Sundholm, *J. Liq. Chromatog.*, 1987, **10**, 2825.
[17] B. Koppenhöfer and E. Bayer, *Chromatographia*, 1984, **19**, 123.
[18] G. Blaschke, *Angew. Chem.*, 1980, **92**, 14.

6

Chiral gas chromatography

For many applications chiral gas chromatography is unsurpassed as an analytical tool, mainly because of the high peak capacity obtained in modern capillary columns. Over the last two decades a rapid development of the technique has taken place and the number of applications continues to grow.

6.1 THE DEVELOPMENT OF CHIRAL STATIONARY PHASES BASED ON HYDROGEN BONDING

6.1.1 Amino-acid and oligopeptide derivatives

In 1966 Gil-Av and his collaborators found that direct optical resolution of a number of N-trifluoroacetyl-D,L-amino-acid esters could be achieved with the use of a glass capillary column of 100 m length, coated with a chiral phase, viz. the lauryl ester of N-trifluoroacetyl-L-isoleucine [1]. It was assumed that hydrogen bond formation between corresponding amide groups and carbonyl oxygen atoms in the CSP and solute, respectively, was the cause of retention and chiral discrimination. These first, really successful, results initiated much research in the field and a large number of CSPs based on amino-acids as chiral synthons were produced and investigated in the following years. Four main structural types (Fig. 6.1) were exhaustively studied. The

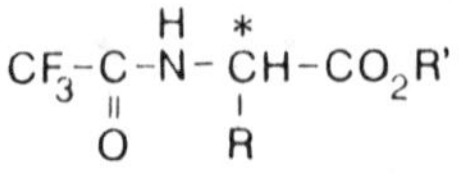

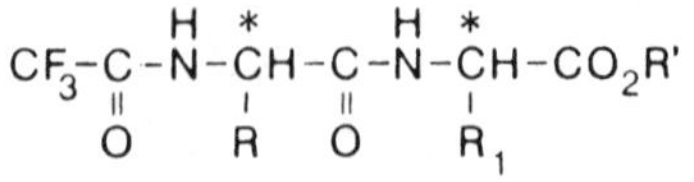

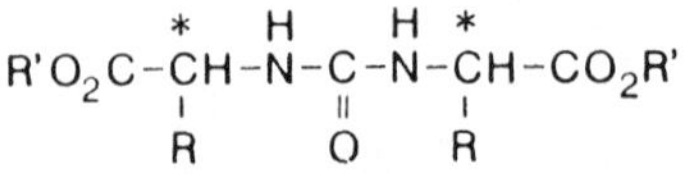

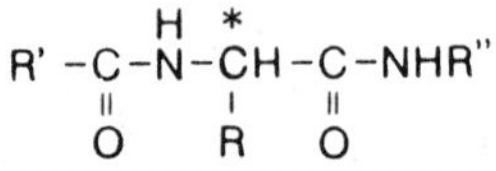

Fig. 6.1 — CSP structures used in early studies of direct optical resolution by GC.

N-TFA-amino-acid esters (**1**) as well as the *N*-TFA-dipeptide esters (**2**) (R′=CF₃) were, in general, found to give higher retention and α-values than the carbonyl-bis-(amino-acid esters) (**3**) at the operating temperatures.

A common problem in GC concerns the volatility and temperature stability of the stationary phase. The CSPs of structure **1** had to be used at relatively low temperatures, i.e. just above the melting point, but the situation was substantially improved by use of the dipeptide phases (**2**). Furthermore, the chiral discrimination could be large enough to permit the use of small packed columns. Thus, Gil-Av and Feibush [2] succeeded in separating the enantiomers of *N*-TFA-D,L-alanine *tert*-butyl ester on a 2-m packed column coated with the dipeptide CSP *N*-TFA-L-valyl-L-valine cyclohexyl ester (**2**, R=R₁=(CH₃)₂CH–, R′=cyclo-C₆H₁₁–).

The first diamide-type (**4**) CSP was reported in 1971 and represented a considerable improvement [3]. It was *N*-lauroyl-L-valine *tert*-butylamide (**4**, R=(CH₃)₂CH–, R′=C₁₁H₂₃, R″=(CH₃)₃C–), which could be used up to 130°C and displayed a high enantioselectivity. The performance of this CSP for a series of amino-acid derivatives is shown in Table 6.1.

Table 6.1 — Enantioselectivity achieved by the use of *N*-lauroyl-L-valine *tert*-butylamide in chiral gas chromatography. (From B. Feibush, *Chem. Commun.*, 1971, 544, reproduced by permission of the author and the copyright holders, Royal Society of Chemistry.)

Compound	α(L/D)	Compound	α(L/D)
Alanine	1.188	Proline	1.057
Valine	1.170	*O*-TFA-serine	1.101
O-TFA-threonine	1.117	Aspartic acid	1.078
t-Leucine	1.084	Glutamic acid	1.170
Alloisoleucine	1.186	Methionine	1.215
Isoleucine	1.159	Phenylalanine	1.198
Leucine	1.280	*O*-TFA-tyrosine	1.262

Capillary column (0.02 in. × 150 ft), temperature 130°C, helium carrier. All compounds were run as their *N*-TFA methyl esters.

The problem of finding an efficient CSP for GC-separation of enantiomers is complicated. First, the thermal properties of the CSP should be adequate, so a combination of low melting point and high boiling point is required. In the case of oligopeptide phases, the use of tri- and higher peptides is often limited, owing to the high melting points of these compounds. On the other hand, many simple amino-acid derivatives have low melting points and possess vapour pressures that lead to severe column bleeding at the desired operating temperatures. Secondly, the stereochemical structure must be adequate for chiral discrimination, i.e. the diastereomeric solvates formed upon dissolution of the racemic solute in the CSP should be different in energy. Thirdly, the column efficiency in terms of plate height should be high, which implies that the mass-transfer rate must not be reduced.

It is obvious that relatively polar solutes are required in order to achieve retention on CSPs based on hydrogen bonding. Most studies have been performed with amino-acids in *N*-acylated and esterified forms. The following section presents some major results and conclusions obtained from systematic investigations in this area.

As a general rule an improvement in the enantiomeric separation factor, α, is obtained with decreasing column temperature. The reason for this is found in the effect of temperature on the difference of the enantiomers with respect to their enthalpy of association with the stationary phase, a difference which decreases with temperature (cf. Section 5.3). This, in turn, means that the derivatives should preferably be as volatile as possible. On the other hand, it has been found that polar compounds are often better resolved than non-polar compounds. This is shown for example by derivatives of 2-hydroxy-acids, where the amides exhibit larger α values than do the corresponding esters. Experimentally, it has been shown that, on dipeptide phases, the N-trifluoroacetyl derivatives of amino-acid esters are better resolved than the corresponding N-pentafluoropropionyl (N-PFP) or N-heptafluoro-butyryl (N-HFB) compounds.

Although not studied in detail, the nature of the ester group is also of fundamental importance. For the same reason as discussed above (column temperature) an increasing chain length generally reduces the α value. It was early found, however, that a certain bulkiness of the ester alkyl group is advantageous. In a comparison of the ethyl and isopropyl esters of N-TFA-D,L-alanine at two different dipeptide phases at 110°C, the isopropyl ester derivative showed significantly higher α values [4]. Accordingly, a major part of the studies of optical resolution of amino-acids by chiral GC has been performed with the N-TFA isopropyl ester derivatives.

An interesting phenomenon was found in connection with investigations of some CSPs of the carbonyl-bis-(amino-acid ester) type. It could be demonstrated that these particular phases were transformed from an isotropic to a smectic state upon a decrease in temperature, and that the α values were significantly increased in the smectic region [5–7].

Research and development of hydrogen-bonding chiral stationary phases for GC has been very intensive and there are many publications in this field. In particular, the number of amide and diamide phases prepared and evaluated is very large. An exhaustive account of these and similar CSPs for GC, and their reported applications, is given in Souter's excellent review, to which the reader is referred for further details. A few of these phases, developed in Japan [8–11], contain a chiral amine component and show broad resolving capacities. They have recently become commercially available as fused silica or glass capillary columns. The structures and properties of these new CSPs are summarized below. Interestingly, the enantioselective hydrogen-bonding properties of these phases are also quite useful in LC with non-polar mobile phases.

N-Lauroyl-L-proline-(S)-1-(α-naphthyl)ethylamide (**5**) exhibits the highest thermal stability and has, like O-lauroyl-(S)-mandelic acid-(S)-1-(α-naphthyl)ethylamide (**6**) been found useful for the resolution of compounds such as N-protected amino-acid esters, O-TFA α-hydroxycarboxylic acid esters, carboxylic acid *tert*-butylamides and -esters, nitriles and N-perfluoroacylamines. A representative result, showing the performance of (**5**), is given in Fig. 6.2.

Two other phases of this kind contain (R,R)-*trans*-chrysanthemic acid as an integral structural element, viz. N-(1R,3R)-*trans*-chrysanthemoyl-(R)-1-(α-naphthyl)ethylamide (**7**) and O-(1R,3R)-*trans*-chrysanthemoyl-(S)-mandelic acid-(R)-1-(α-naphthyl)ethylamide (**8**).

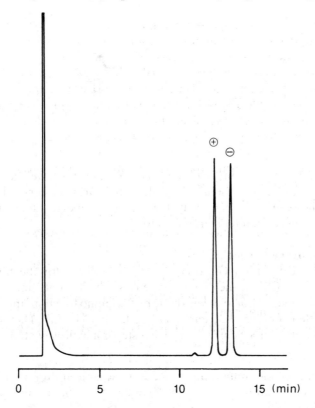

Fig. 6.2 — Separation of the enantiomers of *N*-TFA-1-phenylethylamine on a 0.25 mm × 25 m fused silica capillary coated with *N*-lauroyl-L-proline-(S)-1-(α-naphthyl)ethylamide (Sumipax-CC OA 500). Temperature 130°C, helium (0.8 ml/min), FID. (Courtesy of Sumitomo Chem. Co.).

5   **6**

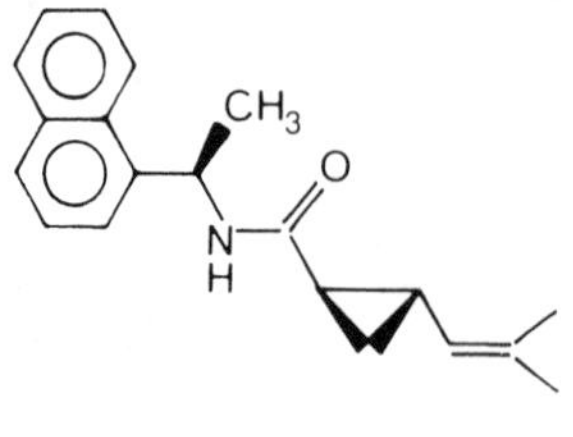

7 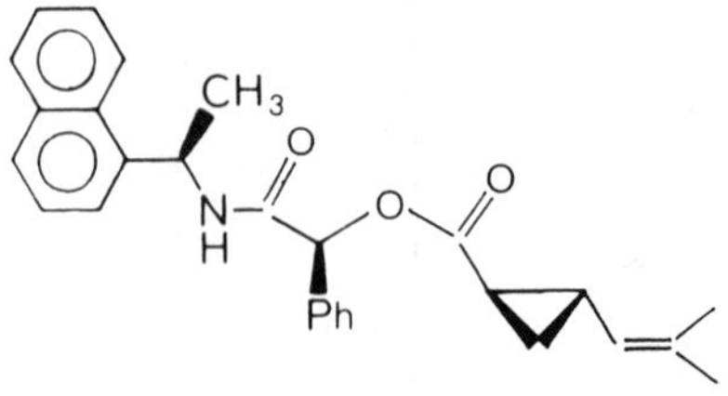   **8**

Although the applicability of (**7**) has so far been the most well documented, both phases appear to be very useful, particularly for various esters. The temperature limits are, however, comparatively low (highest reported temperature, 150°C; recommended maximum temperature, 110°C).

6.1.2 Polysiloxane-bonded phases

A breakthrough in chiral GC occurred in 1977 when, for the first time, a low molecular-weight CSP of the diamide type (L-valine *tert*-butylamide) was attached covalently to a silicone polymer backbone by the amino function, thus yielding a CSP material of then unsurpassed temperature stability [12,13]. The synthesis of the material is outlined in Scheme 6.1. The first step is a hydrosilylation of allyl cyanide

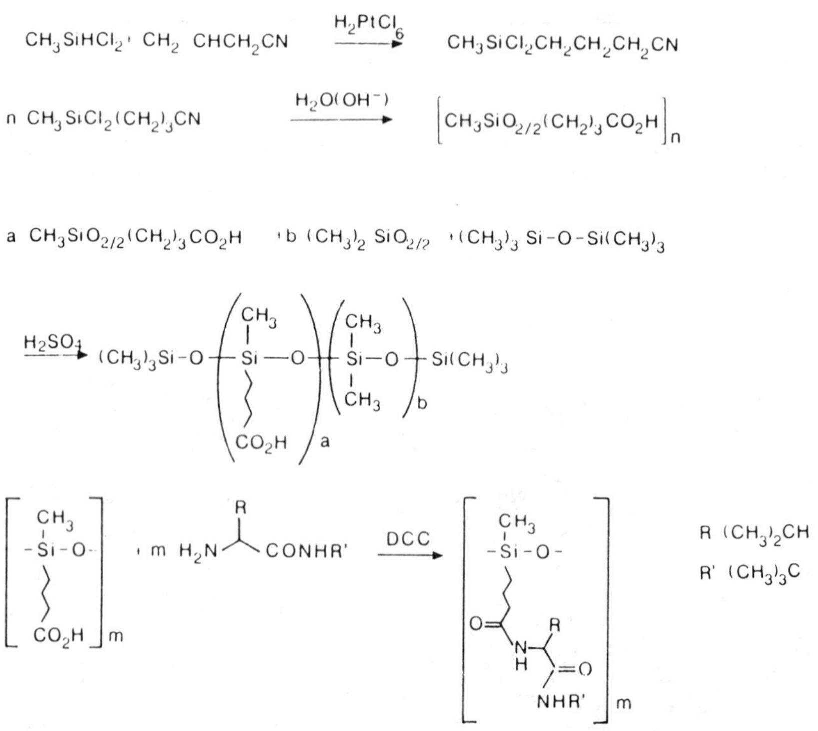

Scheme 6.1 — Reactions used to obtain a polysiloxane-linked L-valine *tert*-butylamide phase.

with methyldichlorosilane. The product is then hydrolysed by treatment with alkali to form a 3-carboxypropyl-substituted polysiloxane. The next step represents a 'dilution' of the carboxyl groups by equilibration with dimethylsiloxane monomer (which actually exists as a cyclic trimer) and hexamethyldisiloxane under conditions of acid catalysis. In the final step, the chiral group is attached by a condensation reaction forming a stable amide bond. It is found experimentally that the 'dilution', i.e. the b/a ratio, should preferably be about 5–7; the main reason for this is the high viscosity found at lower ratios.

The same principle can be applied to achieve coupling of other chiral ligands and a series of different chirally substituted polysiloxanes has been made; polysiloxane-L-valine-*tert*-butylamide has proved to be the most useful. It is marketed under the trade name "Chirasil-Val" as a series of coated capillary columns.

Another approach to the synthesis of chirally substituted polysiloxanes was also made [14,15]. Many commercially available silicones for GC contain cyano functional groups, which can be readily hydrolysed to yield free carboxyl groups for coupling. In this way the same chiral ligand, L-valine *tert*-butylamide, has been covalently bound. These phases differ somewhat in their performance as compared to "Chirasil-Val", owing to the differences in the silicon matrix, notably the presence of phenyl substituents.

In 1981 the same principle of transforming cyanopropyl silicones into suitable materials for chiral derivatization was utilized for the introduction of some new chiral ligands [16–19]. These experiments included hydrolysis of the cyano groups as well as their reduction to form primary amines which permitted the attachment of optically active acids. One of the more interesting phases originating from these studies consists of L-valine-(R)-1-phenylethylamide coupled to a polysiloxane, which has been shown to be useful for optical resolution of a variety of racemates, including *O*-TFA-derivatives of carbohydrates. The procedures used in derivatizing cyanopropylsilicones are outlined in Scheme 6.2

Much research on CSPs has been carried out on polysiloxane-bonded chiral phases. The following section examines the mechanism by which these phases discriminate between the two optical antipodes by means of different hydrogen bond formation.

As already mentioned in the previous chapter, the three-point interaction model has been widely used to construct chiral recognition models, without sufficient experimental support. Thus, the enantioselectivity exhibited by the early GC phases was explained by assuming a three-point hydrogen bonding between the analyte and the chiral sorbent [20]. These assumptions have been found incorrect [21,22] and are not compatible with recent findings [23,24]. Among these results are examples of resolutions of very simple underivatized hydroxy or carbonyl compounds having only one hydrogen-bonding site [25,26].

Of all hydrogen-bonding GC phases, polysiloxane-linked L-valine *tert*-butylamide ("Chirasil-Val") is by far the most exhaustively investigated. Calculations show that L-valine amides will adopt a β-pleated sheet or preferentially an (R)-α-helical structure. In a bonding situation, however, the energetically favoured conformation will, of course, also be dependent on the ligand. Therefore, it has been assumed [27], on the basis of experimental as well as theoretical results, that the

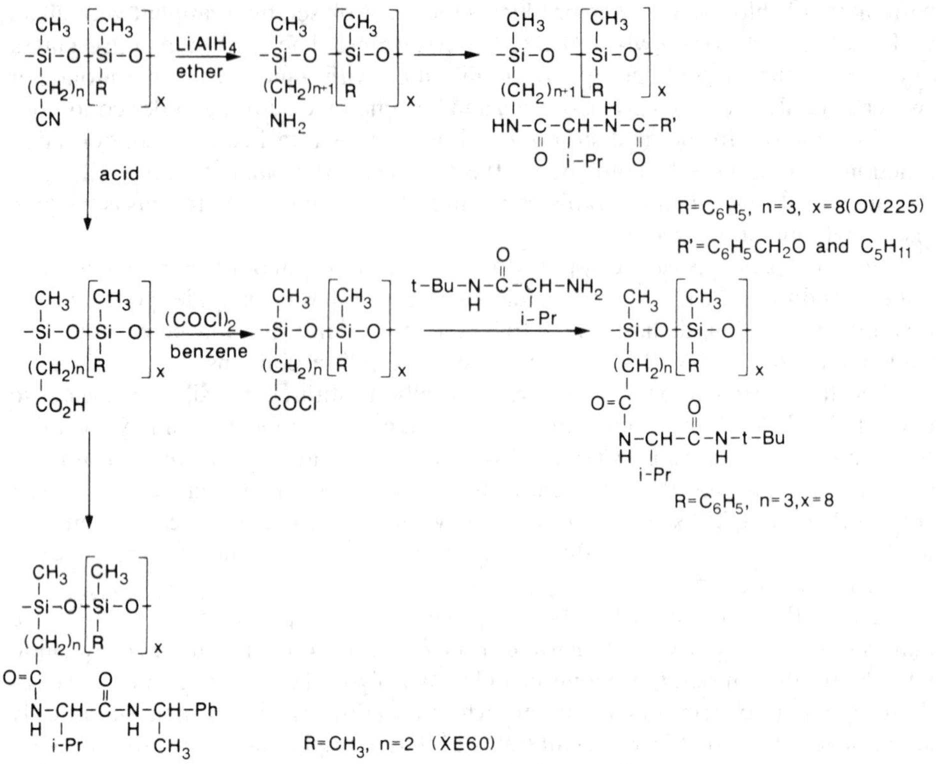

$$
\begin{array}{c}
CH_3 \left[CH_3 \right] \\
-Si-O-Si-O- \\
(CH_2)_n \; R \\
CN
\end{array}
\xrightarrow[\text{ether}]{LiAlH_4}
\begin{array}{c}
CH_3 \left[CH_3 \right] \\
-Si-O-Si-O- \\
(CH_2)_{n+1} \; R \\
NH_2
\end{array}_x
\longrightarrow
\begin{array}{c}
CH_3 \left[CH_3 \right] \\
-Si-O-Si-O- \\
(CH_2)_{n+1} \; R \\
HN-C-CH-N-C-R' \\
\;\;\; O \;\; i\text{-}Pr \;\; O
\end{array}_x
$$

acid

$R=C_6H_5, \; n=3, \; x=8 \,(OV\,225)$

$R'=C_6H_5CH_2O \text{ and } C_5H_{11}$

$$
\begin{array}{c}
CH_3 \left[CH_3 \right] \\
-Si-O-Si-O- \\
(CH_2)_n \; R \\
CO_2H
\end{array}_x
\xrightarrow[\text{benzene}]{(COCl)_2}
\begin{array}{c}
CH_3 \left[CH_3 \right] \\
-Si-O-Si-O- \\
(CH_2)_n \; R \\
COCl
\end{array}_x
\xrightarrow[i\text{-}Pr]{t\text{-}Bu-N-C-CH-NH_2}
\begin{array}{c}
CH_3 \left[CH_3 \right] \\
-Si-O-Si-O- \\
(CH_2)_n \; R \\
O=C \quad O \\
N-CH-C-N-t\text{-}Bu \\
H \; i\text{-}Pr \; H
\end{array}_x
$$

$R=C_6H_5, \; n=3, x=8$

$$
\begin{array}{c}
CH_3 \left[CH_3 \right] \\
-Si-O-Si-O- \\
(CH_2)_n \; R \\
O=C \quad O \\
N-CH-C-N-CH-Ph \\
H \; i\text{-}Pr \; H \; CH_3
\end{array}_x
$$

$R=CH_3, \; n=2 \; (XE\,60)$

Scheme 6.2 — Chemistry involved in the transformation of cyanopropyl silicones into CSPs for GC.

immobilized chiral sectors of "Chirasil-Val" will form association complexes by multiple hydrogen-bonding interaction, either by an intercalation mechanism (Fig. 6.3a) or by complexation in an α-helical conformation (Fig. 6.3b). The first type of mechanism, which is assumed to operate for α-hydroxy- and α-amino-acid derivatives, is characterized by an interaction of two selectors with each ligand in such a way that each selector utilizes one hydrogen-bond acceptor function (valine amide carbonyl group) and two hydrogen-bond donor groups (amide NH groups). In the second mechanism, however, the terminal selector is involved in bonding to the second ligand.

It is obvious that such a model does not apply to ligands lacking donor groups. Examples of such compounds which have been resolved on "Chirasil-Val" are diesters and dicarbonyl compounds. A relevant case is the complete separation of the atropisomeric 2,2'-binaphthol dipentafluoropropionates, where it is found that the (R)-form elutes before the (S)-enantiomer. It is then found that in order for the bound L-valine *tert*-butylamide selector to utilize both NH-donors for hydrogen bonding to the ester carbonyl groups in the substrate, a less favourable conformation has to be adopted for the chiral selector when it binds to the (R)-enantiomer of the ester (cf. Fig. 6.3b).

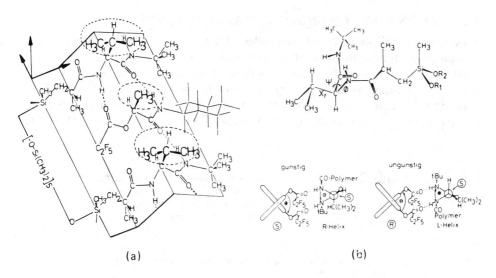

(a)   (b)

Fig. 6.3 — Possible structures of association complexes based on hydrogen bonding between "Chirasil-Val" and different types of ligands. (Reprinted, with permission, from E. Bayer, Z. *Naturforsch.*, 1983, **38b**, 1281. Copyright 1983, Verlag der Zeitschrift für Naturforschung, Tübingen).

The principle of using polysiloxane anchoring of the chiral selector was further developed in 1989 by the introduction of CSPs comprised of polysiloxanes derived from (R,R)-tartramide [28]. In the first version, such CSPs were prepared by reaction of an activated ester of *N*-isopropyltartaric acid monoamide with an aminobutyl polysiloxane; the latter was obtained by lithium aluminium hydride reduction of a commercially available cyanopropyl polysiloxane (cf. Scheme 6.2). By the use of different cyanopropyl silicones as starting materials, CSPs with different properties were obtained, although the immobilized, chiral selector was the same throughout.

By a modification of the synthetic method, which also permitted the introduction of a hydrocarbon spacer, this type of (R,R)-tartramide-based CSP was later improved [29]. The synthetic strategy was based on the use of a hydrosilylation step by which (R,R)-*N*-(10-undecenyl)-*N*'-isopropyldiacetyltartramide was anchored to poly(hydromethylsiloxane). Hydrolytic deacetylation yielded the final product.

This phase was found to possess excellent temperature stability and to be suitable for the coating of fused silica capillaries. Since chiral selectors of this type had previously been shown to interact by dual hydrogen bond association with various diols in apolar solvents, forming diastereomeric complexes, it is perhaps not unexpected to find that an analogous association mechanism operates even in the gas phase. Consequently, these CSPs have proved successful for GC resolutions of a variety of racemic diols. The simplest chiral diol, 1,2-propanediol, was found to give the highest separation factor ($\alpha = 1.084$), with the (S)-enantiomer first eluted. C_2-symmetric diols were also successfully resolved. In the case of 2,4-pentanediol, the *meso*-form eluted before the two enantiomers.

6.2 PHASES BASED ON CHIRAL METAL COMPLEXES

A quite different family of CSPs for capillary GC was introduced in 1977 when it was demonstrated for the first time that an optically active rhodium(I) chelate could be used for the resolution of racemic 3-methylcyclopentene by complexation [30,31]. This new technique, based on the principle of metal co-ordination (cf. Section 2.1.2), was given the name *complexation gas chromatography*. The basic structure used in complexation GC is shown in Fig. 6.4. A chiral β-dicarbonyl ligand, of the same type

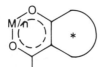

a chiral β–diketone–metal complex; the right-hand part represents the asymmetric cyclic or bicyclic structure element

Fig. 6.4 — Basic structure of the metal–ligand complex CSP.

as used in the NMR shift reagents (cf. Section 3.1.1.2), complexed to a transition metal by the two oxygen functions, constitutes the enantioselective component. The metal complex is used as a solution in squalane $(C_{30}H_{62})$ and coated onto a capillary column. The satisfactory thermal stability and low volatility of the complex and hydrocarbon make such columns useful in a temperature range from below room temperature up to ca. 100°C. As the enantioselective contribution to solute retention is based on electron donation to the metal, the method is well suited to relatively non-polar compounds with π- or lone-pair electrons, such as cyclic alkenes, ethers, thioethers and ketones, but has also been applied to donors such as alcohols and aziridines.

The chiral ligands used in these complexes are derived from readily available optically active natural products, particularly (+)-(1R)-camphor and the related monocyclic (+)-(R)-pulegone. These compounds are transformed into the β-dicarbonyl derivatives by perfluoroacylation at the α-carbon atom. A conversion into the transition metal complex is then easily obtained through the sodium salt [32]. Structures of the chiral metal chelates used so far are given in Fig. 6.5.

6.3 PHASES BASED ON INCLUSION EFFECTS

Quite interestingly, it has recently been shown that it is possible to make use of the inclusion effects that have generated the 'host–guest' solution chemistry (cf. Section 5.2.3), in gas–liquid chromatography. This technique is based on a cyclodextrin (CD) as a chiral selector, mixed with a polar solvent which acts as the liquid matrix. Coating of a Celite support with such a stationary phase gives a chiral GC sorbent for use in packed columns. A remarkable enantioselectivity can be obtained for certain hydrocarbons. However, for GC the columns must be used at exceptionally low temperatures (<70°C).

Some of the first reports on the use of this type of GC showed that a mixture of isomeric xylenes was well separated on a column where β-CD in formamide was used

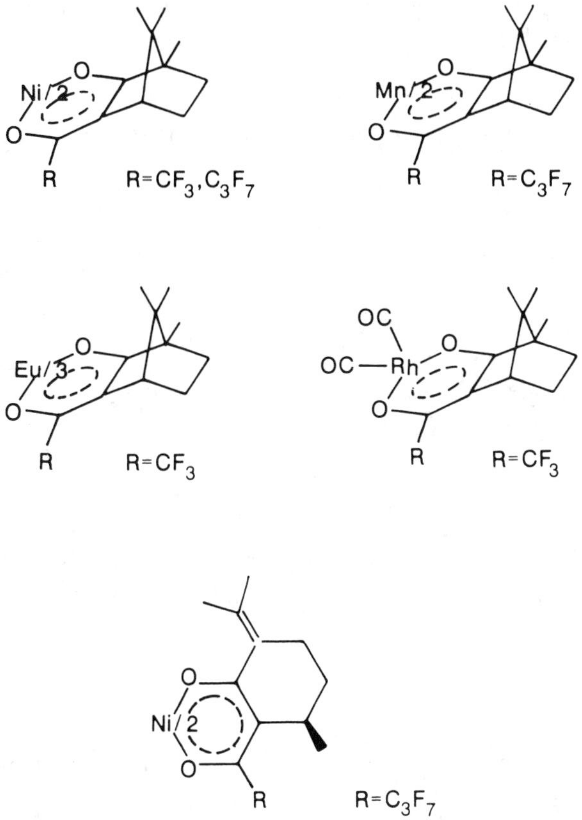

Fig. 6.5 — Chiral metal chelates employed for optical resolution by complexation GC.

as CSP [33]. Next, it was found that when α-CD and β-CD were compared as chiral selectors for optical resolution of racemic α- and β-pinenes, only the α-CD phase gave rise to separation of the optical antipodes [34]. However, the β-CD phase caused a much larger retention of the compounds, indicating a more stable β-CD–solution complex.

These new chiral sorbents show very great promise for separation of hydrocarbon enantiomers by GC. The technique is treated in further detail in Section 8.2.4.

Further progress was made in 1988, when it was first discovered that by substitution of all the free hydroxyl groups of α-cyclodextrin, a highly temperature-stable liquid, with great potential use as a chiral stationary phase in capillary gas chromatography, was obtained [35,36]. It was found that this phase permitted baseline resolution of a variety of trifluoroacetylated carbohydrates, methyl glycosides, 1,4- and 1,5-anhydroalditols, polyols and polyhydroxy-acid methyl esters. The mechanism behind the observed enantioselectivity in this case is not clear, although in most cases dipole–dipole interactions were found to be more probable than inclusion effects. However, since CSPs of this type have all been developed from typical inclusion-complex forming cyclodextrins, and since partial operation of such effects cannot be excluded, it seems reasonable to treat this subject under the present heading.

Several CSPs based on both α- and β-cyclodextrin soon followed [37–44]; their structures are given in Fig. 6.6. It should be noted that the structure variation

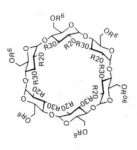

Lipodex A $R^2=R^3=R^6=$ n–pentyl   Lipodex C $R^2=R^3=R^6=$ n–pentyl

Lipodex B $R^2=R^6=$n–pentyl, $R^3=$ acetyl   Lipodex D $R^2=R^6=$ n–pentyl,$R^3=$ acetyl

Fig. 6.6 — Cyclodextrin derivatives used as CSPs in capillary GC (Lipodex columns).

obtained reflects the great difference in reactivity between the 3-hydroxyl group and the two others. Thus, it is possible to obtain the 2,6-di-O-alkylated cyclodextrin in a first step, leaving a free 3-hydroxyl group for subsequent modification, either by alkylation or esterification. The four basic types of derivatives shown in Fig. 6.6, called Lipodex A-D, have been extensively investigated and found to differ significantly in enantioselective properties. The types of compounds preferentially separated on the various columns are listed in Table 6.2. An example, showing the performance of capillary columns packed with this type of material, is given in Fig. 6.7.

Table 6.2 — Compounds preferentially separated on the various Lipodex-phases. (Reprinted by courtesy of Macherey-Nagel GmbH)

Lipodex A	Carbohydrates, polyols, diols, alcohols
	Epoxyalcohols, hydroxy acid esters (all as O-TFA derivatives)
	Spiroacetals, ketones
Lipodex B	Carbohydrates, aldols (both as O-TFA-derivatives)
	Amino alcohols (as N,O-TFA derivatives)
	Diols, glycerol derivatives (both as cyclic carbonates)
	Succinimides, lactones
Lipodex C	Alcohols, cyanhydrins, hydroxy acid esters (all as O-TFA-derivatives), olefins
Lipodex D	Amines, amino alcohols, amino acid esters, *trans*-cycloalkane-1,2- and 1,3-diols (all as TFA-derivatives), lactones

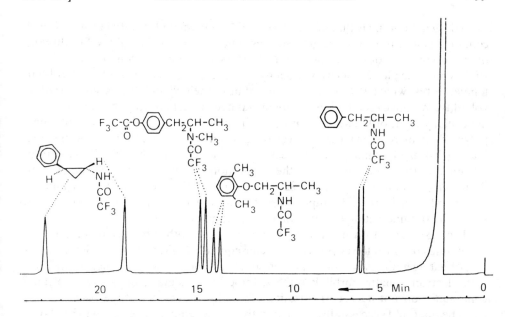

Fig. 6.7 — Enantiomer separation of a series of basic drugs, as trifluoroacetyl derivatives, on a
45-m Lipodex D capillary column at 175°C. (Reprinted, with permission, from W. A. König, S.
Lutz, G. Wenz and E. von der Bey, *J. High Res. Chromatog., Chromatog. Commun.*, 1988, **11**,
506. Copyright 1988, Dr. Alfred Huethig Publishers.)

Although the role of the size of the cyclodextrin cavity in these phases is still not
quite clear, an interesting result was obtained from studies of a series of trifluoroacetylated α-hydroxy acid methyl esters. Resolution experiments with these esters on
perpentylated α- and β-cyclodextrin columns, respectively, showed that while on
both columns only the lower homologues can be separated, the optimum of
enantioselective interaction differs. On the α-CD-based CSP the highest separation
factor was found for the lactic acid derivative, whereas for the other CSP a maximum
was obtained for the 2-hydroxyisovaleric acid derivatives. The two columns were
operated under identical conditions (column temperature 50°C). This dependence
on the size of the CD cavity indicates a retention mechanism with contributions from
stereoselective inclusion phenomena.

By analogy with Chirasil-Val, a CSP consisting of permethyl-β-cyclodextrin
anchored to a polysiloxane backbone has also been prepared [45]. A number of
hydroxy-containing compounds (diols, hydroxy acid esters, carboxylic acids) were
resolved on this phase without any derivatization.

6.4   RELATIVE MERITS OF THE VARIOUS MODES OF CHIRAL GAS
CHROMATOGRAPHY

As we have learnt from the previous section, the techniques of chiral GC available
today cover a broad range of compounds. A common feature of all GC-phases based

on hydrogen bonding is that they are, generally speaking, best suited to compounds containing polar functional groups, such as amides, esters and alcohols. For this and other reasons various *N*-acylamino-acid esters have been favoured subjects for studies on these phases. Such compounds, however, often require quite high column temperatures, which will cause severe bleeding of the CSP unless this has a very low volatility. We have seen that this problem has been attacked either by increasing the molecular weight of the selector or by linking it to an essentially non-volatile inert polymer. An increase in molecular weight has been made either by incorporation of a long hydrocarbon chain or by increasing the number of polar bonds. A certain dilemma is that quite generally the separation factor obtained increases with decreasing column temperature [22], which means that some of the advantages associated with increased temperature stability are neutralized by the decreased selectivity obtained at higher operating temperatures.

The CSPs based on metal complexation, on the other hand, are suitable for compounds of much lower polarity and consequently higher volatility. Since metal-co-ordinating abilities are found even in simple alkenes, as well as other compounds with electron-donating orbitals (ethers, thioethers, esters, etc), many resolvable compounds do not need derivatization. This also means that these columns may often be operated successfully at comparatively low temperatures. As will be shown later, capillary columns coated with CSPs of this kind are excellent for head-space analysis (e.g. studies of chiral alkene production). Furthermore, they have proved useful in the investigations of various chiral pheromones.

Finally, the technique of using chiral selectors which might form inclusion complexes with the analytes appears to be highly versatile. Columns containing α-cyclodextrin dissolved in the stationary phase are thus far unique in being capable of separating enantiomers of saturated hydrocarbons. The alkylated cyclodextrin phases (Lipodex) show enantioselectivity towards a wide spectrum of analytes. Interestingly, this includes hydrocarbons [43], although no resolutions of completely saturated hydrocarbons have yet been performed on these phases.

A summary of the different types of CSPs used and their main applications and properties is given in Table 6.3.

Table 6.3 — Types of compounds used as stationary phase materials in chiral gas chromatography (an asterisk denotes a commercially available phase or column; see Appendix)

1. Carbonyl-bis (amino-acid esters)

Amino-acid	Esters used	Main types of analytes studied	References
L-Valine	Methyl	N-Perfluoroacyl amines and amino-acids	[7]
	Ethyl	N-Perfluoroacyl amines and amino-acids	[7]
	Isopropyl	N-Perfluoroacyl amines and amino-acids	[7, 21, 46–52]
	tert-Butyl	N-Perfluoroacyl amines and amino-acids	[7]
L-Leucine	Isopropyl	N-Perfluoroacyl amines	[21]
D-Leucine	Isopropyl	N-Perfluoroacyl amines	[6, 7]
L-Proline	Isopropyl	N-Perfluoroacyl amines	[47]
L-Valine (1-) + glycine (1'-)	Isopropyl	N-Perfluoroacyl amines	[50]

2. N-Acylpeptide esters

Peptide	Acyl group	Esters used	Main types of analytes studied	References
L-Valyl-L-valine	TFA	Isopropyl	N-Perfluoroacyl amino-acid-esters	[20]
		Cyclohexyl	N-Perfluoroacyl amino-acid esters	[22,48,53,54]
		o-Carboranyl isopropyl	N-Perfluoroacyl amino-acid esters	[55]
	Acetyl	Isopropyl	N-Perfluoroacyl amino-acid esters	[20]
	Lauroyl	Cyclohexyl	N-Perfluoroacyl amino-acid esters	[56]
		Lauryl	N-Perfluoroacyl amino-acid esters	[57]
L-Valyl-L-leucine	TFA	Cyclohexyl	N-Perfluoroacyl amino-acid esters	[54,58]
	PFP	Cyclohexyl	N-Perfluoroacyl amino-acid esters	[54]
	Lauroyl	Cyclohexyl	N-Perfluoroacyl amino-acid esters	[56]
L-Leucyl-L-valine	TFA	Cyclohexyl	N-Perfluoroacyl amino-acid esters	[56]
L-Leucyl-L-leucine	TFA	Cyclohexyl	N-Perfluoroacyl amino-acid esters	[56]
L-Norvalyl-L-norvaline	TFA	Cyclohexyl	N-Perfluoroacyl amino-acid esters	[59–62]
L-Norleucyl-L-norleucine	TFA	Cyclohexyl	N-Perfluoroacyl amino-acid esters	[59]
L-Isoleucyl-L-isoleucine	TFA	Cyclohexyl	N-Perfluoroacyl amino-acid esters	[62]
L-Alanyl-L-alanine	TFA	Cyclohexyl	N-Perfluoroacyl amino-acid esters	[59,60]
L-Phenylalanyl-L-leucine	TFA	Cyclohexyl	N-Perfluoroacyl amino-acid esters	[63–66]
L-Methionyl-L-methionine	TFA	Cyclohexyl	N-Perfluoroacyl amino-acid esters	[67]

(continued p. 102)

Table 6.3 (*contd.*)

2. *N*-Acylpeptide esters (contd.)

Peptide	Acyl group	Esters used	Main types of analytes studied	References
L-α-Aminobutyryl-L-aminobutyric acid	TFA	Cyclohexyl	*N*-Perfluoroacyl amino-acid esters	[59,62,68]
	TFA	*o*-Carboranyl propyl	*N*-Perfluoroacyl amino-acid esters	[55]
L-Leucyl-L-leucyl-L-leucine	TFA	Cyclohexyl	*N*-Perfluoroacyl amino-acid esters	[54]
L-Valyl-L-valyl L-valine	TFA	Isopropyl	*N*-Perfluoroacyl amino-acid esters	[20]

3. Substituted triazines

Compound	Main types of analytes studied	References
N,N'-[2,4-(6-Ethoxy-1,3,5-triazine)diyl]-bis(L-valyl-L-valyl-L-valine isopropyl ester)*	Wide variety of applications reported	[69–73]
N,N'-[2,4-(6)-Ethoxy-1,3,5-triazine)diyl]-bis(L-valyl-L-valine isopropyl ester)*	Wide variety of applications reported	[69]
N,N',N''-[2,4,6-(1,3,5-triazine)triyl]-tris(*N*α-lauroyl-L-lysine-*tert*-butylamide)*	Wide variety of applications reported	[69,71]

4. Menthyl esters

Alcohol	Acid	Main types of analytes studied	References
(−)-Menthol	(+)-Tartaric (diester)	Perfluoroacyl amines and amino-acids, other amides	[74]
rac.-Menthol	(−)-Malic (diester)	Perfluoroacyl amines and amino-acids, other amides	[74]

5. Substituted amides, diamides and related compounds

A. Compounds not polymer-bound

(1) *N*-Acylamino-acid amides

Amino-acid	Acyl group	Amide used	References
L-Valine	Lauroyl	Isopropyl	[22]
		Cyclohexyl-, -heptyl -octyl and others	[22]
		tert-Butyl	[3,22,75–77]
	Isobutyryl, pivaloyl, *tert* butylacetyl	Dodecyl	[22]
	Caproyl	Hexyl	[66]
	Docosanoyl	*tert*-Butyl	[12,77]
L-Leucine	Docosanoyl	*tert*-Butyl	[76]
	Lauroyl	*tert*-Butyl	[75,76]
L-Alanine	Lauroyl	*tert*-Butyl	[75,76]
L-Penylalanine	Lauroyl	*tert*-Butyl	[75,76]
D-Phenylglycine	Lauroyl	*tert*-Butyl	[75,76]
L-Proline	Lauroyl	(S) or (R)-1-(α-Naphthyl)-ethylamide*	[9]

(continued p. 103)

Table 6.3 (*contd.*)

(2) *O*-Acyloxy acid amides

Hydroxy-acid	Acyl group	Amide used	References
(S)-3-Phenyl-lactic	BOC	*tert*-Butyl	[17]
(S)-Mandelic	Lauroyl	(S) or (R)-1-(α-Naphthyl)ethylamide*	[9]
	(1R,3R)-*trans*-Chrysanthemoyl	(S) or (R)-1-(α-Naphthyl)ethylamide*	[11,78]

(3) Other amides

Acyl group	Amide used	References
Lauroyl	(S) or (R)-1-(α-Naphthyl)-ethylamide	[8,79]
	(S)-α-Phenylethylamide	[8]
(1R,3R)-*trans*-Chrysanthemoyl	(R)-1-(α-Naphthyl)ethyl-amide*	[10]

B. Polysiloxane-bonded compounds

Phase	Main types of analytes studied	References
L-Valine *tert*-butylamide/polysiloxane ("Chirasil-Val")	A very large number of different analytes have been optically resolved	[24,80–92]
L-Valine *tert*-butylamide/modified OV-225		[14]
L-Valine *tert*-butylamide/modified XE-60		[93]
L-Valine (S) or (R)-α-phenylethylamide/modified XE-60 or OV-225		[17,19,94–99]
O-BOC-L-valine or -L-leucine/modified OV-225		[16]
(R,R)-*N*-Alkyl-*N'*-isopropyltartramide linked to polysiloxane	Diols	[28,29]

6. Metal complexes

Ligand	Metal	Main types of analytes studied	References
(1R)-3-HFB-camphorate	Cu	A wide variety of volatile	[99]
	Mn	compounds resolvable, e.g.	[100,101]
	Ni	alcohols, ketones, esters, oxiranes, ethers, spiroketals	[32,102]
(1R)-3-TFA-camphorate	Mn		[101]
	Ni		[103]
	Rh(I)(dicarbonyl mono-complex)		[103]

Copper–Schiff's base complexes (binuclear):

Ligand components		References
Aldehyde	Amine	
Salicylaldehyde	(S)-2-Amino-1,1-di-phenylpropan-1-ol*	[104]
	(R)-2-Amino-1,1-bis-(5-*tert*-butyl-2-octyl-	[104,105]

(continued p. 104)

Table 6.3 (*contd.*)

oxyphenyl)-propan-1-ol*	
(S)-2-Amino-1,1-bis-	[105]
(5-*tert*-butyl-2-heptyl-	
oxyphenyl)-3-phenyl-	
propan-1-ol*	

7. Cyclodextrin derivatives

Cyclodextrin	Substitution	References
α- and β-CD	Methyl	[106,107]
α-CD	Perpentyl (Lipodex A)	[35,36,39]
α-CD	Hexakis(3-*O*-acetyl-2,6-di-*O*-pentyl) (Lipodex B)	[38]
β-CD	Perpentyl (Lipodex C)	[43]
β-CD	Heptakis(3-O-acetyl-2,6-di-*O*-pentyl) (Lipodex D)	[37]

BIBLIOGRAPHY

C. H. Lochmüller and R. W. Souter, Chromatographic Resolution of Enantiomers: Selective Review, *J. Chromatog.*, 1975, **113**, 283; *Chromatog. Rev.*, 1975, **19**, 283.

C. H. Lochmüller, Gas Chromatographic Separation of Enantiomers, *Sep. Purif. Meth.*, 1979, **8**, 21.

R. H. Liu and W. W. Ku, Chiral Stationary Phases for the Gas-Liquid Chromatographic Separation of Enantiomers, *J.Chromatog.*, 1983, **271**, 309; *Chromatog. Rev.*, 1983, **27**, 309.

V. Schurig, Gas Chromatographic Methods, in *Asymmetric Synthesis*, Vol. I, J. D. Morrison (ed.), Academic Press, New York, 1983.

S. Allenmark, Recent Advances in Methods of Direct Optical Resolution, *J. Biochem. Biophys. Methods*, 1984, **9**, 1.

B. Koppenhöfer and E. Bayer, Chiral Recognition in Gas Chromatographic Analysis of Enantiomers on Chiral Polysiloxanes, in *The Science of Chromatography*, F. Bruner (ed.), Elsevier, Amsterdam, 1985, p. 1.

R. W. Souter, *Chromatographic Separation of Stereoisomers*, CRC Press, Boca Raton, 1985.

W. A. König, *The Practice of Enantiomer Separation by Capillary Gas Chromatography*, Hüthig Verlag, Heidelberg, 1987.

W. A. König, A new generation of chiral phases for gas chromatography (GC). *Nachr. Chem. Tech. Lab.*, 1989, **37**, 471.

REFERENCES

[1] E. Gil-Av, B. Feibush and R. Charles-Siegler, *Tetrahedron Lett.*, 1966, 1009.
[2] E. Gil-Av and B. Feibush, *Tetrahedron Lett.*, 1967, 3345.
[3] B. Feibush, *Chem. Commun.*, 1971, 544.
[4] J. A. Corbin, J. E. Rhoad, and L. B. Rogers, *Anal. Chem.*, 1971, **43**, 327.
[5] C. H. Lochmüller and R. W. Souter, *J. Phys. Chem.*, 1973, **77**, 3016.
[6] C. H. Lochmüller and R. W. Souter, *J. Chromatog.*, 1973, **87**, 243.
[7] C. H. Lochmüller and R. W. Souter, *J. Chromatog.*, 1974, **88**, 41.
[8] N. Oi, H. Kitahara, Y. Inda and T. Doi, *J. Chromatog.*, 1981, **213**, 137.
[9] N. Oi, H. Kitahara, Y. Inda and T. Doi, *J. Chromatog.*, 1982, **237**, 297.
[10] N. Oi, T. Doi, H. Kitahara and Y. Inda, *J. Chromatog.*, 1982, **239**, 493.
[11] N. Oi, H. Kitahara and T. Doi, *J. Chromatog.*, 1983, **254**, 282.
[12] H. Frank, G. J. Nicholson and E. Bayer, *J. Chromatog. Sci.*, 1977, **15**, 174.
[13] H. Frank, G. J. Nicholson and E. Bayer, *Angew. Chem.*, 1978, **90**, 396. (*Int. Ed.*, 1978, **17**, 363).
[14] T. Saeed, P. Sandra and M. Verzele, *J. Chromatog.*, 1979, **186**, 611.
[15] T. Saeed, P. Sandra and M. Verzele, *J. High Resol. Chromatog.*, *Chromatog. Commun.*, 1980, **3**, 35.

[16] W. A. König and I. Benecke, *J. Chromatog.*, 1981, **209**, 91.
[17] W. A. König, I. Benecke and S. Sievers, *J. Chromatog.*, 1981, **217**, 71.
[18] W. A. König, I. Benecke and H. Bretting, *Angew. Chem.*, 1981, **93**, 688.
[19] I. Benecke, E. Schmidt and W. A. König, *J. High Resol. Chromatog., Chromatog. Commun.*, 1981, **4**, 553.
[20] B. Feibush and E. Gil-Av, *Tetrahedron*, 1970, **26**, 1361.
[21] C. H. Lochmüller, J. M. Harris and R. W. Souter, *J. Chromatog.*, 1972, **71**, 405.
[22] U. Beitler and B. Feibush, *J. Chromatog.*, 1976, **123**, 149.
[23] H. Frank, G. J. Nicholson and E. Bayer, *Angew. Chem.*, 1978, **90**, 396 (*Int. Ed.*, 1978, **17**, 363).
[24] B. Koppenhöfer and E. Bayer, *Chromatographia*, 1984, **19**, 123.
[25] B. Koppenhöfer, H. Allmendinger and G. Nicholson, *Angew. Chem.*, 1985, **97**, 46.
[26] B. Koppenhöfer and H. Allmendinger, *Chromatographia*, 1986, **21**, 503.
[27] E. Bayer, *Z. Naturforsch.*, 1983, **38b**, 1281.
[28] K. Nakamura, S. Hara and Y. Dobashi, *Anal. Chem.*, 1989, **61**, 2121.
[29] K. Nakamura, T. Saeki, M. Matsuo, S. Hara and Y. Dobashi, *Anal. Chem.*, 1990, **62**, 539.
[30] V. Schurig and E. Gil-Av, *Isr. J. Chem.*, 1976–77, **15**, 96.
[31] V. Schurig, *Angew. Chem.*, 1977, **89**, 419 (*Int. Ed.*, 1977, **16**, 110).
[32] V. Schurig and W. Bürkle, *J. Am. Chem. Soc.*, 1982, **104**, 7573.
[33] D. Sybilska and T. Koscielski, *J. Chromatog.*, 1983, **261**, 357.
[34] T. Koscielski, D. Sybilska and J. Jurczak, *J. Chromatog.*, 1983, **280**, 131.
[35] W. A. König, P. Mischnick-Lübbecke, B. Brassat, S. Lutz and G. Wenz, *Carbohydrate Res.*, 1988, **183**, 11.
[36] W. A. König, S. Lutz, P. Mischnick-Lübbecke, B. Brassat and G. Wenz, *J. Chromatog.*, 1988, **447**, 193.
[37] W. A. König, S. Lutz, C. Colberg, N. Schmidt, G. Wenz, E. von der Bey, A. Mosandl, C. Günther and A. Kustermann, *J. High Res. Chromatog., Chromatog. Commun.*, 1988, **11**, 621.
[39] W. A. König, S. Lutz and G. Wenz, *Angew. Chem.*, 1988, **100**, 989; (*Int. Ed.*, 1988, **27**, 979).
[40] W. A. König, S. Lutz and G. Wenz, *Proc. 4th Internat. Symp. on Cyclodextrins* (München 1988), Kluwer Academic Publishers, Dordrecht, 1988.
[41] W. A. König, S. Lutz, P. Mischnick-Lübbecke, B. Brassat, E. von der Bey and G. Wenz, *Starch/Stärke*, 1988, **40**, 472.
[42] W. A. König, S. Lutz, M. Hagen, R. Krebber, G. Wenz, K. Baldenius, J. Ehlers and H. tom Dieck, *J. High Res. Chromatog., Chromatog. Commun.*, 1989, **12**, 11.
[43] J. Ehlers, W. A. König, S. Lutz, G. Wenz and H. tom Dieck, *Angew. Chem.*, 1988, **100**, 1614; (*Int. Ed.*, 1988, **27**, 1556).
[44] W. A. König, S. Lutz, G. Wenz, G. Görgen, C. Neumann, A. Gäbler and W. Boland, *Angew. Chem.*, 1989, **101**, 180; (*Int. Ed.*, 1989, **28**, 178).
[45] P. Fischer, R. Aichholz, U. Bölz, M. Juza and S. Krimmer, *Angew. Chem.*, 1990, **102**, 439.
[46] J. A. Corbin and L. B. Rogers, *Anal. Chem.*, 1970, **42**, 974.
[47] R. W. Souter, *J. Chromatog.*, 1975, **114**, 307.
[48] J. A. Corbin and L. B. Rogers, *Anal. Chem.*, 1970, **42**, 1786.
[49] B. Feibush and E. Gil-Av, *J. Gas Chromatog.*, 1967, **5**, 257.
[50] C. H. Lochmüller and J. V. Hinshaw, Jr., *J. Chromatog.*, 1979, **178**, 411.
[51] B. Feibush, E. Gil-Av and T. Tamari, *J. Chem. Soc. Perkin Trans. II*, 1972, 1197.
[52] H. Rubinstein, B. Feibush and E. Gil-Av, *J. Chem. Soc. Perkin Trans. II*, 1973, 2094.
[53] S. Nakaparksin, P. Birrell, E. Gil-Av and J. Oro, *J. Chromatog. Sci.*, 1970, **8**, 177.
[54] J. A. Corbin, J. E. Rhoad and L. B. Rogers, *Anal. Chem.*, 1971, **43**, 327.
[55] R. Brazell, W. Parr, F. Andrawes and A. Zlatkis, *Chromatographia*, 1976, 9, 57.
[56] I. Abe and S. Musha, *J. Chromatog.*, 1980, **200**, 195.
[57] H. Iwase, *Chem. Pharm. Bull.*, 1975, **23**, 1608.
[58] W. Parr and P. Howard, *Chromatographia*, 1971, **4**, 162.
[59] W. Parr and P. Howard, *Anal. Chem.*, 1973, **45**, 711.
[60] W. Parr and P. Howard, *J. Chromatog.*, 1972, **66**, 141.
[61] W. Parr and P. Howard, *J. Chromatogr.*, 1972, **67**, 227.
[62] P. Howard and W. Parr, *Chromatographia*, 1974, **7**, 283.
[63] W. Parr, C. Yang, E. Bayer and E. Gil-Av, *J. Chromatog. Sci.*, 1970, **8**, 591.
[64] W. Parr, C. Yang, J. Pleterski and E. Bayer, *J. Chromatog.*, 1970, **50**, 510.
[65] W. König, W. Parr, H. A. Lichtenstein, E. Bayer and J. Oro, *J. Chromatog. Sci.*, 1970, **8**, 183.
[66] K. Grohmann and W. Parr, *Chromatographia*, 1972, **5**, 18.
[67] F. Andrawes, R. Brazell, W. Parr and A. Zlatkis, *J. Chromatog.*, 1975, **112**, 197.
[68] W. Parr and P. Y. Howard, *J. Chromatog.*, 1972, **71**, 193.

[69] N. Oi, T. Doi, H. Kitahara and Y. Inda, *J. Chromatog.*, 1981, **208**, 404.
[70] N. Oi, M. Horiba and H. Kitahara, *Agric. Biol. Chem.*, 1979, **43**, 2403.
[71] N. Oi, M. Horiba and H. Kitahara, *J. Chromatog.*, 1980, **202**, 299.
[72] N. Oi, M. Horiba and H. Kitahara, *Agric. Biol. Chem.*, 1981, **45**, 1509.
[73] M. Horiba, S. Kida, S. Yamamoto and N. Oi, *Agric. Biol. Chem.*, 1982, **46**, 281.
[74] N. Oi, H. Kitahara and T. Doi, *J. Chromatog.*, 1981, **207**, 252.
[75] S.-C. Chang, R. Charles and E. Gil-Av, *J. Chromatog.*, 1982, **235**, 87.
[76] S.-C. Chang, R. Charles and E. Gil-Av, *J. Chromatog.*, 1982, **238**, 29.
[77] R. Charles, U. Beitler, B. Feibush and E. Gil-Av, *J. Chromatog.*, 1975, **112**, 121.
[78] N. Oi, R. Takai and H. Kitahara, *J. Chromatog.*, 1983, **256**, 154.
[79] S. Weinstein, B. Feibush and E. Gil-Av, *J. Chromatog.*, 1976, **126**, 97.
[80] H. Frank, A. Rettenmeier, H. Weicker, G. J. Nicholson and E. Bayer, *Anal. Chem.*, 1982, **54**, 715.
[81] H. Frank, G. J. Nicholson and E. Bayer, *J. Chromatog.*, 1978, **146**, 197.
[82] A. L. Leavitt and W. R. Sherman, *Methods Enzymol.*, 1982, **89**, 3.
[83] E. Bailey, P. B. Farmer and J. H. Lamb, *J. Chromatog.*, 1980, **200**, 145.
[84] H. Frank, A. Rettenmeier, H. Weicker, G. J. Nicholson and E. Bayer, *Clin. Chim. Acta*, 1980, **105**, 201.
[85] G. J. Nicholson, H. Frank and E. Bayer, *J. High Resol. Chromatog.*, *Chromatog. Commun.*, 1979, **2**, 411.
[86] W. Chinghai, H. Frank, W. Guanghua, Z. Liangmo, E. Bayer and L. Peichang, *J. Chromatog.*, 1983, **262**, 352.
[87] H. Frank, J. Gerhardt, G. J. Nicholson and E. Bayer, *J. Chromatog.*, 1983, **270**, 159.
[88] K. M. McErlane and G. K. Pillai, *J. Chromatog.*, 1983, **274**, 129.
[89] K. D. Haegele, J. Schoun, R. G. Alken and N. D. Huebert, *J. Chromatog.*, 1983, **274**, 103.
[90] J. H. Liu and W. W. Ku, *Anal. Chem.*, 1981, **53**, 2180.
[91] I. Abe, S. Kuramoto and S. Musha, *J. High Resol. Chromatog.*, *Chromatog. Commun.*, 1983, **6**, 366.
[92] H. Frank, W. Woiwode, G. Nicholson and E. Bayer, *Liebigs Ann. Chem.*, 1981, 354.
[93] I. Abe, S. Kuramoto and S. Musha, *J. Chromatog.*, 1983, **258**, 35.
[94] W. A. König and I. Benecke, *J. Chromatog.*, 1983, **269**, 19.
[95] W. A. König, I. Benecke and K. Ernst, *J. Chromatog.*, 1982, **253**, 267.
[96] W. A. König, I. Benecke and S. Sievers, *J. Chromatog.*, 1982, **238**, 427.
[97] W. A. König, W. Franke and I. Benecke, *J. Chromatog.*, 1982, **239**, 227.
[98] W. A. König, I. Benecke and H. Bretting, *Angew. Chem.*, 1981, **93**, 688 (*Int. Ed.*, 1981, **20**, 693).
[99] R. Weber and V. Schurig, *Naturwiss.*, 1981, **68**, 330.
[100] R. Weber, K. Hintzer and V. Schurig, *Naturwiss.*, 1980, **67**, 453.
[101] V. Schurig and R. Weber, *J. Chromatog.*, 1981, **217**, 51.
[102] V. Schurig, W. Bürkle, A. Zlatkis and C. Poole, *Naturwiss.*, 1979, **66**, 423.
[103] V. Schurig, *Chromatographia*, 1980, **13**, 263.
[104] N. Oi, K. Shiba, T. Tani, H. Kitahara and T. Doi, *J. Chromatog.*, 1981, **211**, 274.
[105] N. Oi, M. Horiba, H. Kitahara, T. Doi, T. Tani and T. Sakakibara, *J. Chromatog.*, 1980, **202**, 305.
[106] T. Koscielski, D. Sybilska, S. Belniak and J. Jurczak, *Chromatographia*, 1986, **21**, 413.
[107] V. Schurig and H.-P. Novotny, *J. Chromatog.*, 1988, **441**, 155.

7

Chiral liquid chromatography

Though the concept of chiral GC is quite clear, viz. that enantiomers are separable with a chiral stationary phase, the situation is more complex for chiral LC. Here we can distinguish between two fundamentally different cases, depending upon whether enantiodifferentiation takes place through a chiral recognition effect by the stationary phase or by a chiral constituent of the mobile phase forming a diastereomeric complex *in situ* during the chromatographic process. From an *experimental* point of view, a clear distinction can be made between the use of a chiral column (i.e. a column containing a CSP) together with an achiral mobile phase, and the use of an achiral column together with a chiral mobile phase. In the latter case, however, the actual mode of chiral separation will depend on the relative affinity of the chiral constituent for the stationary phase and the analyte, respectively. In one extreme case, the chiral constituent may become strongly adsorbed on the stationary phase, thereby converting it into a CSP, and the separation process would then be regarded as a chiral recognition by the CSP thus generated. In the opposite case, the chiral constituent has a much lower affinity for the stationary phase than for the analyte. This means that diastereomeric complexes are generated in the mobile phase and the separation takes place as a normal LC separation of diastereomers. Between these two extreme cases there is probably a range of mixed retention modes. In other words: are diastereomeric complexes formed at the surface of the chromatographic sorbent, or in the mobile phase, or both?

Chiral sorbents for LC may further be classified with respect to their general structural types. Some are based on synthetic or natural polymers and are totally and intrinsically chiral. Others consist of chiral selectors of low molecular weight which are bound to a hard, incompressible matrix, usually silica. There are also sorbents consisting of polymers anchored to silica in order to give improved column performance.

A rough classification of the various modes of chiral LC is given in Table 7.1. This is mainly based on the general nature of the chromatographic sorbent, without regard to the type of retention process involved, which is often quite complex and difficult to evaluate in any detail.

Table 7.1 — Summary of the main methods used for direct optical resolution by liquid chromatography

Site of chiral selector	Basic principle	Capacity	Column efficiency
Stationary phase (CSP)	Use of an intrinsically chiral, polymeric stationary phase, either of natural origin (polysaccharides and derivatives, proteins) or synthetic (synthetic polymers with chiral substituents or 'grafted' chiral cavities)	Analytical to preparative	Low to moderate, depending on whether a support material is used or not
Stationary phase (CSP)	Use of bonded synthetic chiral selectors	Analytical to preparative	Moderate to fairly high
Mobile phase (CMP)	Addition of chiral constituents to the mobile phase system used. Column achiral, usually an alkyl-silica used in reversed-phase mode	Analytical	Moderate to fairly high

7.1 CHIRAL STATIONARY PHASES BASED ON NATURALLY OCCURRING AND SYNTHETIC POLYMERS

Owing to the early recognition of the chiral nature and ready availability of many natural products, particularly carbohydrates, such compounds were among the first to be tried as sorbents for optical resolution by LC. As early as 1938 a partial resolution of a racemic camphor derivative was obtained on a column packed with lactose [1]. Lactose remained a column material of interest for some years and was used with success in the first nearly complete chromatographic chiral resolution described in the literature, which took place in 1944, when Tröger's base was resolved on a 0.9 m long lactose column [2]. The resolving capacity of a polysaccharide, viz. cellulose, was first realized by the observation that a racemic amino-acid could occasionally give two spots in paper chromatography [3–5]. Dalgliesh advanced his three-point interaction theory in 1952 on the basis of results from paper chromatography of racemic amino-acids [6]. Other early findings on direct optical resolution of amino-acids by means of paper chromatography [7] and cellulose thin-layer chromatography (TLC) [8] were reported. This led to further use of cellulose and cellulose derivatives, as well as investigations of starch and cyclodextrins for the purpose of chiral LC. At present, derivatives of a large number of natural polysaccharides have been investigated or are under investigation as potential chiral sorbents.

The principle of using chiral polymers has also been exploited with many different types of totally synthetic materials, by various approaches, and the results seem to be very promising.

Further, the enantioselectivity of proteins, first observed by studies of binding equilibria in solution (for reviews see [9,10]), has been successfully utilized for analytical chiral LC.

7.1.1 Polysaccharides and derivatives
7.1.1.1 Underivatized polysaccharides
A. Cellulose

The linear polysaccharide cellulose represents the most common organic compound of all. Its chemical constitution is that of a linear poly-β-D-1,4-glucoside (Fig. 7.1a).

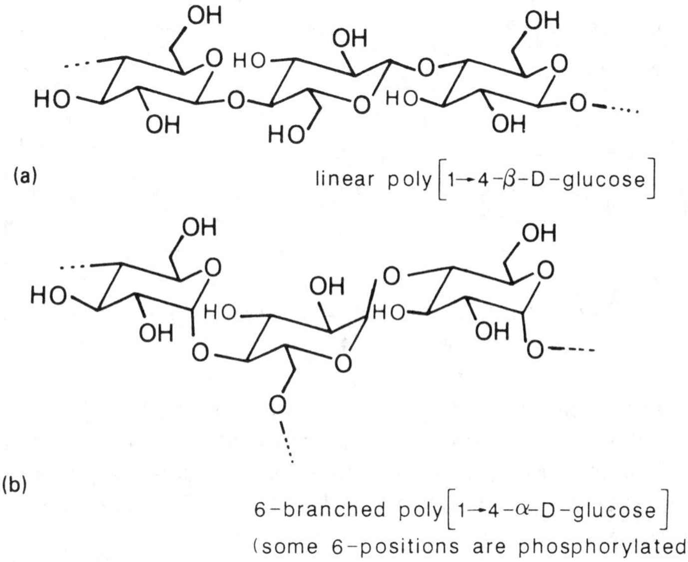

(a)

linear poly $\left[1\rightarrow4-\beta-D-glucose\right]$

(b)

6-branched poly $\left[1\rightarrow4-\alpha-D-glucose\right]$

(some 6-positions are phosphorylated)

Fig. 7.1 — The chemical structure of (a) cellulose and (b) amylopectin, the main component of starch.

It forms very long chains containing at least 1500 (+)-D-glucose units per molecule. The molecular weight of cellulose ranges from 2.5×10^5 to 1×10^6 or more. In a cellulose fibre these long molecules are arranged in parallel bundles and held together by numerous hydrogen bonds between the hydroxyl groups. In the native state cellulose is therefore built up from partially crystalline regions. These are not regenerated on precipitation of cellulose from solution. As seen from Fig. 7.1a the (+)-D-glucose repetitive unit contains five chiral centres and three hydroxyl groups. All the ring substituents are equatorial.

It has been found that partial hydrolysis of natural cellulose with dilute mineral acid can yield a material with a high degree of crystallinity because hydrolytic cleavage will take place preferentially in the amorphous regions. Such a material contains ca. 200 glucose units per chain and is usually called 'microcrystalline cellulose' [11]. It is marketed as 'Avicel' by several chemical companies.

Although derivatives of cellulose have been used in many successful resolutions, very good results have also been obtained by the use of unmodified cellulose and are therefore worth mentioning. The compounds resolved are, without exception, highly polar with multiple sites for hydrogen bond formation. Some typical results are summarized in Table 7.2.

It has been found [22] that on treatment with dilute alkali, cellulose loses its enantioselective properties owing to a transformation of the native, metastable form into a rearranged and stable amorphous form.

B. Starch

The other widespread polysaccharide, also built from (+)-D-glucose units, is starch. Its structure is more complex than that of cellulose. It is composed of ca. 20% amylose and 80% amylopectin, the latter being an insoluble fraction. Both are

Table 7.2 — Examples of optical resolutions performed by liquid chromatography on cellulose

Type of compound	LC mode	References
Amino-acids, amino-acid	Paper	[3–5,7,12]
derivatives	Thin layer	[8,13,14]
	Column	[15–17]
Diaminodicarboxylic acids	Column	[18]
Synthetic alkaloids	Column	[19,20]
Cathecins	Column	[21]

entirely composed of (+)-D-glucose units, linked by α-glucoside bonds. Whereas amylose is a linear polymer, amylopectin is branched by C_1–C_6 connections (Fig. 7.1b).

Depending upon the source, there are different particle sizes of starch available. The material obtained from potatoes is relatively coarse (60–100 μm) and has been favoured for column chromatography. Despite its ready availability and non-swellable properties in aqueous media, which give good flow properties, it has so far found very little use.

As in the case of cellulose, starch appears to be most suitable for polar aromatic compounds. Its use for resolution of atropisomers with structures containing polar substituents has been particularly well documented [23–27]. These separations show a very pronounced dependence on the nature of the mobile phase, and are especially influenced by the ionic strength. Figure 7.2 illustrates the chromatographic behav-

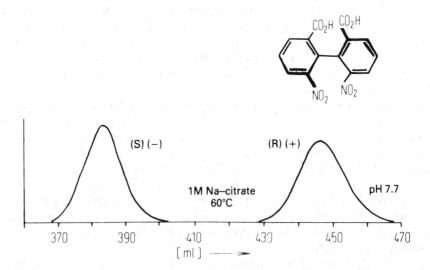

Fig. 7.2 — Optical resolution of atropisomers on a starch column. (Reprinted, with permission, from H. Hess, G. Burger and H. Musso, *Angew. Chem.*, 1978, **90**, 645. Copyright 1978, Verlag Chemie GmbH).

iour of a starch column after application of racemic 2,2′-dinitrodiphenic acid and elution with 1*M* sodium citrate buffer, pH 7.7, at 60°C. The separation factors obtained are quite satisfactory but the column efficiency is modest.

C. Cyclodextrins

As early as 1908 it was discovered by Schardinger [28] that new crystalline carbohydrates, so-called dextrins, were formed if starch was subjected to degradation by a
micro-organism, *Bacillus macerans* [29]. These compounds were found to be normal
β-1,4-D-glucosides, but cyclized to rings of 6–12 units. Those with the three smallest
rings (6–8 units) have been called α-, β-, and γ-cyclodextrins (CD), respectively, and
form inclusion complexes with various compounds of the correct size. The diameter
of a β-CD ring is 8Å and its volume is ca. 350Å^3. The stability of the inclusion
complex is largely dependent on the hydrophobic and steric character of the guest.
These phenomena make CDs, particularly β-CD, which is easily available, highly
promising for use in chiral LC.

The major development in chiral LC with cyclodextrins started with the technique of using them as mobile phase additives in TLC experiments [30–32]. This
technique has also been applied with success to column chromatography and will be
treated in Section 7.3. Earlier some efforts to use cross-linked cyclodextrin gels for
chromatographic purposes had also been made [33–36]. The first attempts to
immobilize CDs on solid supports were made quite recently [37,38]. By an improvement in coupling techniques, a highly efficient β-CD silica-bonded phase column is
now available [39,40].

Since the formation of inclusion complexes with CDs in aqueous systems is based
mainly on hydrophobic interaction, it is logical that a CD column operates entirely in
a reversed-phase mode. Consequently, the mobile phase systems normally used are
the same as those used in ordinary reversed-phase LC, usually methanol/water or
acetonitrile/water. This also means that buffers can be used to control the pH and
possibly affect the retention of charged solutes.

The rather special type of solute–CSP interaction present in the case of immobilized CDs deserves particular attention. The inclusion complexes formed are of great
interest, not only from a theoretical point of view. This field, which belongs to 'host–
guest' chemistry (like the crown-ether complexes which will be treated in Section
7.2.1), is important in achieving better understanding of the role and function of
ordered molecular complexes in biological systems.

The conformation of a cyclodextrin in an aqueous system is generally assumed to
approximate to a truncated cone, Fig. 7.3, possessing a hydrophobic internal surface.
Hydrophobic molecules such as benzene or hexane, which can diffuse in and out of
the cavity, are reversibly adsorbed on this surface. Retention of a hydrophobic solute
should largely be dependent on the efficiency of the contact with the interior of the
cavity. Enantioselectivity is likely to be associated with the chiral structure at the
entrance of the cavity, caused by the exposed 2- and 3-hydroxy groups in the glucose
units. If the solute is of a suitable size, allowing good contact with the internal surface
and hence restricted in movement, then a different interaction between the chiral
cavity entrance and the substituents of the two enantiomers may cause a difference in
both the complexation constants and the chromatographic k' values.

The effect of the mobile phase on the enantioselectivity appears to be large. In
general, both the k' and α values tend to decrease with increasing content of organic
modifier. In most cases, methanol as a retention modifier will decrease α to a lesser
degree than acetonitrile. Retention on CD columns is markedly affected by the

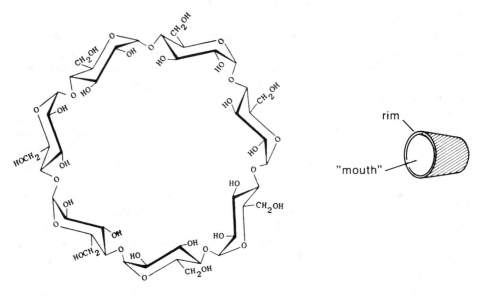

Fig. 7.3 — The chemical structure of β-cyclodextrin and its assumed conformation in an aqueous solution.

temperature, decreasing rapidly to zero between 60 and 80°C in most cases. This might be due to an increased conformational mobility of the ring with increased temperature.

7.1.1.2   Polysaccharide derivatives

A. Cellulose triacetate

In 1966 Lüttringhaus and co-workers [41,42] found that a partially acetylated cellulose (described as a 2.5 acetate) could be used with ethanol to achieve optical resolution in column chromatography. Some years later another German research group carefully investigated the heterogeneous acetylation of native (microcrystalline) cellulose and found that a triacetate could be prepared with almost complete preservation of the microcrystallinity and excellent resolving properties [43]. They pointed out that the microcrystallinity was essential for the enantioselective properties of the material, since the optical resolution power was totally lost on dissolution and reprecipitation. The metastable state of the material was evident from these experiments as the change was found to be irreversible. On the basis of the results from these investigations it was concluded that microcrystalline cellulose triacetate (MCTA) gave retention by means of inclusion of the solute into molecular cavities in the chiral matrix. Therefore the term 'inclusion chromatography' was used [44,45].

Owing to the availability and low cost of 'microcrystalline' cellulose (Avicel), the well described technique of its acetylation and the interesting properties of the product, MCTA has been the subject of extensive research during the last decades. Large columns can be packed with this cheap material and relatively large quantities of sample can be used, permitting preparative work. A typical example of a resolution on MCTA is shown in Fig. 7.4.

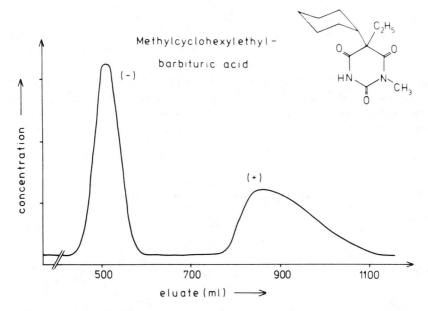

Fig. 7.4 — Separation of the enantiomers of 205 mg of (±)-methylcyclohexylethylbarbituric acid on 210 g of MCTA. (Column 85 × 2.5 cm, ethanol 96%, flow-rate 50 ml/hr). (Reprinted from G. Blaschke, *J. Liquid Chromatog.*, 1986, **9**, 341 by courtesy of Marcel Dekker, Inc.).

It is very important that MCTA is allowed to swell in boiling ethanol before being packed into a column. Ethanol (95%) is a good medium for swelling, and does not dissolve MCTA.

The inclusion mode of retention of solutes on MCTA is consistent with the very different chromatographic behaviour shown by benzene and mesitylene (1,3,5-trimethylbenzene), the first being much more strongly retained owing to better permeation into the cavities. 1,3,5-Tri-*tert*-butylbenzene is totally excluded and therefore of use for void volume determinations [46].

Further evidence for the inclusion model of retention is the fact that very non-polar solutes, lacking functional groups, can be resolved. Thus, racemic *trans*-1,2-diphenylcyclopropane is easily resolved into its antipodes on MCTA [47]. Optical resolutions of a great number of structurally quite different compounds on MCTA have been described to date, many on a preparative or semipreparative scale. This will be treated further in Chapter 9.

The main drawbacks of MCTA are its compressibility and relatively large, irregular and inhomogeneous particle size. This means that preparative columns can only be run at very low linear flow-rates (cf. the conditions used in Fig. 7.4). The latter problem can be partially solved by grinding and careful fan-sieving of the material. However, since MCTA requires a swollen state to function well, reduction of the compressibility is a more difficult problem.

An exhaustive study of the influence of the supramolecular structure of cellulose triacetate on the chromatographic optical resolution of several racemates has been made by Francotte *et al.* [48]. Their results confirmed the original ideas by Hesse and Hagel [43–45] that the inclusion of low molecular-weight chiral molecules into a specific spatial arrangement of the glucose units of the polysaccharide chains is of fundamental importance for the chiral discrimination process.

During experiments with deposition of CTA into silica gel particles, it was found by a Japanese research group [49,50] that, although the dissolved and reprecipitated polymer had apparently lost most of its microcrystalline structure, the new material still possessed some resolving capacity. The results presented in Table 7.3 are

Table 7.3 — A comparison between microcrystalline (I) and reprecipitated (II) cellulose triacetate sorbents in optical resolutions under identical conditions

Compound	Sorbent	k_1'	k_2'	α
(structure: quinoxaline-type diamine with two aryl groups)	I	2.61 (−)	5.36 (+)	2.05
	II	0.59 (+)	0.91 (−)	1.53
(structure: stilbene oxide, Ph–epoxide–Ph)	I	7.82 (+)	11.3 (−)	1.45
	II	0.94 (−)	1.23 (+)	1.31
Ph–CH–CONH₂ with OH	I	2.08	3.08	1.48
	II		0.80	1.00
(structure: benzene ring)	I		10.3	
	II		0.46	

Column: 250×4.6 mm. Eluent: ethanol. The signs denote the optical rotation of the enantiomer eluted. (Reprinted from T. Shibata, I. Okamoto and K. Ishii, *J. Liquid Chromatog.*, 1986, **9**, 313 by courtesy of Marcel Dekker, Inc.).

illustrative. As expected, all the k' values found are much lower than in MCTA, particularly for benzene. Further, the α values are diminished and a reversal of the elution order appears. However, a much higher column efficiency is found, which often more than compensates for the lower α values. This new material also permits use of solvents other than ethanol, and of higher flow-rates for a faster chromatographic run.

The mechanism of retention by these CSPs appears to be highly complex and not yet satisfactorily elucidated. Many factors have been found to play important roles, such as the average molecular weight of the polymer, the molecular weight distribution, the solvent used for deposition onto the support and the nature of the support [51]. Nevertheless, the substantial improvements obtained by the presence of a rigid support and the wider choice of mobile phases are evident. This is illustrated by Fig. 7.5 which shows chromatograms of racemic *trans*-stilbene oxide under four

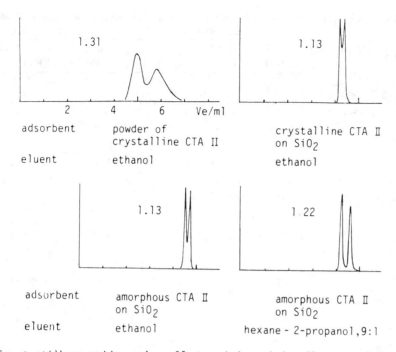

Fig. 7.5 — Optical resolution of racemic *trans*-stilbene oxide by two different types of reprecipitated CTA under various conditions. (Reprinted from T. Shibata, I. Okamoto and K. Ishii, *J. Liquid Chromatog.*, 1986, **9**, 313 by courtesy of Marcel Dekker, Inc.).

different conditions. The effect of the silica support on column efficiency, as well as that of the mobile phase on α, is quite clear. The degree of crystallinity of the CTA obtained by reprecipitation was established by means of X-ray powder diffraction.

These results, which demonstrated that the microcrystallinity of CTA is not an absolute requirement for efficient chiral recognition, led to renewed interest in the use of carbohydrate-based polymers for optical resolution, with vigorous research activity, particularly in Japan. Derivatives of cellulose and other polysaccharides have been the subject of extensive investigations.

B. Other derivatives of cellulose

There are essentially four types of derivatives that are easily prepared from cellulose by modification of the free hydroxyl groups, viz. organic esters, nitrates, carbamates (obtained by reaction with isocyanates) and ethers. Of these the esters and carbamates have been shown to be potentially the most useful. Table 7.4 shows the derivatives investigated so far, with the α values for some racemic compounds resolved. Note that the elution order of the enantiomers on CTA is the reverse of that on cellulose tribenzoate (CTB).

It seems to be quite clear that the better properties exerted by the ester and carbamate derivatives, compared to the ethers, are associated with the polar

Table 7.4 — Examples of separation factors (α) obtained in optical resolutions on various derivatives of cellulose

Compound	Substituent on cellulose					
	COCH$_3$	NO$_2$	COPh	CH$_2$Ph	CONHPh	COCH=CHPh
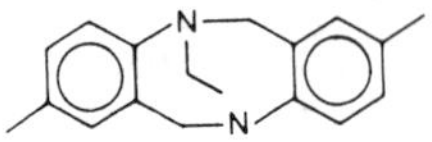	1.31 (+)	1.33 (+)	1.0 (−)	1.34 (+)	1.32 (+)	2.82 (+)
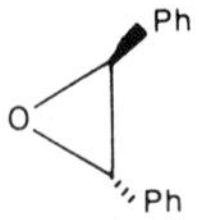	1.22 (−)	1.61 (−)	1.47 (+)	1.0	1.32 (+)	1.15 (+)
Ph–CH–CONH$_2$ OH	1.08	1.10	1.0	1.0	1.0	1.0
Ph–CH–C–Ph OH O	1.05 (−)	1.0 (−)	1.12 (+)	1.0	1.0 (+)	1.08 (−)
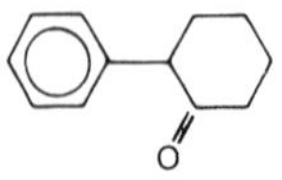	1.07 (+)	1.14 (+)	1.47 (−)	1.0	1.14 (−)	1.26 (−)
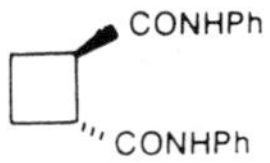	1.13 (−)	1.22 (−)	2.06 (+)	1.0	1.25 (+)	1.52 (−)
	1.39 (+)		1.17 (−)	1.0		1.07 (−)

Column: 250 × 4.6 mm. The cellulose derivatives were coated (20–22%) on LiChrospher Si-1000. The α values were not all obtained under identical conditions. The signs denote the optical rotation of the first-eluted enantiomer. (Reprinted from T. Shibata, I. Okamoto and K. Ishii, *J. Liquid Chromatog.*, 1986, **9**, 313 by courtesy of Marcel Dekker, Inc.).

carbonyl groups, which cause increased retention of polar solutes. An investigation of the k' values obtained for a series of solutes of increasing polarity, by the use of a CTB-column and a tribenzylcellulose column gave very consistent results. If the k' ratio [with k'(CTB) as the denominator] were to be calculated for such a series, it would be found to range from <2 for saturated and chlorinated hydrocarbons, through 2 for aromatic hydrocarbons with non-polar substituents, to >3 for amides, alcohols, lactones, sulphoxides and aliphatic nitro compounds [52].

An important property of the CSPs based on cellulose derivatives is their usefulness for optical resolution of chiral aliphatic compounds. Quite often, an aromatic substituent will promote separation, but it is by no means essential. In Table 7.5 some non-aromatic compounds resolved on cellulose derivative columns are given together with α values obtained.

Table 7.5 — Various non-aromatic compounds resolved on derivatives of cellulose

Compound	α (on triacetate)	Compound	α (on tribenzoate)
	1.22		1.21
	1.31		1.44
	1.23		1.41
	1.61		1.80
	1.21		

Conditions as described under Table 7.4. Elution orders were not established. (Reprinted from T. Shibata, I. Okamoto and K. Ishii, *J. Liquid Chromatog.*, 1986, **9**, 313 by courtesy of Marcel Dekker, Inc.).

In summary, the five most useful cellulose derivatives possess properties characterized [53] below.

Triacetate—for many racemates, especially effective for substrates with a phosphorus atom as a stereogenic centre. In general low separation factors.

Tribenzoate—for racemates with carbonyl group(s) in the neighbourhood of a stereogenic centre.

Trisphenylcarbamate—for polar racemates. Sensitive to the molecular geometry of the substrates.

Tribenzyl ether—effective with protic solvents as mobile phases.

Tricinnamate—for many aromatic racemates and barbiturates. High retention times.

These silica-coated CSPs have been commercialized and columns ("Chiralcel") are marketed by Daicel Chem. Co. (see Appendix for details).

C. Derivatives obtained from other polysaccharides

Derivatives of a variety of polysaccharides other than cellulose have been prepared and many of them show interesting properties. Of the polysaccharides shown in

Scheme 7.1 pullulan was the least promising, as its benzoate failed to resolve any of the compounds tested. This was taken as an indication of the importance of a certain structural regularity. Pullulan has the 1→6 linkage randomly distributed over every 3–4 α-1→4 linkages.

It has now been demonstrated that many readily available polysaccharide derivatives can be used as CSPs for optical resolution by LC. Recently, a series of phenylcarbamate derivatives of various carbohydrate polymers was investigated [54]. Many of these new CSPs were found to give better resolution of certain racemates than the corresponding cellulose derivatives. The data collected in Table 7.6 give some insight into the resolving capacities of the particular chiral sorbents investigated.

As already pointed out, very little is known about the mechanisms of enantio-selection by the polysaccharide derivatives. The finding that aliphatic hydrocarbons, such as *cis,trans*-1,3-cyclo-octadiene (**1**), can be optically resolved on

1

MCTA [55] is hard to interpret without the assumption of an inclusion effect, i.e. a steric discrimination caused by a difference in the permeability of the enantiomers into chiral cavities, leading to non-identical retention times on the column. For compounds bearing polar substituents, however, there appear to be contributions from hydrogen bonding and dipole–dipole interactions. Contrary to what might have been expected, it has turned out that conformationally rigid solutes are not better resolved than more flexible ones [56]. This result, exemplified in Table 7.7, illustrates the complexity of the chiral recognition mechanism in these CSPs. It is reasonable to assume that a certain conformation of a substrate molecule is necessary for a good steric fit causing enantioselection. If such a conformation cannot be attained, there will be no resolution. It is also conceivable that great differences in these steric requirements will be present among the various polysaccharide derivatives [51].

7.1.2 Derivatives of polyacrylamide and similar synthetic polymers
As we have seen in Section 7.1.1, carbohydrate biopolymers are very useful and readily available chiral starting materials, which require only simple derivatization procedures to yield selective column materials for enantioseparation. Synthetic chiral polymers, however, cannot be produced without the use of a chiral reagent or a chiral catalyst. In the first case, a chiral derivatization of a suitable monomer is performed and the product is then polymerized to form a polymer network having chiral substituents (Fig. 7.6a). In the second case, the monomer is polymerized under the influence of a chiral catalyst, which will produce an optically active polymer since

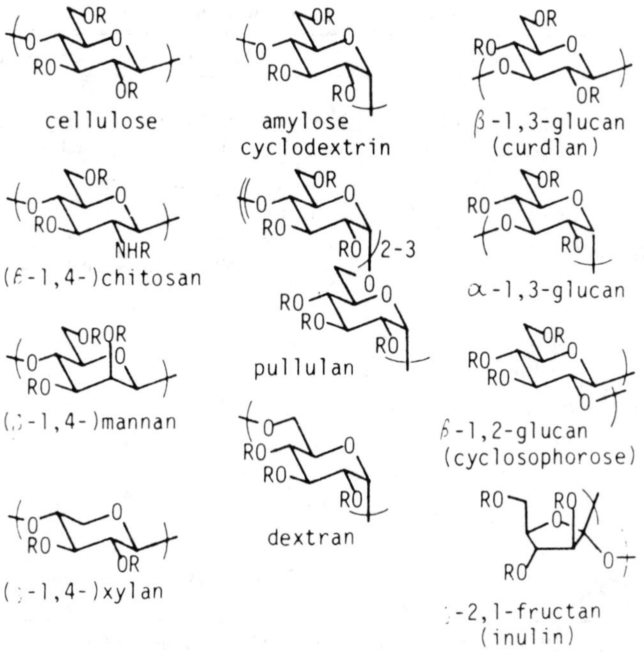

Scheme 7.1 — Polysaccharides used as ester and phenylcarbamate derivatives to produce new CSPs.

the stereoregulatory influence of the catalyst will yield an isotactic polymeric structure of a certain preferred helicity (Fig. 7.6b). Here, the chirality of the polymer is inherent, i.e. it is caused only by the helical structure.

7.1.2.1 *Sorbents based on chirally substituted synthetic polymers*
This type of sorbent has been successfully developed by Blaschke and co-workers [56–62]. So far the work has been concentrated on polyacrylamide and polymeth-acrylamide derivatives, where the chiral substituents originate from an optically active amine or amino-acid component. By the use of a suspension polymerization technique, polymer particles of desired mean diameter and acceptable size homoge-neity can be obtained. Free-radical initiation is used and the porosity of the gel particles is regulated by the relative amount of cross-linking agent added. The particles swell in organic solvents and the material is only used in low-pressure LC systems.

The resolving capacity of these sorbents is highly dependent on a variety of factors. Apart from the substituents on the polyacrylamide backbone, the degree of cross-linking, the nature of the cross-link and the mobile phase compositions are the most essential considerations. Systematic investigations of these factors began in 1973 when partial resolution of doubly radiolabelled (^{3}H and ^{14}C) mandelic acid and mandelamide was studied by analysis of the enantiomer composition of eluted

Table 7.6 — Examples of compounds which have been optically resolved on phenylcarbamate sorbents derived from various polysaccharides

Compound		Sorbent derived from
CONHPh —Ph		cellulose amylose
CH₃ CH₃	R = OH R = NO₂	cellulose, amylose cellulose
HO–CH–CF₃		chitosan xylan
Ph–C–CH–Ph, O OH		xylan
M(CH₃COCHCOCH₃)₃	M = Co, Cr	inulin

Data obtained from Y. Okamoto, M. Kawashima and K. Hatada, *Polymer Preprints*, 1984, **33**, 1607. With permission.

Table 7.7 — Separation factors obtained from optical resolution of a series of benzoates and dibenzoates on silica coated with cellulose tribenzoate

Compound	α	Compound	n	α
OBz	1.16	OBz (CH₂)ₙ–OBz	1	1.14
OBz	1.00		2	1.74
OBz	1.21		3	1.44
			4	1.83
(Bz = –C–C₆H₅) O				

Data obtained from T. Shibata, I. Okamoto and K. Ishii, *J. Liquid Chromatog.*, 1986, **9**, 313. (Reprinted by courtesy of Marcel Dekker, Inc.).

(a) Use of chiral *substituent*:

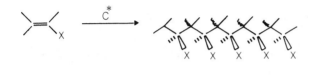

polymerization
cross linking

chiral monomer

chiral
polymer network

(b) Use of chiral *catalyst*:

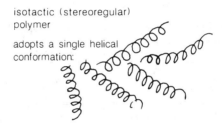

isotactic (stereoregular)
polymer

adopts a single helical
conformation:

Fig. 7.6 — Two different routes to chiral sorbents based on synthetic polymers. (a) From chiral monomer; (b) from achiral monomer.

fractions by liquid scintillation counting [56]. Since the $^{3}H/^{14}C$ activity ratio is proportional to the enantiomer composition, this detection technique permits a rather precise determination of α even at very low resolution.

The method is, of course, limited to those cases where it is possible to radiolabel the two enantiomers of a compound with the appropriate radioisotope. Accordingly, the elution profiles of both enantiomers are obtained in a single chromatographic experiment.

The preferred route to these polymers was shown to be that given in Scheme 7.2. The great variation of performance with substituents is evident from Table 7.8, which indicates that two of the most useful polymers are those corresponding to (1) $R = H$, $R_1 = CO_2R'$, $R_2 = C_6H_5CH_2$, $R' = C_2H_5$ and (2) $R = CH_3$, $R_1 = CH_3$, $R_2 = cyclo\text{-}C_6H_{11}$.

These substituted polyacrylamides have been found particularly well suited to polar compounds with functional groups capable of hydrogen bonding. It is therefore

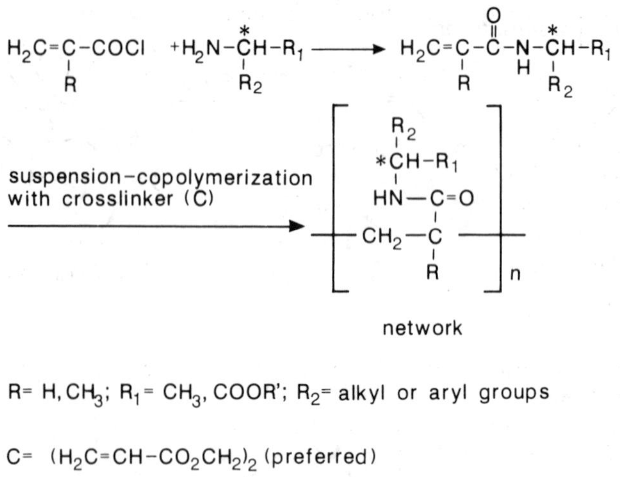
H2C=C-COCl +H2N-CH-R1 → H2C=C-C-N-CH-R1
suspension-copolymerization
with crosslinker (C)
network
R= H, CH3; R1= CH3, COOR'; R2= alkyl or aryl groups
C= (H2C=CH-CO2CH2)2 (preferred)

Scheme 7.2 — Synthetic routes to the chiral polyacrylamides and polymethacrylamides.

Table 7.8 — Optical resolution ability of a series of variously substituted polyacrylamide sorbents. (Reprinted, with permission, from G. Blaschke, *Angew. Chem.*, 1980, **92**, 14. Copyright 1980, Verlag Chemie GmbH).

R	R_1	R_2	R′	Optical yield (%)	
				Mandelic acid	Mandelamide
H	CH_3	C_6H_5	—	28	35
CH_3	CH_3	C_6H_5	—	8	81
H	CH_3	*cyclo*-C_6H_{11}	—	12	35
CH_3	CH_3	*cyclo*-C_6H_{11}	—	58	92
CH_3	CH_3	1-naphthyl	—	0	97
H	CH_3	*p*-I-C_6H_4	—	0	0
H	COOR′	CH_3	C_2H_5	14	18
H	COOR′	C_6H_5	C_2H_5	46	34
H	COOR′	$C_6H_5CH_2$	C_2H_5	51	96
H	COOR′	*p*-OH-$C_6H_4CH_2$	C_2H_5	0	0
H	COOR′	$C_6H_5CH_2$	*tert*-C_4H_9	89	80

logical that relatively non-polar mobile phases have been found most useful. Combinations of hydrocarbons, ethers and possibly small amounts of an alcohol are typical; examples include toluene–dioxan, hexane–dioxan and toluene–dioxan–methanol mixtures. Protic co-solvents such as methanol strongly decrease the retention and therefore normally form less than 10% of the mobile phase composition.

Although hydrogen bonding is apparently the major binding contribution to retention of the solute, the mechanism of enantiomer discrimination appears to be quite complex. As in the case of polysaccharide-derived sorbents, enantioselection is assumed to be caused by inclusion phenomena, i.e. the binding groups in the asymmetric cavities into which the solute enantiomers are thought tó diffuse are more favourably located for one of the antipodes, which therefore will be preferentially retained.

Columns packed with sorbents of this kind have mainly been used for semipreparative work (sample amounts between ca. 1 and 250 mg) on racemic pharmaceuticals. These applications will be treated in Chapter 9. An interesting modification of the sorbent consists of a silica-bonded non-cross-linked polyacryloyl (S)-phenylalanine ethyl ester. This has been used for analytical HPLC and gives considerably improved column efficiency [63].

7.1.2.2 Sorbents based on isotactic linear polymethacrylates of helical conformation

A vinyl polymer with chirality caused only by its helicity was first prepared in 1979 by Okamoto and co-workers [64,65]. Optically active poly(triphenylmethyl methacrylate) was then obtained according to Scheme 7.3 by asymmetric anionic polymeriza-

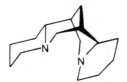

$$R=CH_3; \quad C= (-)-\text{sparteine:}$$
$$Ar, Ar_1 = C_6H_5, C_5H_4N$$

linear, isotactic,

single–handed helical

Scheme 7.3 — Asymmetric polymerization reaction used to produce the right-handed helical poly(triarylmethyl methacrylates).

tion of triphenylmethyl methacrylate under the influence of a chiral initiator in toluene at low temperature. The success of the reaction is strongly dependent on the initiator used, which is based on a complex between an optically active diamine and butyllithium or lithium amide. Both (−)-sparteine–butyllithium, and (+)-(2S,3S)-dimethoxy-1,4-bis(dimethylamino)butane–lithium amide, were found to give the

(+)-polymer in good yield [66]. At a degree of polymerization greater than ca. 70 the polymer is insoluble in most common organic solvents.

The material can be used as such after grinding and sieving to a particle size averaging ca. 30 μm [67]. However, a more efficient and durable chromatographic sorbent is obtained by adsorption of the low molecular-weight soluble fraction of the polymer on silanized silica (10 μm and 1000 or 4000 Å) [68].

It is perhaps not surprising that this CSP turned out to be excellent for optical resolution of racemic aromatic hydrocarbons of linear and planar chirality [69]. The compounds in Fig. 7.7 are examples of such hydrocarbons that are all resolved with high α values. They are characterized by high hydrophobicity and a rigid molecular geometry. A difference in the extent to which the enantiomers would interact with the helical CSP (in which the triphenylmethyl group is assumed to attain a propeller-like conformation) is therefore very reasonable on intuitive grounds. This is particularly true in the case of hexahelicene, for which the highest α value (>13) was reported. The enantiomer most strongly retained on the (+)-CSP is the (+)-form , which has P- (right-handed) helicity. Since it was found that the (+)-CSP interacts very strongly with itself, but only weakly with the (−)-polymer, it is very likely that the (+)-CSP also has P-helicity [70,71]. The same P-helicity of the most strongly retained enantiomer was also found for all other compounds investigated that possessed this type of chirality. The situation is shown in Fig. 7.7. This correlation means that chromatographic retention data can be used for determination of absolute configuration of these types of compound.

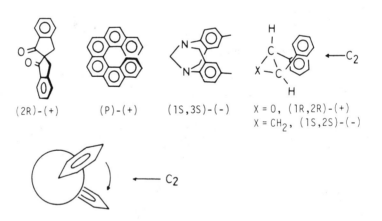

(2R)-(+)   (P)-(+)   (1S,3S)-(−)   X = O, (1R,2R)-(+)
                                                        X = CH$_2$, (1S,2S)-(−)

Fig. 7.7 — Schematic representation of the helical CSP-structure and absolute configurations of most strongly retained enantiomers, showing their resemblance to (P)-(+)-hexahelicene. (Reprinted from Y. Okamoto and K. Hatada, *J. Liquid Chromatog.*, 1986, **9**, 369 by courtesy of Marcel Dekker, Inc.).

Because the structure of this CSP is not cross-linked and the coated-silica version of the sorbent is based only on physical adsorption, some limitations are put on the choice of mobile phase, for solubility reasons. Thus, aromatic hydrocarbons, chloroform and tetrahydrofuran (which dissolves the polymer) should be avoided. To date, methanol has been the preferred mobile phase [51] and there is generally a tendency towards increase in α value with increasing polarity of the solvent, but retention times may be unacceptably long in many cases.

The use of these sorbents together with protic mobile phase systems is disadvantageous owing to the solvolytic instability of the ester bond. Thus in methanol the CSP gradually undergoes solvolysis to yield methyl triphenylmethyl ether as a byproduct. (It should be remembered that the trityl group is commonly used for protection of carboxyl groups in peptide synthesis and is easily removed under weakly acidic conditions). It is therefore advisable to work at low temperature with alcohols as eluents.

Partly in order to reduce the ester lability, a series of similar polymers, (2), has been prepared. In general all of these polymers gave inferior resolution although the 2- and 4-pyridyl analogues, when used with hexane–2-propanol eluent mixtures, gave interesting results for some polar solutes [72]. It has been suggested that hydrogen bonds involving the pyridyl groups may play a role in these cases.

2

It is further of interest that compounds without aromatic substituents have been resolved on the (+)-poly(triphenylmethyl methacrylate) CSP, viz. the trisacetylacetonates of cobalt, chromium and aluminium [73,74].

The usefulness of this CSP is evident from Scheme 7.4, in which various compounds resolved are given.

Columns containing this type of optically active synthetic polymers coated onto 10-μm silica are available (as "Chiralpak") from Daicel (see Appendix for details).

7.1.3   Synthetic polymers containing 'grafted' chiral cavities

As early as 1949 Pauling presented the idea of constructing, by synthetic means, polymer network cavities which should fit only one of two enantiomers [75]. Basically, the principle rests on an imitation of an enzyme's binding site, which can usually be regarded as a chiral cavity or cleft in the protein, often highly specific with respect to binding of substrate enantiomers because of the precise steric requirements for multiple bond attachment. Because the experimental technique can be compared with making a plaster cast from an original template it has also been called 'molecular imprinting'. Thus, the molecules of a particular compound act as templates around which a rigid polymeric network is cast. This procedure, which is

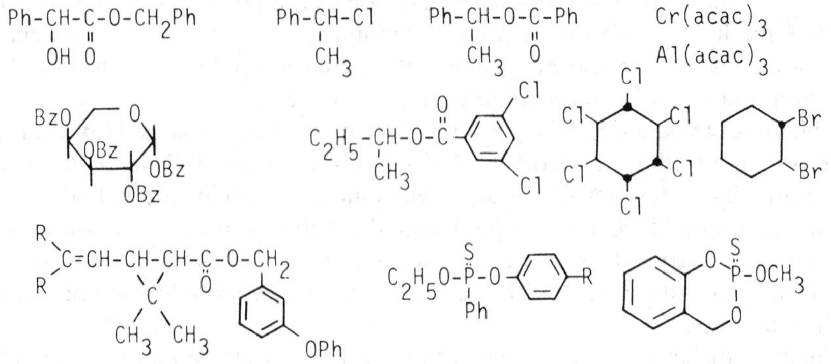

Ph-CH-C-O-CH₂Ph   Ph-CH-Cl   Ph-CH-O-C-Ph   Cr(acac)₃
| ‖ | | ‖
OH O   CH₃   CH₃ O   Al(acac)₃

BzO

C₂H₅-CH-O-C-
        |
        CH₃

R
  \
   C=CH-CH-CH-C-O-CH₂
  /   \ / ‖
R   C O
        / \
      CH₃ CH₃

C₂H₅O-P-O-⟨ ⟩-R
       |
       Ph

O-P-OCH₃

Scheme 7.4 — Various compounds optically resolved on (+)-poly(triphenylmethyl methacry-late) sorbents.

conceptually straightforward but most delicate to carry out in practice, can be summarized in three distinct steps.

(1) Formation of a complex between the (chiral) compound used as template and a polymerizable monomer.

(2) Polymerization with cross-linking to form a rigid matrix.

(3) Removal of the template, either simply by some washing technique, or by means of a hydrolytic or similar reaction, which has to be used in the case of covalent attachment to the template. These steps are visualized in Scheme 7.5.

Extensive research on this technique has been carried out since 1977 when it was shown that a polystyrene sorbent, prepared with the use of an optically active template, could be used for partial resolution of the corresponding racemate [76–80]. The preferred method has been to use the rapid and reversible formation of boronic acid ester bonds between a carbohydrate structure and a vinyl-substituted phenyl-boronic acid (Fig. 7.8). In this way the monomer units are covalently attached to the template (cf. Scheme 7.5, step 1). Polymerization and cross-linking with a divinyl compound, followed by hydrolysis and washing out of the template will then yield the chiral sorbent.

The chromatographic performance of this kind of sorbent is shown in Fig. 7.9. The chromatogram illustrates one of the difficulties associated with the technique, viz. the necessity of fast mass-transfer to reduce band broadening. Although boronic ester formation is a very fast process, it is still somewhat too slow for chromato-graphic purposes. It is important to keep in mind that chromatographic sorption–desorption equilibria are always based on non-covalent interactions, except for some protein separations by affinity chromatography where thiol–disulphide interchange may contribute.

In accordance with these facts, it was found that the column efficiency could be improved by increasing the temperature and also by acid–base catalysis, induced by the addition of ammonia to the mobile phase [81].

The choice of the mobile phase is complicated by the requirement that the polymer should be resistant to swelling in the medium, as deformation of the cavities

Complexation with monomer units:

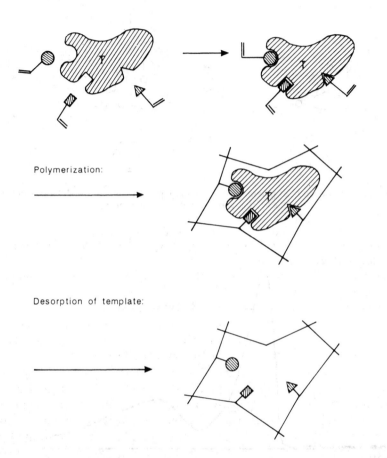

Polymerization:

Desorption of template:

Scheme 7.5 — The principle of 'molecular imprinting' with formation of a defined arrangement
of binding groups within the microcavity.

will otherwise occur, with loss of selectivity. A mixture of acetonitrile with 4–6% of
water and 2–8% of concentrated ammonia solution has been found to be very useful.
The flow-rate has a pronounced influence on the result and very low flow-rates are
usually necessary.

Interestingly, the k' values increase with increasing temperature. The increase is
larger for the most strongly retained enantiomer, leading to an improved α-value.

Another "molecular imprinting" approach has been taken by Mosbach and
collaborators. This is based on non-covalent interactions, usually electrostatic or
hydrogen bonding interactions, between the monomer units and the template
molecules. This technique was originally intended to produce polymeric materials
with a particular substrate selectivity [82], but has been further developed towards
polymers containing enantiomer-differentiating chiral cavities from the use of

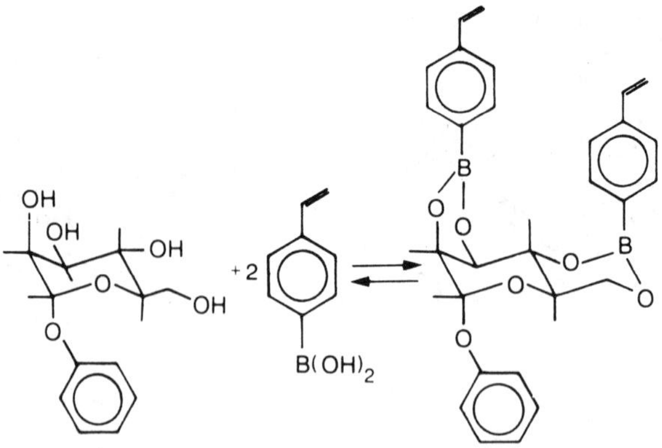

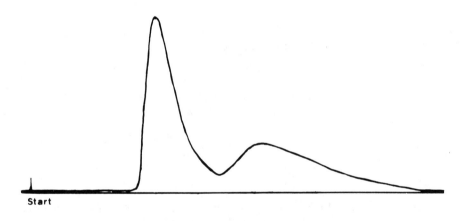

Fig. 7.8 — Introduction of polymerizable vinyl groups into phenyl-β-D-mannopyranoside by esterification with 4-vinylbenzeneboronic acid.

Fig. 7.9 — Elution profile from the chromatography of phenyl-β-D.L-mannopyranoside on macroporous polymer imprinted with the D-enantiomer. (Mobile phase: acetonitrile with 4% concentrated ammonia solution and 5% water, flow-rate 0.1 ml/min. Sample amount: 200 μg). (Reprinted, with permission, from G. Wulff, H.-G. Poll and M. Minarik, *J. Liquid Chromatography*, 1986, **9**, 385 by courtesy of Marcel Dekker, Inc.).

optically active templates [83–86]. The best results so far have been obtained by the use of amides of aromatic amino acids as templates, or "print molecules", and by carrying out the polymerization at a relatively low temperature [86]. For these purposes methacrylic acid (MAA) was used as monomer, with ethyleneglycol dimethacrylate (EDMA) added as a crosslinker. Although some separation factors reported from chromatographic experiments with columns packed with polymer particles of this kind are amazingly large, the plate numbers of the columns are very low. Table 7.9 gives a summary of the templates used in the synthetic procedure and

Table 7.9 — Resolution on chiral polymer sorbents obtained by "molecular imprinting". (Reproduced from B. Sellergren, *Chirality*, 1989, **1**, 63, with permission. Copyright 1989, Wiley-Liss, Inc.)

Polymer and print molecule	Particle diameter, μm	Column temperature, °C	k_2'(config.)	α	R_s
A,L-PheNHPh	10–30	80–85	4.3 (L)	3.2	1.2
C,L-*p*-H₂NPheOEt	10–30	80–85	2.7 (L)	1.8	0.8
E,L-PheNHPh	32–45	80–85	5.1 (L)	4.9	1.0
G, L-TrpOEt	45–65	80–85	1.2 (L)	1.8	0.5
H,L-PheNHPh	32–45	23	1.2 (L)	3.7	0.9
I,D-*p*-H₂NPheNHPh	32–45	23	1.2 (D)	5.7	0.9

the final chromatographic results. The polymer designation A–I refers to differences in the MAA/EDMA ratio, and also in temperature used during polymerization. The importance of using template molecules having a multitude of interaction sites has been pointed out recently [87].

7.1.4 Proteins

The complicated molecular structure of proteins makes them very interesting for binding studies. The technique of affinity chromatography developed from the knowledge of the ability of certain protein–ligand pairs to form very strong complexes. Such pairs could be derived from natural systems, such as enzyme–substrate analogue, enzyme–cofactor, hormone–receptor protein, etc., but it was soon realized that many 'unnatural' synthetic compounds could also show very strong binding power (high affinity) for proteins.

The availability and importance of serum proteins, particularly serum albumins, have made them preferred models for binding studies. It was known from Scatchard analyses that binding to a protein involves multiple equilibria, i.e. a number of binding sites, some of which have different affinity for the ligand. It therefore seemed quite probable that the net result, the overall binding constant, could be different for the two enantiomers in a racemate. Further, from the knowledge concerning the substrate enantioselectivity often shown by enzymes, the presence of high affinity sites with enantiomer-differentiating ability would be expected in other proteins as well.

A number of studies of solution equilibria between serum proteins and various ligands, particularly pharmacologically active compounds, have been made, which demonstrated significantly different binding constants for the respective enantiomers [88,89]. Such effects could, however, be demonstrated more clearly by chromatographic techniques. Thus, in 1973 the previously known higher affinity of L-tryptophan for bovine serum albumin (BSA) was used to separate the enantiomers on a column packed with a BSA–Sepharose gel [90]. Elution of the D-form was performed with a borate buffer (pH 9), then the L-form was desorbed by changing to dilute acetic acid. The technique was later used for determination of the enantioselectivity of certain drugs for serum albumins [91,92]. In the following years analytical chiral LC-methods based on the use of immobilized proteins as stationary phase materials were rapidly developed and shown to be useful for a variety of resolution problems.

7.1.4.1 *Immobilized albumin*

Early work with the use of low-pressure LC-systems and isocratic elution with
phosphate and borate buffers from columns packed with BSA coupled to Sepharose,
demonstrated that the optical resolution of charged solutes such as the free amino-
acids tryptophan, kynurenine [3-(2-aminobenzoyl)alanine] and their 5- and 3-
hydroxy derivatives, respectively, is extremely dependent on the pH of the mobile
phase [93]. It was further shown that compounds with chirality at a sulphur atom,
such as methyl *o*-carboxyphenyl sulphoxide, could be resolved and that, in addition
to pH, the ionic strength of the mobile phase had a great influence on retention and
resolution [94].

A significant improvement in column performance and ease of operation was
obtained with the introduction of a silica support (spherical 7 or 10 μm particles, 300
Å average pore diameter) for the immobilized albumin [95]. Analytical columns
packed with such albumin-silica sorbents ("Resolvosil") are useful in combination
with aqueous buffered eluents for optical resolution of a variety of racemates.

As with the previously described polymeric CSPs, the mechanism of chiral
recognition is complex and not yet known in any detail. However, the major
contributions to the overall retention have been elucidated from systematic investigations of the effects of solute structure and mobile phase composition. In many
respects the albumin-silica sorbents act like alkyl-silica reversed-phase materials.
Alkanols, preferably l-propanol, can be used as retention modifiers, causing faster
elution due to a decrease in hydrophobic interaction with the sorbent. There are
essentially three variables by means of which the mobile phase system can be
optimized for a particular resolution problem, viz. the pH, ionic strength and organic
solvent-modifier [96]. There appear to exist simultaneous contributions mainly from
charge and dipole interactions on the one hand and hydrophobic interactions on the
other. Further, effects from hydrogen bonding and charge-transfer interaction are
likely to be present. The large effects of the mobile phase on the k' values of the
solute enantiomers are understandable in view of the dynamic properties of a protein
with respect to charge distribution and conformation. BSA consists of no less than
581 amino-acids in a single chain (m.w. 6.6×10^4) and its higher structure is to a large
extent determined by the 17 disulphide bridges present in the molecule. At pH 7.0 its
net charge is -18 and its isoelectric point is 4.7. Thus, as is well-known from enzyme
chemistry, a solvent change may cause changes in a binding site of a protein by charge
as well as conformational effects.

Depending on the charge of the solute, a pH change of the mobile phase may
cause either increased or decreased k' values. Whereas amines and free amino-acids
bearing a positive charge are more strongly retained at higher pH, the reverse is true
for negatively charged solutes. Systematic studies of a series of *N*-benzoyl-D,L-amino
acids have given some insight into the mechanism of interaction between a solute and
the protein. The effect of various mobile phase parameters on the k' and α values is
shown in Fig. 7.10. First, it is evident that retention is increased to a large extent by
increasing the hydrophobic character of the amino-acid (SER < ALA < PHE).
Secondly, the increasing negative net charge of the protein with increasing pH will
cause a decreased k' for all six species (due to the effect on ionic interaction). Next,
the effect of buffer concentration could be interpreted as an increased ionic

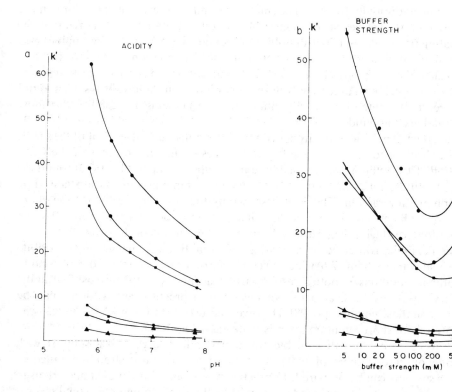

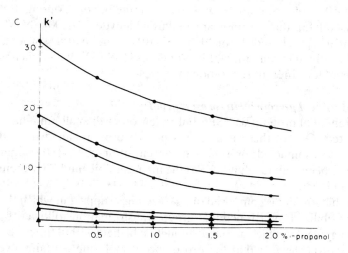

Fig. 7.10 — Retention (*k'*) of the enantiomers of *N*-benzoyl-D.L-serine (▲), -alanine (■) and -phenylalanine (●) as a function of: (a) pH, (b) buffer strength, and (c) organic modifier (1-propanol). (Reprinted, with permission, from S. Allenmark, B. Bomgren and H. Borén, *J. Chromatog.*, 1984, **316**, 617. Copyright 1984, Elsevier Science Publishers B.V.).

adsorption at low ionic strength. The small but significant increase in k' for the most strongly retained species at the highest buffer concentrations is probably an effect originating from increased hydrophobic interaction. Because ionic (coulombic) and hydrophobic interactions are oppositely influenced by the ionic strength, the two effects should overlap, giving a minimum in solute adsorption (k') at a certain point. Finally, the effect of an organic solvent-modifier is obvious; it has always been found to decrease the retention of the solute, the effect being largest for the most hydrophobic compounds.

The effect of pH and ionic strength on the retention of uncharged solutes is small, but significant, and attributable solely to changes in the binding sites of the CSP. Addition of 1-propanol causes a decrease in retention comparable to that found for charged solutes, indicating that it affects mainly the binding contribution caused by hydrophobic interaction. This is also shown by the very large effect of the chain length of 1-alkanols on retention, the higher alkanols being much more effective competitors for the binding sites, and causing faster elution of the solute.

The flexibility conferred by regulating retention through changes in the mobile phase is shown in Scheme 7.6, which gives an idea of the procedure used in the search for optimal conditions for optical resolution of a racemate on a "Resolvosil" column.

Very large α values, actually not desirable in practical applications, can be obtained on these columns [97,98]. However, adjustment to give adequate baseline resolution with short retention times is quite easy (Fig. 7.11).

By using techniques that enable covalent bonding to the silica support with simultaneous crosslinking of the protein, sorbents with increased resistance to organic solvents can be prepared. In practice this means that the columns are used under essentially the same conditions as the alkylsilica columns in conventional reversed-phase liquid chromatography. Thus, such columns are compatible with buffers containing 0–50% of acetonitrile or 0–60% of methanol. Among the crosslinkers found useful for this purpose are glutardialdehyde [99,100], N,N'-disuccinimidyl carbonate [101] and formaldehyde [101]. The chiral sorbents obtained differ significantly in chromatographic properties [101,102], which means that they can be selected for different separation purposes.

7.1.4.2 *Immobilized α_1-acid glycoprotein (orosomucoid)*

Despite the great potential of proteins as CSPs, only a few other than albumin have so far been investigated. One of these is a human plasma protein called α_1-acid glycoprotein (AGP) or orosomucoid, present in a concentration of 55–140 mg per 100 ml of plasma. It has been claimed that AGP is the main cationic binding protein in the human organism [103].

The chromatographic sorbent is composed of a 300-Å silica support in which the protein has been immobilized by a functional group transformation followed by cross-linking to form aggregates which are large enough to be held within the pores. The procedure is based on the fact that the orosomucoid molecule contains five carbohydrate units, which together constitute ca. 45% of the molecular weight. Oxidation with sodium metaperiodate converts some alcohol functions of these sugar units into aldehyde groups. Incorporation of this modified protein into the silica, followed by raising of the pH of the buffer solvent, yields cross-linking by

Chiral stationary phases

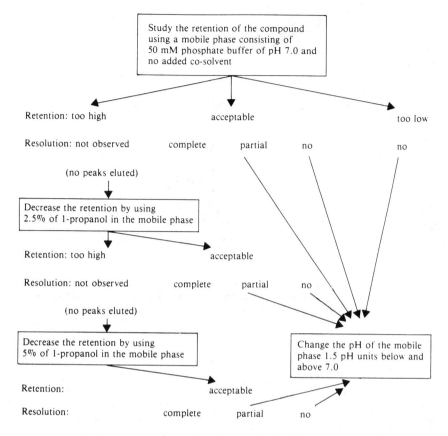

Scheme 7.6 — Optimization procedure in optical resolution by systematic changes in mobile phase composition. (Reprinted, with permission, from S. Allenmark and B. Bomgren, in *Affinity Chromatography — A Practical Approach*, P. G. D. Dean, W. S. Johnson and F. A. Middle (eds.), p. 108. Copyright 1985, IRL Press, Oxford.).

Schiff-base formation. Hydrolytically stable bonds are then formed by reduction of the imino functions with sodium cyanoborohydride [104]. The immobilization chemistry is outlined in Scheme 7.7. Columns packed with this sorbent are manufactured under the name "EnantioPac" and "Chiral AGP" (later version).

In the unbound native form AGP has an isoelectric point of 2.7, i.e. it is ca. 2 pI-units more acidic than albumin. Its molecular weight is 4.1×10^4. Sialic acid (14 residues in all) present in the sugar units is presumed to be involved in the binding of ammonium-type compounds at neutral pH [105], a process thought to be associated with some enantioselectivity.

Basically, the solute retention is governed by the same types of primary interaction processes as in albumin-silica and the methods used for regulation of retention and resolution are essentially the same, although 2-propanol has been the preferred alkanol modifier [106]. Further, cationic as well as anionic modifiers have been used. An analysis of their effects on the retention of charged solutes has shown

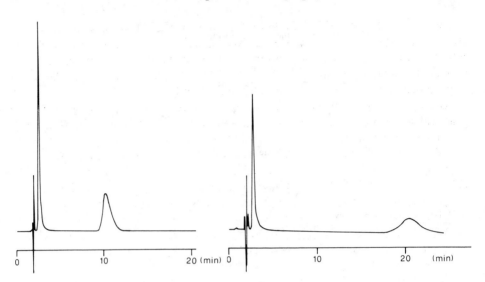

Fig. 7.11 — An example of the effect of the mobile phase on resolutions; chromatogram of D,L-kynurenine under two slightly different conditions: (a) pH 7.1, (b) pH 7.6. (S. Allenmark and S. Andersson, unpublished work).

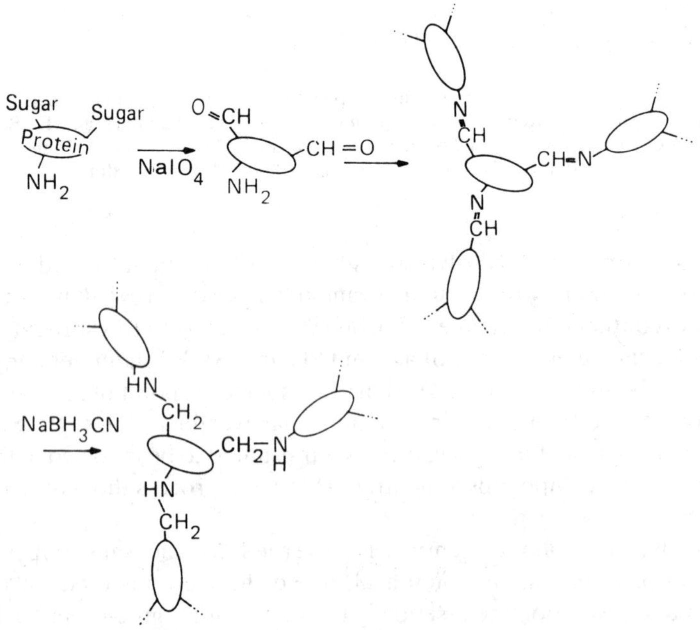

Scheme 7.7 — Reactions used for immobilization of the protein by cross-linking.

consistency with an ion-pair equilibrium model for the solute–sorbent interaction [107]. Thus, the same principles which are widely used [108] for the regulation of retention of charged solutes in ordinary reversed-phase chromatography can be applied to optical resolutions on "EnantioPac" columns. This again shows the importance of hydrophobic binding sites in protein CSPs.

Because the use of such ion-pairing retention modifiers is of great importance for practical applications such as drug analysis, to which we will return in Chapter 8, let us briefly outline the theoretical background of ion-pair chromatography [109].

We assume that the solute, S, is a basic compound which exists in the protonated form, HS^+, at the pH used for the separation. A charged ion-pairing compound Q^+X^- is added to the mobile phase. The distribution of the solute between the mobile and the stationary phase will then be influenced by the concentration of the added compound.

We know from Chapter 4 [Eq. (4.3)] that the capacity ratio, k', can be expressed as $k' = Kq$, where K is the distribution constant (C_s/C_m) of the solute and q denotes the phase ratio (V_s/V_m).

It can be shown [109] by consideration of the equilibria involved, that the capacity ratio of the solute of interest, k'_{HS^+}, will take the form:

$$k'_{HS^+} = \frac{qK°K_{HSX}[X^-]}{1 + (K_{QX}[Q^+][X^-])} \tag{7.1}$$

where $K°$ is the monolayer capacity.

The validity of this expression can be experimentally verified. Rearrangement will yield the expression:

$$\frac{[X^-]}{k'_{HS^-}} = \frac{(1 + K_{QX}[Q^+][X^-])}{qK°K_{HSX}} \tag{7.2}$$

Thus, a plot of $[X^-]/k'_{HS^+}$ against $[Q^+][X^-]$ should be linear if the model is correct.

An example of the effect of sodium octanoate ($X^- = C_8H_{17}CO_2^-$) on the optical resolution of atropine (3), and homoatropine (4), on an "EnantioPac" column is shown in Fig. 7.12. Note that the linear relationships mean that the α values are unchanged, as predicted from the assumption of a single binding site.

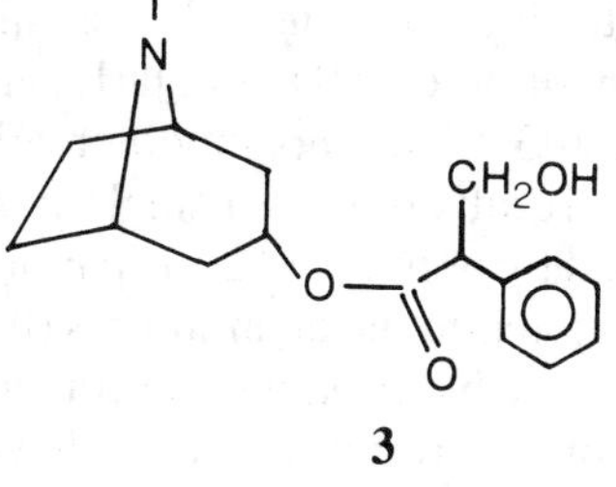

3 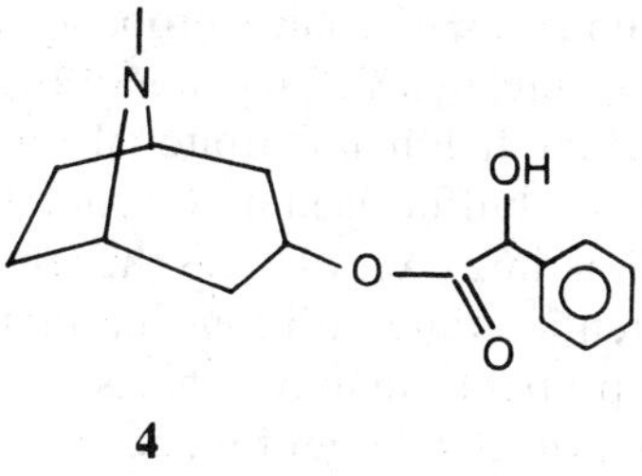   **4**

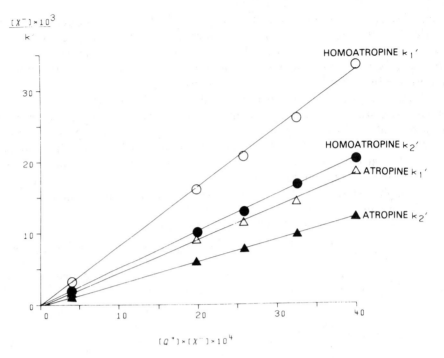

Fig. 7.12 — Application of the ion-pair distribution model to the retention of the enantiomers of atropine and homoatropine. Mobile phase: octanoic acid in 0.02M phosphate buffer, pH 7.0. (Reprinted from G. Schill, I. W. Wainer and S. A. Barkan, *J. Liquid Chromatog.*, 1986, **9**, 641 by courtesy of Marcel Dekker, Inc.).

The effect of hydrophobic charged modifiers, as well as the effects of pH and various alkanols, on a variety of ionic solutes have been extensively studied by Schill and co-workers [107,110] and by Hermansson and Eriksson [111].

The properties of the second version of AGP-based columns, the "Chiral-AGP" column, have been extensively discussed by Hermansson [112]. The fact that very small changes in the mobile phase composition can dramatically change the retention behaviour of a chiral solute on a protein column is well illustrated by two experiments performed with the Chiral-AGP column, shown in Fig. 7.13. First, the choice of the organic modifier is highly important. Under otherwise identical mobile-phase conditions, the use of acetonitrile (10%) as a modifier gives baseline resolution of verapamil (A), whereas 1-propanol, added to give approximately the same retention (4%), results in a single peak for the two enantiomers (B). Similarly, an attempt to resolve methylphenobarbitonal on Chiral-AGP with no modifier added to the phosphate buffer used as the mobile phase results only in a single peak at a long retention time (A), whereas the effect caused by only 2% of 2-propanol is to reduce the retention time with concomittant separation into the enantiomers (b). The latter effect probably mainly reflects a reduction of hydrophobic interaction with the protein, which is larger for one of the enantiomers, resulting in a smaller k'-value for that enantiomer.

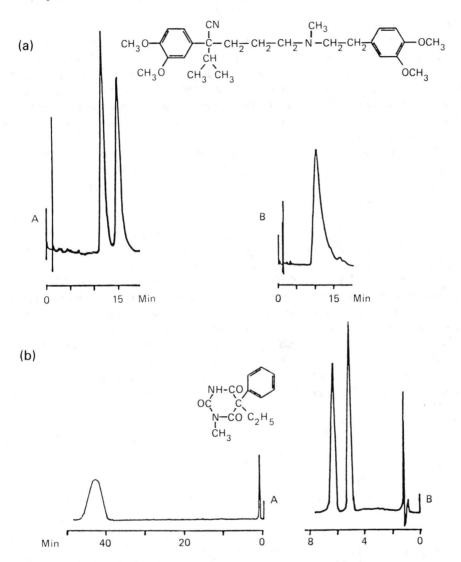

Fig. 7.13 — Chromatograms illustrating the drastic improvement in chromatographic performance with often small changes of the mobile phase composition. (*a*) Influence of the *nature* of
the organic modifier: Left: 10% acetonitrile, right: 4% 1-propanol. Both chromatograms:
Chiral-AGP column, 10m*M* phosphate buffer, pH 7.0+modifier; solute: verapamil.
(*b*) Influence of the *concentration* of the organic modifier: Left: 0% 2-propanol, right: 2% 2-
propanol. Both chromatograms: Chiral-AGP column, 10m*M* phosphate buffer, pH 7.0+modi-
fier; solute: methylphenobarbital. (Reprinted, with permission, from J. Hermansson, *Tr.
Anal. Chem.*, 1989, **8**, 251. Copyright 1989, Elsevier Science Publishers BV).

From studies of the effect of column temperature (in the range 25–77°C) on the
separation [112], it was found that although column efficiency is improved by
increased temperature, the net result is a decreased resolution due to a rapid
decrease of α with increasing temperature.

7.1.4.3 *Immobilized ovomucoid*

Another acid glycoprotein, ovomucoid (OVM), was introduced in 1987 as a chiral stationary phase for liquid chromatography [113,114]. The protein, which is present in hen's egg white and is isolated relatively easily, has a molecular weight of 28 000 and an isoelectric point of *ca.* 4.5. It is composed of 186 amino acids in a single chain that occupies three homologous tandem domains and incorporates 9 disulphide bonds. In all 4–5 asparagine residues are glycosylated and sialic acid is present at 0.5–1.0% of the total weight.

The first method reported for coupling of OVM to silica was based on the use of *N,N'*-disuccinimidyl carbonate and 3-aminopropylsilica. The resulting chiral sorbent was found to be stable under the mobile phase conditions used, and to be useful for enantiomer separation.

These columns are now marketed under the name Ultron ES-OVM and have been used for optical resolution of a variety of biologically active compounds [115,116]. Like the AGP-based columns they have been of particular interest for enantiomer separation of cationic analytes, such as basic drugs.

Since ovomucoid, like AGP, contains sialic acid in its carbohydrate region, it was found to be of interest to study the role of the sialic acid on the retention properties of ovomucoid [115]. Therefore, an ovomucoid column was cycled with a neuraminidase-containing buffer for several hours, a procedure which effected liberation of the sialic acid from the glycoprotein. Then, this neuraminidase-treated column (NT-OVM) was compared, under identical conditions, with an untreated (OVM) column with respect to chromatographic properties. A racemic chlorpheniramine maleate sample was used as the analyte and the results are shown in Fig. 7.14. Obviously, the

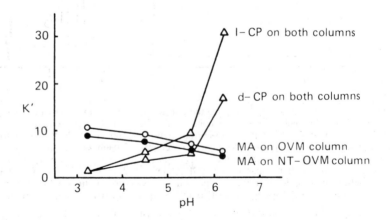

Fig. 7.14 — pH–retention profiles for *rac.* chlorpheniramine (CP) maleate (MA) on an OVM column and a neuraminidase-treated OVM (NT-OVM) column, respectively. (Reprinted, with permission from T. Miwa, H. Kuroda, S. Sakashita, N. Asakawa and Y. Miyake, *J. Chromatog.*, 1990, **511**, 89. Copyright 1990, Elsevier Science Publishers BV.)

sialic acid residue has only a minor influence on the retention of maleic acid and does not influence the retention and optical resolution of the chiral, cationic analyte.

When a chemical deglycosylation was performed prior to immobilization, it was found, however, that the enantioselective properties were completely lost. Further, the pH–capacity ratio profile was found to be the opposite of that of the non-deglycosylated OVM column. These results suggested that the carbohydrate moiety in OVM is essential for preservation of the chiral recognition.

7.1.4.4 Immobilized avidin
Another protein from egg white, avidin, has long been of interest to biochemists because of its remarkably strong binding of a particular ligand, viz. biotin. The binding constant has been determined to be $10^{15} M^{-1}$, which corresponds almost to the strength of a covalent bond. It is therefore a highly efficient inhibitor of biotin-dependent enzymes. This very basic protein has a molecular weight of 65 600 and consists of four identical subunits, each of which binds one molecule of biotin and has a glycosidic chain linked to an asparagine. The isoelectric point is 9.5–10. Avidin has recently been immobilized onto silica and the columns thus produced have shown good enantiomer-differentiating properties particularly for anionic analytes, as expected from the significant positive net charge of the protein at neutral pH [117]. Among the compounds found to be resolvable on this type of column are some 2-arylpropionic acids (the "profens": ibuprofen, ketoprofen and flurbiprofen; see Section 8.3.2.3). No enantioselectivity and very low retention were found on chromatography of a series of amine compounds on the column, further demonstrating the effect of the net charge of this protein phase. It was also shown that biotin completely eliminates the chiral recognition ability of this CSP, as judged from the results obtained from chromatography of the selected profens.

7.1.4.5 immobilized cellobiohydrolase
While the proteins discussed thus far have been of animal origin, the most recent protein used in chiral chromatography is derived from a microbial source. A mould, *Trichoderma reesei*, which can be easily cultivated, is a producer of a group of cellulose-degrading enzymes. These cellulases can readily be isolated from the culture broth and subsequently purified. One of the enzymes, cellobiohydrolase-I (CBH-I), has recently been immobilized onto silica and found to be highly interesting as a tool for liquid chromatographic chiral separations [118,119]. The purification of CBH-I is a relatively simple process, involving gel filtration followed by anion-exchange chromatography at two different pH-values. This procedure removes the two endoglucanases present in the culture broth, as well as a second cellobiohydrolase, CBH-II. CBH-I has an isoelectric point between 3.5 and 3.6 [119]. It is the dominating cellulase produced by *Trichoderma reesei*, which is capable of yielding up to *ca.* 20 g of total cellulase per litre of culture medium. An interesting aspect of the technique of using microbially produced proteins as chiral selectors in liquid chromatography, is the possible application of genetic engineering to enhance expression of desirable proteins. This is likely to become an important route in the future. A schematic drawing of the structural organisation of CBH-I is given in Fig. 7.15. Here C denotes the catalytically active "core" region of the protein, B the

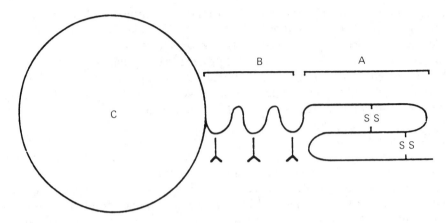

Fig. 7.15 — General structure of the enzyme cellobiohydrolase-I obtained from the mould *Trichoderma reesei.*

glycosylated interconnecting region and A the terminal, 36 amino-acids long, cellulose-binding region.

In summary, although generalizations are difficult to make in the field of chiral separations with proteins, some observations, common to all present types of columns described, are noteworthy.

(1) Alkanol modifiers generally decrease retention of both enantiomers. Most probably, this effect is due to their tendency to reduce hydrophobic interaction between the protein surface and the solute. Since hydrophobic interaction is an essential part of the sorption equilibrium, any reduction will cause faster elution from the column. The effect on the enantiomeric separation factor, however, will depend on the relative decrease in retention of the two enantiomers. Often, but not always, α will either be reduced or essentially unchanged.

(2) The effect of pH on retention is markedly larger for charged than for uncharged solutes. Generally, a decrease in pH gives reduced retention for cationic solutes and an increased retention for anionic solutes. The effects on α are often quite large but difficult to predict or systematize.

(3) Solutes added to the mobile phase can act in different ways depending on whether they bind to the same or a different site as the analyte. If it acts by competition with the analyte for a binding site, the additive will decrease retention. On binding to a different site, however, the additive may cause an allosteric interaction which may change the binding of the analyte at the other site. This can either increase or decrease the retention. The two situations are indicated schematically in Scheme 7.8. Since the two enantiomers of a chiral analyte can bind to the same or to different sites in the protein, the net result of such interactions is hard to predict and sometimes unexpected and large effects are found experimentally. Thus, increasing the concentration of the ion-pairing modifier dimethyloctylamine (DMOA) in the mobile phase gave a dramatic increase in α for the racemic acid 2-(5-methoxy-2-naphthyl)propanoic acid (racemic Naproxen) when this analyte was

competitive interaction   allosteric interaction

Scheme 7.8 — Competitive *vs.* allosteric effect of a mobile phase additive.

run on an AGP-based column at pH 7.0. It was found that the effect was almost entirely due to an increase in k_2' caused by the addition, and that it levelled off at about 5 mM DMOA in the buffer which also contained 1% of 2-propanol. A similar effect, although to a lesser degree, was found for 2-phenoxypropionic acid. These results indicate that the two enantiomers do not compete for the same binding site; instead they are interacting with the protein independently. The increased affinity for the second enantiomer created by the additive could well be the result of an allosteric interaction, i.e. a conformational change in the protein enhancing the binding of this enantiomer.

Such effects have been reported from resolutions on immobilized HSA [120–122]. Domenici and co-workers found during studies of some benzodiazepinones that the enantiomers of a chiral additive could have quite different effects. The results were interpreted in terms of mechanisms involving simple competition for the same binding site as well as involving effects from allosteric interaction as indicated by Scheme 7.8. Some experimental results are shown by Fig. 7.16. In the first case, the additive (ibuprofen) competes with the analyte for the same binding site, thus causing decreased retention. The effect is substantial for the (S)-form of the analyte and (R)-ibuprofen is the most efficient competitor. In the second case, however, an increased retention for the last eluted analyte enantiomer is obtained by the additive (S)-warfarin. This seems to indicate that the analyte and the additive bind to different, but mutually dependent, sites and that the warfarin binding causes a conformational change of the benzodiazepine binding site, leading to a higher affinity for the (S)-form of the analyte.

(4) The complexity of the molecular interactions with protein phases and the presence of a variety of binding sites are factors that cause retention data to vary with the analyte concentration, unless the latter is kept very low. The reason is that as binding sites become saturated and no longer contribute to retention, a decrease will occur in the k'-values of both enantiomers and to a larger extent in the one most strongly bound, resulting also in a reduction of α. The overall retention can therefore be regarded as consisting of one stereoselective and one non-stereoselective part, and the observed k' can be written as a sum: $k' = k_n' + k_s'$, where the subscripts n and s denote the contributions from the non-stereoselective and stereoselective retention mechanisms, respectively. These views have been experimentally verified by studies of the elution profiles from a BSA-based column under non-linear chromatographic conditions [123]. The behaviour of each enantiomer of a selected, resolvable analyte

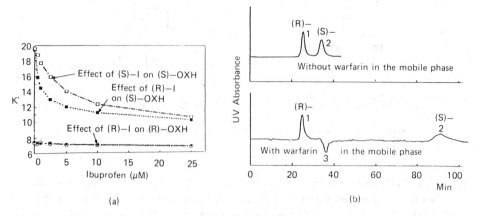

(a)   (b)

Fig. 7.16(a) — Illustration of the effect of mobile phase additives of different configuration on the capacity ratios of (R) and (S)-oxazepam hemisuccinate. HSA column, (R) and (S)-enantiomers of ibuprofen (I) added. (Reproduced from E. Domenici, C. Bertucci, P. Salvadori, S. Motellier and I. W. Wainer, *Chirality*, 1990, **2**, 263 with permission. Copyright 1990, Wiley-Liss, Inc.). (b) Chromatograms showing the effect of (S)-warfarin (40μ*M*) added to the mobile phase on the capacity ratios of (R) and (S)-lorazepam hemisuccinate. (HSA column (4.6×150 mm), mobile phase: 50m*M* phosphate buffer pH 7.0 with 6% of 1-propanol, flow rate 0.8 ml/min). (From E. Domenici, C. Bertucci, P. Salvadori and I. W. Wainer, *J. Pharm. Sci.*, 1991, **80**, 164, reproduced with permission of the copyright owner, the American Pharmaceutical Association.)

under these conditions was well in agreement with a two-site Langmuir adsorption isotherm model. Further, the saturation capacity of the non-selective sites was found to be 10 times higher than that of the non-selective ones.

7.2 BONDED SYNTHETIC CHIRAL SELECTORS

The CSPs described under this heading are all characterized by their well-defined molecular structures bonded to some solid support, usually silica. These low molecular-weight chiral compounds, here called selectors, have often been chosen on a more rational basis because their enantioselective properties can often be evaluated from NMR studies on their solutions. This also means that the elution order from use of sorbents based on such chiral selectors is often predictable from a chiral recognition mechanism.

7.2.1 Crown ethers ('host–guest' complexation)

Macrocyclic polyethers are known under the name crown ethers because molecular models of them often resemble a crown in shape. Their ability to form strong complexes with metal cations as well as substituted ammonium ions has led to very extensive research [124]. One interesting field is their use as phase-transfer agents, where the formation of lipophilic alkali-metal ion inclusion complexes is utilized for the transfer of alkali-metal salts into organic solvents [125]. The field is now known as a branch of 'host–guest' complexation chemistry, mimicking Nature's principle for structural recognition, so common in biological regulatory systems.

The first successful synthesis of optically active crown ethers [126,127] led Cram and his collaborators to investigate their use for optical resolution purposes [128]. This resolution principle was then transferred into an LC separation technique by the use of a chiral crown ether in the mobile phase or covalently bound to a silica support [129]. Such a chiral host is able to discriminate between enantiomeric ammonium compounds, such as esters of D,L-amino-acids, because the multiple hydrogen bonds formed between the ammonium group and the ether oxygens will, for steric reasons, lead to a less stable complex with one of the enantiomers.

The situation is shown by Fig. 7.17. The immobilized selector, the optically active crown ether of (R,R)-configuration, acts as the host, incorporating a guest, represented by the (S)-form of methyl phenylalaninate (as hydrochloride). All three protons of the ammonium group are hydrogen-bonded to crown ether oxygen atoms.

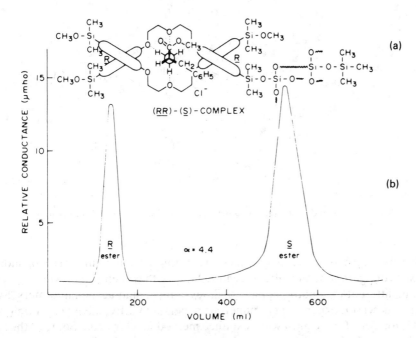

Fig. 7.17 — (a) Structure of the optically active crown ether used for optical resolution. (b) Chromatographic resolution of methyl phenylalaninate hydrochloride. (Reproduced from L. R. Sousa, G. D. Y. Sogah, D. H. Hoffman and D. J. Cram, *J. Am. Chem. Soc.*, 1978, **100**, 4569, with permission. Copyright 1978, American Chemical Society).

The conformational freedom within the complex is rather restricted and the (S)-enantiomer will permit an energetically more favourable conformation in the complexed form. Note that the crown ether, despite its apparent symmetry, can exist in four optically active forms because it is derived from the atropisomeric binaphthol units, which are resolvable owing to hindered rotation. The structure shown is obtained from the (R)-enantiomer.

7.2.2   Metal complexes (chiral ligand exchange)

The ability of transition metals to participate in complex formation was early exploited for the purpose of enantiomer separation. The pioneering work was carried out by Davankov, who published his first papers on the new technique, chiral ligand exchange chromatography (CLEC) as early as 1970 [130–132]. The method used was to immobilize L-proline on a chloromethylated styrene–divinylbenzene copolymer and to utilize the ternary complex formation in the presence of copper(II) ions and amino-acid anions (Scheme 7.9).

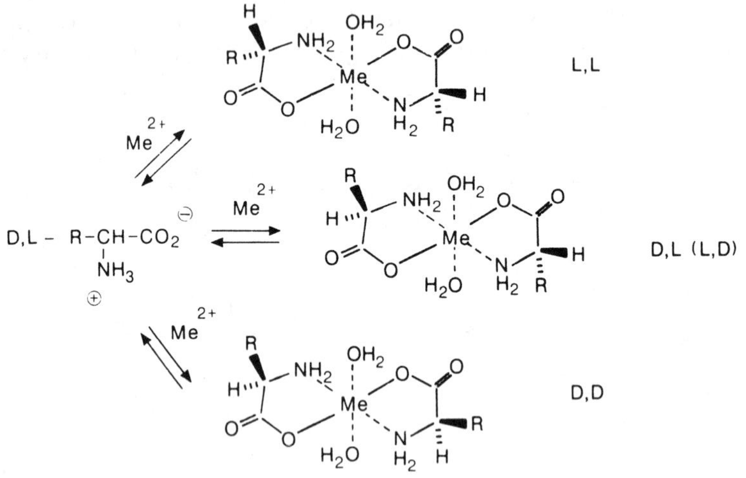

Scheme 7.9 — Bidentate transition-metal amino-acid ternary complex formation.

Because diastereomeric complexes are formed from amino-acid ligands of opposite configuration, any difference in stability of these complexes will, of course, result in different chromatographic mobilities of the amino-acid enantiomers. Since the first successful experiments, numerous papers on CLEC have been published, and it is today by far the most widely studied method for chiral LC. Some of the most important results from these investigations are summarized below.

(1) Of all chelating metal ions studied [Cu(II), Ni(II), Zn(II), Hg(II), Co(III), Fe(III), etc.], copper(II) forms the most stable complexes and seems to have the greatest potential for liquid-chromatography purposes.

(2) Cyclic amino-acids, such as L-proline and L-hydroxyproline, form, together with copper(II), the most enantiodifferentiating chiral selectors.

(3) The method of immobilization and the nature of the matrix are highly important.

(4) In general, an unchanged enantiomeric elution order is obtained for a series of different bidentate amino-acids from a given column.

The requirement of sufficient stability of the ternary complex means that this condition can only be fulfilled by a very limited number of racemic compounds. Since five-membered ring formation is favoured, compounds such as α-amino- and α-hydroxy-acids are the most suitable. Not surprisingly, it has also been found that β-amino-acids (where formation of a six-membered ring is necessary) are difficult to resolve by CLEC [133]. The number of co-ordinating groups influences the relative complex stability. Bidentate ligands, such as the neutral amino-acids lacking other polar substituents, therefore show an elution order (on a polystyrene L-Pro or L-HO-Pro sorbent) which is opposite to that obtained with, for example, the acidic amino-acids (Asp, Glu), which are terdentate.

A way to rationalize these and other results is to consider the various possibilities of stabilizing the sorption complex by electron donation to the metal atom through solvation or other types of ligand participation along the axial directions in the complex. It is assumed that water molecules normally stabilize the complex by co-ordination in an axial position. Therefore, complex stability will be highly dependent on substituent effects causing changes in this solvent participation. Such effects are very likely to take place, owing to the fixed geometry and reduced conformational mobility in the sorption complex. This also explains the importance of the matrix used. The situation is illustrated in Fig. 7.18.

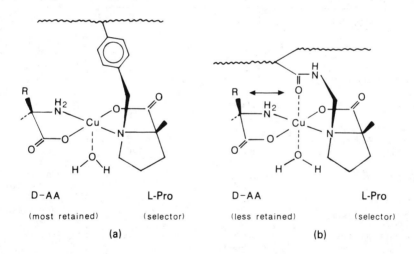

D−AA   L-Pro   D−AA   L-Pro

(most retained)   (selector)   (less retained)   (selector)

(a)   (b)

Fig. 7.18 — Steric effect on solvent participation in ternary complexes formed with an immobilized ligand.

If we start by looking at the situation for the polystyrene–L-Pro sorbent (a), we find that in the two possible diastereomeric sorption complexes formed by a bidentate ligand, the L–L complex will be the least stable because of steric hindrance to solvation. Therefore, the elution order in this case will be L prior to D. With a terdentate ligand, on the other hand, a participation by a neighbouring group, in this case a carboxylate group, will produce a more efficient stabilization than the water molecule. As this participation is only possible in the L–L complex, this will be the most stable and the elution order will consequently be D prior to L. Now, considering the polyacrylamide–L-Pro sorbent (b), the effect of the polar functions in this matrix has to be taken into account. Here, the strongly electron-donating carbonyl groups of the polyacrylamide will participate [134]. In this case it is assumed that the effect of the participation of the amide carbonyl group in complex stabilization (the polyacryl-amide–L-Pro behaves as a terdentate ligand) will be reduced in the L–D complex owing to steric interference with the amino-acid side-chain R. This view is further supported by experimental stability data for analogous low molecular-weight model systems [135]. The elution order of bidentate α-amino-acids, D prior to L, is fulfilled by the compounds investigated, with proline as an exception. The separation factor is strongly increased with the size and branching of the substituent R [136].

The intensive research in this field since 1970 has resulted in a variety of sorbents for CLEC, based on various matrices (polystyrene, polyacrylamide, polymethacry-late, silica) and chiral selectors (where *N*-anchored L-Pro, L-allo-HO-Pro and L-HO-Pro have been among the most useful). A summary is given in Table 7.10.

Table 7.10 — Various types of sorbents used for CLEC

Immobilized ligand	Matrix	Metal ion	Racemates resolved
Various L (and D)-AA	polystyrene	Cu^{2+}, Ni^{2+}, Zn^{2+}	Various D.L-AA, 2-amino-alcohols, mandelic acid
L-Pro	polystyrene	Cu^{2+}	D.L-Pro
N-Carboxymethyl-L-Val	polystyrene	Cu^{2+}	D.L-AA
(R)-*N*,*N*'-dibenzyl-1,2-pro-pandiamine	polystyrene	Cu^{2+}	D.L-AA
L-Pro, L-AA	polyacrylamide	Cu^{2+}	D.L-AA
L-Pro	silica	Cu^{2+}	D.L-AA
L (and D)-Pro, L-HO-Pro, L-Val, L-His	silica	Cu^{2+}, Ni^{2+}, Co^{3+}, Zn^{2+}	D.L-AA
L-Pro-NH₂	silica	Cu^{2+}	D.L-Trp, D.L-Tyr, D.L-Phe
L-Pro-NH₂	silica	Cd^{2+}	Dns-D.L-AA, barbi-turates, hydantoins
t-BOC-L-Pro-NH₂ or	silica	Cu^{2+}	D.L-AA, Dns-D.L-AA
L-Val-NH₂			
Linear polyacrylamide–L-Pro-NH₂	silica (adsorption)	Cu^{2+}	D.L-AA
N-Acyl-L-Val-NH₂	silica		*N*-acyl-D.L-AA ester

Use of silica as a support has significantly improved the normally quite low column efficiency and facilitated the application of modern LC-technology. Despite this, the effective plate height is still comparatively large. This is undoubtedly due to a slow exchange process, on the chromatographic time-scale, in the sorption complex. Therefore, a gain in resolution is usually obtained by an increase in column temperature to around 50°C [137].

The mobile phase is always an aqueous, usually buffered, system containing a low concentration (ca. 0.1–1.0 m*M*) of the complexing metal ion. The use of a water-miscible organic solvent such as methanol or acetonitrile as a modifier is sometimes advantageous. Use of sorbents comprised of a linear polyacrylamide/Cu(II) phase adsorbed onto silica was also found to give slow ligand exchange kinetics [136].

Techniques for covalent binding of amino-acids to silica have been developed [138–151]. A series of silica-bound amino-acids as fixed ligands was investigated by Gübitz and co-workers [143–145] who found that on complexation with Cu(II) ions cyclic amino-acids show higher enantioselectivity than aliphatic ones, and phenylalanine as a fixed ligand gives an elution order for all amino-acids (i.e. L- prior to D-forms) that is the reverse of that obtained with the other ligands investigated.

Methods utilizing non-covalent immobilization of the amino-acid/metal complex by hydrophobic interaction with a reversed-phase (alkyl-silica) sorbent, have also been developed and extensively investigated. Even though some of these techniques actually do not require any chiral additive in the mobile phase [152], they represent borderline cases, and because of their similarity to other methods based on the combination of a reversed-phase non-chiral column and a mobile phase containing a chiral additive, they are presented in Section 7.3.

7.2.3 Selectors based on charge-transfer complexation
The chiral selectors described in this section are characterized by the operation of an aromatic $\pi-\pi$ bonding interaction as an essential element of the retention process. Interactions of this type are well known and are found to take place between so-called π-donor and π-acceptor molecules. A π-donor will have a tendency to lose an electron because the resulting positive charge will be accommodated by the π-system. Conversely, a π-acceptor can readily stabilize a negative charge and has therefore a tendency to accept an additional electron in its π-system. In this way a π-donor/acceptor pair will form a complex when a charge can be transferred from the donor to the acceptor molecule. The strength of such complexes may be considerable. Often the donor molecule is called a π-base and the acceptor a π-acid.

One of the earliest applications of charge-transfer (CT) complexation to the problem of optical resolution by LC was carried out in the 1960s by Klemm and co-workers [153,154]. Polycyclic aromatic compounds, including the helicenes, were known to form CT-complexes with acceptors such as nitroaromatics. An interesting resolving agent based on π-acidity had been designed and synthesized by Newman and co-workers [155–157]. This compound, optically active α-(2,4,5,7-tetranitro-fluorenylideneamino-oxy)propionic acid (TAPA, **5**) was shown to be capable of resolving a variety of aromatic π-bases by CT-complex formation [153–155].

5

By the use of a TAPA-impregnated chromatographic support, Klemm succeeded in partial resolution of some aromatic ethers and hydrocarbons. The ability of TAPA to resolve chiral aromatic hydrocarbons lacking functional groups attracted much interest, and resolutions of a variety of helicenes by LC on TAPA bound to silica or physically adsorbed on silica or alumina have been published, notably by Gil-Av and co-workers [158,159] and by Wynberg and co-workers [160,161]. Other resolutions based on this principle have also been reported [162–164].

Among other CT-complexing chiral selectors suggested are N-dinitrophenyl-L-alanine [165] and binaphthylphosphoric acid (**6**) [166,167], which have also been found useful for resolutions of helicenes.

6

A breakthrough in the use of CT-adsorption for optical resolution by LC was made by the introduction of N-(3,5-dinitrobenzoyl)amino-acids (**7**) as chiral, immo-bilized CT-acceptor ligands. This approach, taken by Pirkle and co-workers [168–170], was preceded by the use of an optically active anthryl carbinol, viz. (R)-(−)-2,2,2-trifluoro-1-(9-anthryl)ethanol ("Pirkle's alcohol") (**8**), for the resolution of racemic 2,4-dinitrophenyl methyl sulphoxide by recycling on a silica column satur-ated with the anthryl carbinol selector† [171]. The successful results led to a chiral sorbent based on this optically active π-base covalently bound to a silica support. This sorbent was used for optical resolutions of a variety of π-accepting racemic solutes, including sulphoxides, amines, amino-acids, hydroxy-acids, lactones, alco-hols, amino-alcohols and thiols [172].

7   **8**

The great success of the technique was interpreted by Pirkle, in the framework of Dalgliesh's three-point interaction theory, as being a result of a combination of simultaneous π–π interaction and hydrogen bonding in the non-polar solvent used as the mobile phase [173]. The solute–sorbent interaction proposed as a chiral recogni-tion model is shown in Fig. 7.19.

† These early experiments were carried out in a glass column, which permitted direct observation of the red charge-transfer complex formed between the alcohol and the sulphoxide. Moreover, the intensity of the colour varied inversely with temperature, demonstrating reversible complexation.

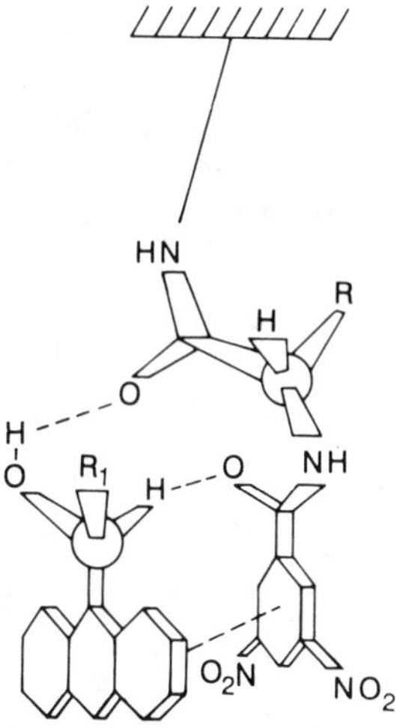

Fig. 7.19 — Pirkle's chiral recognition model. The charge-transfer complex formation causes simultaneous additional enantiomer-dependent contacts between the partners.

Since this chiral sorbent had shown excellent properties with respect to 3,5-dinitrobenzoylated racemic compounds such as amino-acids, application of the principle of 'reciprocal' behaviour (meaning that if optically active A resolves the enantiomers of B, then optically active B resolves the enantiomers of A) led to the synthesis of (R)-N-(3,5-dinitrobenzoyl)phenylglycine as a π-acidic chiral selector. This could be very conveniently used when ionically bound to 3-aminopropyl-silica in combination with 0–20% of 2-propanol in hexane as a mobile phase. Studies of various substituted anthryl carbinols on this CSP contributed to a large extent to a better understanding of the mechanism behind the enantioselective adsorption, which often resulted in high α values [168].

It was found that substituents, R_1, in the anthryl carbinol system, that changed the π-basicity also changed the α values in the same direction. Further, α appeared to increase with the size of R_1. An additional functional group present in R_1 had a minor influence, unless it could competitively interact with the CSP or essentially alter the conformational behaviour of the molecule.

In order to obtain a stable, covalent bond to the aminopropyl-silica support and hence a wider choice of mobile phase, the DNB-amino-acid was reacted with the amino terminal of the silica with EEDQ as condensing agent. Interestingly, the amide-bonded phases differed significantly from their ionically bound analogues when used under identical conditions.

The two most successful CSPs of type (**7**), also commercialized, are those based on phenylglycine and leucine [R = $CH_2C_6H_5$ and $CH_2CH(CH_3)_2$] [174].

The reciprocal aspect of chiral recognition was further utilized as a strategy for the design of another series of CSPs. It had been observed [175] that a long-chain ester of N-(2-naphthyl)-D,L-alanine was resolved on an (S)-N-(3,5-dinitrobenzoyl)-leucine column with a very large α value (10.5). Therefore, new CSPs, comprised of a naphthyl-amino-acid anchored to the silica by a long-chain ester function, were prepared [176,177]. It may be instructive to take a brief look at the synthetic route leading to these sorbents, as exemplified by the case of valine [R = CH(CH$_3$)$_2$] in Scheme 7.10.

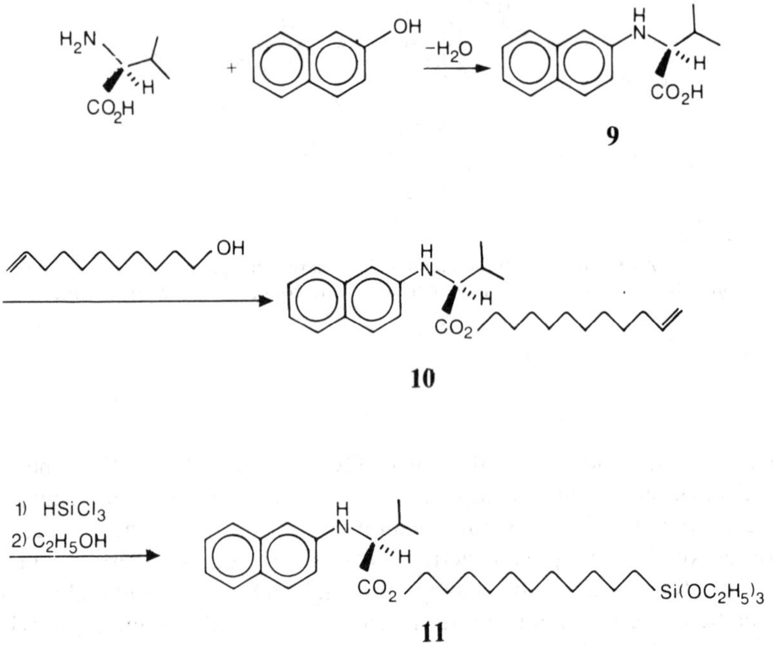

9

10

11

Scheme 7.10 — Synthetic route to immobilized, π-donating chiral selectors.

First, L-valine is reacted with 2-naphthol in a modified Bucherer reaction, yielding (S)-(−)-N-(2-naphthyl)valine (**9**). This product is then esterified with 10-undecen-1-ol by acid catalysis, yielding (**10**). The next step involves a hydrosilylation reaction, meaning that trichlorosilane is added to the terminal double bond, with chloroplatinic acid as catalyst. The chiral trichlorosilane obtained is then converted (without purification) into the corresponding triethoxysilane (**11**). The latter is then bound to silica.

The enantioselective properties of these sorbents are shown in Table 7.11 for a series of derivatives of amines and alcohols. In order to make full use of the π-donating capacity of the naphthylamino function in the CSP, the 'selectand' (the analyte) must carry a good π-acceptor such as the 3,5-dinitrophenyl group. This can

Table 7.11 — Chromatographic data of 3,5-dinitrophenylated derivatives of some amino-acids, amines and alcohols. (Reprinted from W. H. Pirkle and T. C. Pochapsky, *J. Am. Chem. Soc.*, 1986, **108**, 352, with permission. Copyright 1986, American Chemical Society).

Compound	Derivative	k'_1 (enantiomer)	α	2-Propanol in hexane (%)
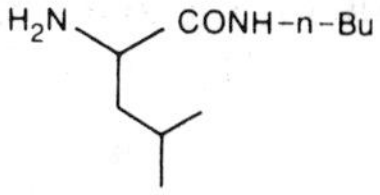 H_2N—...—CONH–n–Bu	DNB	0.38 (R)	17.66	10
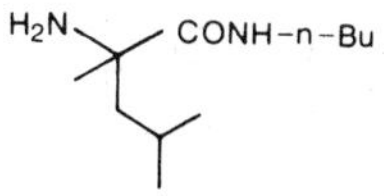 H_2N—...—CONH–n–Bu	DNB	7.97	1.45	5
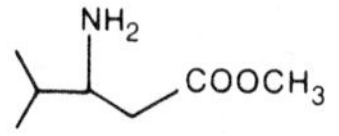 NH_2 ...COOCH$_3$	DNB	0.93 (S)	1.97	5
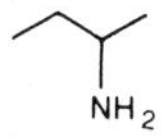 NH_2	DNAn	5.87 (S)	1.19	5
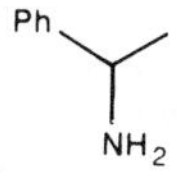 Ph ... NH_2	DNAn	3.27 (R)	1.33	20
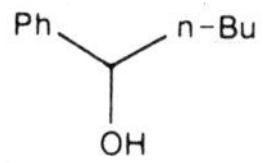 Ph ... n–Bu, OH	DNAn	3.19 (R)	1.24	5
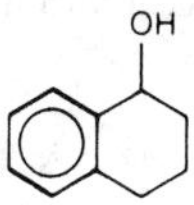 OH (tetralin)	DNAn	5.35 (R)	1.22	5
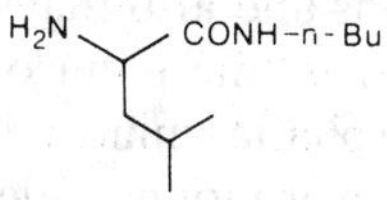 H_2N—...—CONH–n–Bu	DNB	9.0 (R)	2.61	Reversed-phase conditions: 50% methanol–water

DNB = 3,5-dinitrobenzoyl-, DNAn = 3,5-dinitroanilido-

CSP: $-\overset{|}{Si}-(CH_2)_{11}O-C\overset{O}{\underset{\|}{}}\cdots\overset{H}{\underset{i-Pr}{C}}\cdots N\text{—naphthyl}$

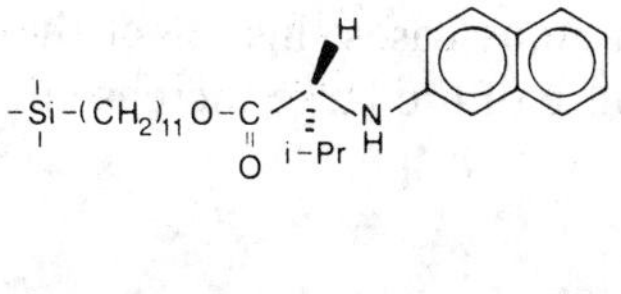

be introduced rather easily by acylation (forming amides and esters) or by carbamoylation with an isocyanate (forming urea derivatives and carbamates). These key reactions are shown in Scheme 7.11.

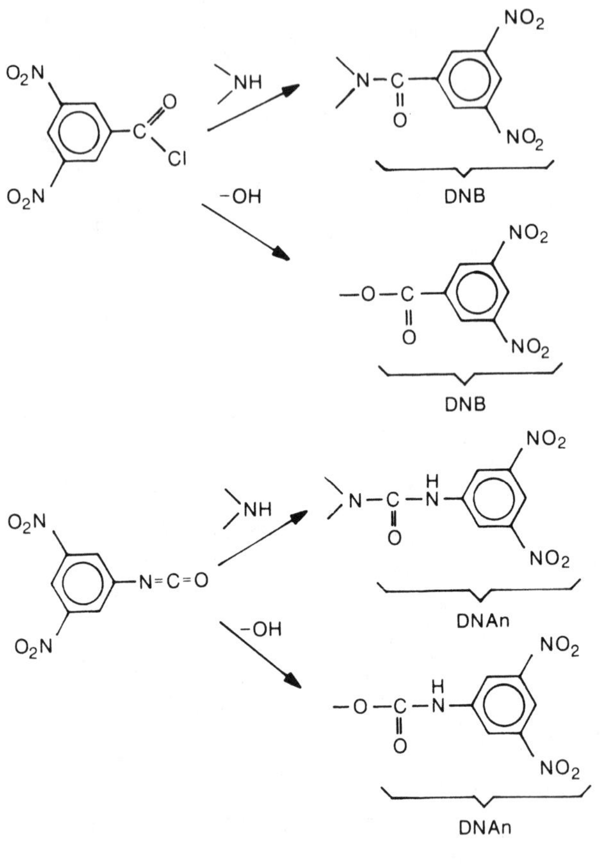

Scheme 7.11 — Derivatization reactions used for introduction of the π-acceptor function into the analyte.

The data presented in Table 7.11 give some indications that derivatives of α-amino-acids are preferred over the β-isomers as selectands and also that α-methyl substitution in the α-amino-acid is highly unfavourable from the point of view of chiral recognition. It is also noteworthy that very simple aliphatic amines and amino-alcohols are well resolvable as DNAn derivatives. Further, a pronounced decrease in separability is found in more polar eluents. Although many factors are involved here, it is reasonable that in a reversed-phase system a significant contribution to overall retention stems from hydrophobic interaction. In many cases this effect should be rather unselective. This is most likely the main cause of the large increase in the k'

values of the α-amino-acid derivatives with increased mobile phase polarity, leading to a drastic reduction in α but no change in elution order.

Other applications of the reciprocity principle have led to various new chiral bonded selectors acting as π-donors. These include the hydantoin- [178], arylalkyl-amine- [179–184] and phthalide-based [185] CSPs. The reciprocity principle and the importance of competing (opposite sense) chiral recognition processes has recently been reviewed and discussed [186].

An interesting new chiral stationary phase, based on an α-amino phosphonate and belonging to the π-acidic category of phases, has been prepared by Pirkle and his collaborators [187]. The synthetic route, given in Scheme 7.12, shows the elegant

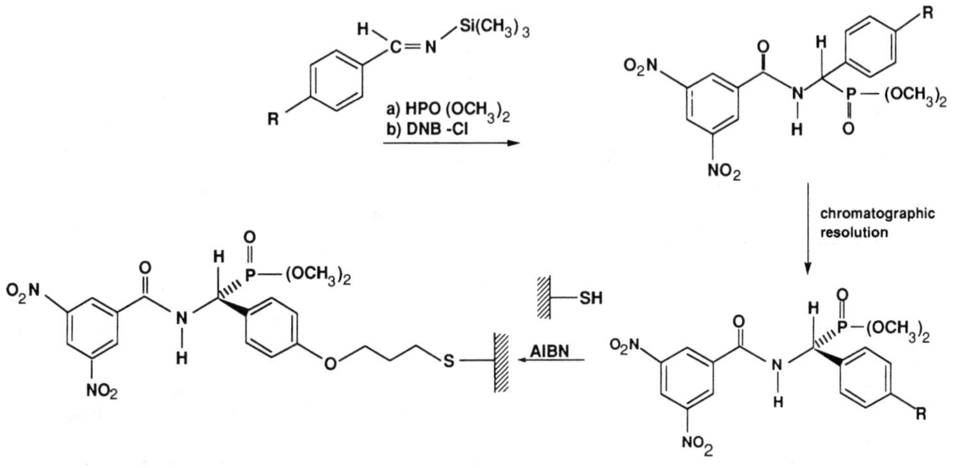

Scheme 7.12 — Synthetic route to a silica-bound dimethyl
N-(3,5-dinitrobenzoyl)-α-aminobenzylphosphonate.

strategy used and is well worth some additional comments. First, hexamethyldisila-zane is reacted with butyllithium in hexane, yielding trimethylsilylamine (as the lithium bis(trimethylsilyl)amide complex). Addition of 4-allyloxybenzaldehyde then gives the imine. In the next step dimethylphosphite is added to the imine double bond. Then, reaction with 3,5-dinitrobenzoyl chloride gives the racemic dimethyl N-(3,5-dinitrobenzoyl)-α-amino-(4-allyloxy)benzylphosphonate. By preparative column chromatography (bed *ca.* 25×750 mm), with a (+)-(R)-N-(2-naphthyl)alanine undecyl ester CSP bonded to a 60-μm irregular silica, and elution with 2% 2-propanol in hexane, the two enantiomers are then separated. The first eluted (−)-(R)-form is used in the final step for the synthesis of the chiral sorbent. By the use of 3-mercaptopropyl-functionalized silica and 2,2′-azobis(isobutyronitrile) (AIBN) as radical initiator, the thiol group is added to the allylic double bond in the phospho-nate, yielding the chiral sorbent.

Chromatographic experiments showed this CSP to behave rather similarly to the previously described *N*-(3,5-dinitrobenzoyl)amino acid-based phases although some differences in selectivity exist. It was, however, superior for the resolution of a series of α-amino phosphonates as 3,5-dinitrophenyl derivatives, giving α-values as high as 3.35.

An analogous coupling technique, exploiting L-tyrosine as the chiral starting material, has been used for the synthesis of a series of charge-transfer-based chiral sorbents [188–190] (Scheme 7.13). The three functional groups present in tyrosine

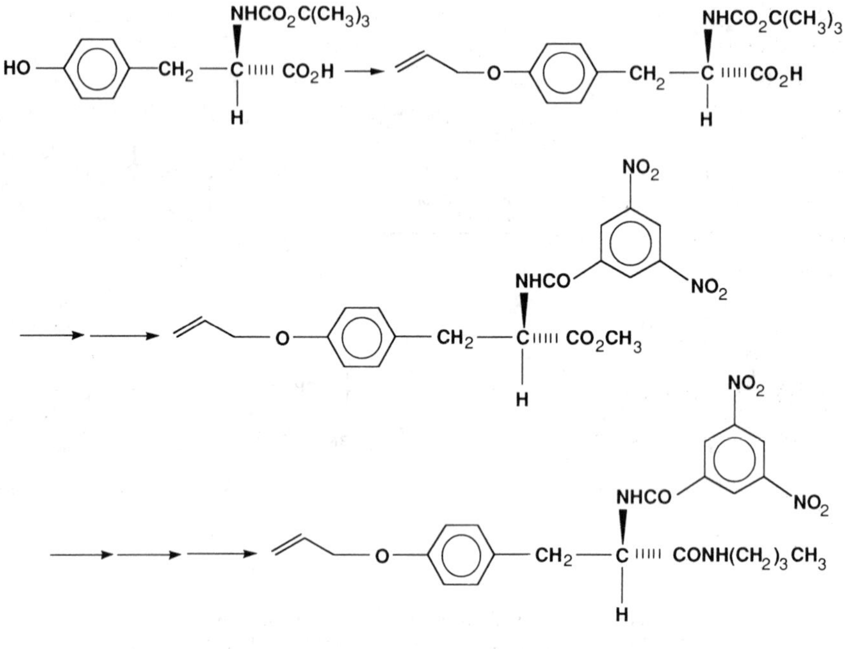

Scheme 7.13 — Chiral sorbents derived from L-tyrosine.

make it possible to transform it into a variety of interesting chiral derivatives which can be covalently immobilized onto a suitably functionalized silica. This has also led to the synthesis of "mixed" π-acidic and π-basic CSPs; the idea being that this should give rise to a broader applicability [191]. This concept of a "mixed" CSP has also previously been used in other sorbents based on synthetic selectors [192,193].

All these tyrosine-derived phases are useful for resolution of a variety of aromatic racemates in normal-phase (hexane-based) conditions. One of them, commercially available as ChyRoSine-A (see Appendix), has also been used successfully in the reversed-phase mode. The preparation of the phases shown in Scheme 7.13 will be described in Chapter 11.

Another promising class of selectors, also based on charge-transfer interaction, has been introduced by Gasparrini and collaborators [194]. These selectors are based on optically active *trans*-1,2-diaminocyclohexane (DACH), which is attached through one of the amino groups to the silica matrix, while the second amino group is used for the attachment of an aromatic substituent like the π-accepting 3,5-dinitrobenzoyl group (DACH-DNB CSP). Owing to the ease of preparation, stability, and good performance, this type of chiral sorbent is likely to be of particular interest in preparative organic chemistry for small-scale optical resolution by LC. The loading capacity for a 20-mm i.d. column has been estimated to be between 10 and 500 mg, depending on the separation factor [195]. Columns packed with the DACH-DNB CSP have been found to resolve a similar spectrum of racemates as the π-accepting Pirkle-phases. They are also favourably operated in a normal-phase mode, with hexane as the major mobile phase component and with cosolvents like dioxan, dichloromethane and 2-propanol as retention modifiers [196].

A similar kind of strategy was used by Uray and Lindner [197], who used a non-cyclic diamine, (S,S)-diphenylethanediamine (DPEDA), as the chiral starting material. The racemic diamine is readily available from benzil [198] and is relatively easy to resolve by tartaric acid (cf. Section 11.3.14). As in the former case, one of the primary amino groups is used for attachment to the silica surface via a hydrocarbon spacer, while the other is 3,5-dinitrobenzoylated. The reactions are outlined in Scheme 7.14.

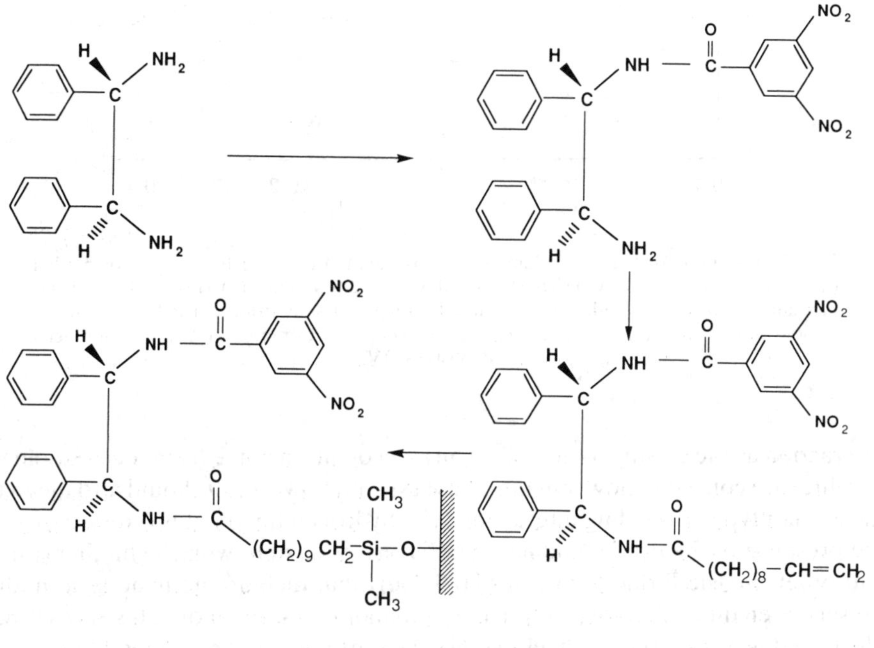

Scheme 7.14 — Synthesis of silica-bound (S,S)-DPEDA.

A comprehensive study of the DACH-DNB CSP has been made with the use of a variety of racemic methyl 2-arylpropanoates as analytes [199]. Two organic modifiers were used with hexane as the major component of the mobile phase, viz.

dichloromethane and ethanol. It was found that a mobile phase with 1% of ethanol gave a much better correlation between log α and Hammett's σ-values than a phase with 15% of dichloromethane (Fig. 7.20). This was taken as an indication of a π–π

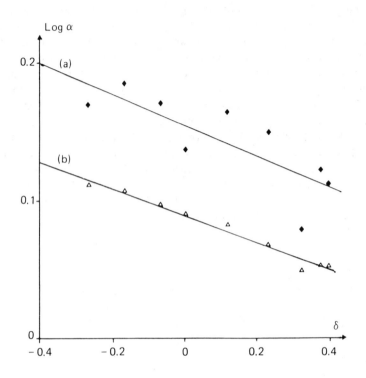

Fig. 7.20 — Plots of log α *vs.* Hammett's substituent constant σ for a series of methyl 2-(aryloxy)propanoates. DACH-DNB CSP, flow rate 2.0 ml/min. (*a*) Hexane/15% dichloromethane, (*b*) hexane/1% ethanol. (Reprinted, with permission, from A. Tambuté, L. Siret, M. Caude and R. Rosset, *J. Chromatog.*, 1991, **541**, 349. Copyright 1991, Elsevier Science Publishers BV.)

interaction as the driving force of the chiral recognition in the former case. Since the α-values are constantly lower in this phase system, however, it should also mean that the second type of bonding interaction, viz. hydrogen bonding, is reduced owing to the presence of the protic modifier. This is also consistent with the finding (for this and other π-basic Pirkle-type CSPs [189,200]) that dichloromethane as a modifier preserves enantioselectivity better than ethanol or 2-propanol. This fact is nicely illustrated in Fig. 7.21, which also shows the efficiency of the two modifiers used in reducing k'_2.

7.2.4   Selectors based on hydrogen bonding
By an extension of the principle used in gas chromatography on chiral amide stationary phases, viz. multiple hydrogen bonding, Hara and associates constructed

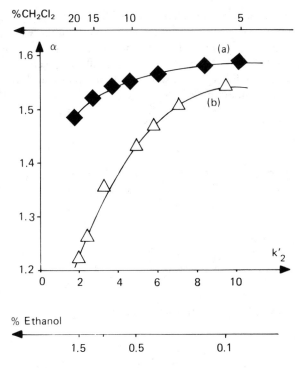

Fig. 7.21 — Plot of α vs. k_2' for the analyte methyl 2-(tolyloxy)propanoate. (a) Hexane/
dichloromethane, (b) hexane/ethanol. Other conditions as in Fig. 7.20. (Reprinted, with
permission, from A. Tambuté, L. Siret, M. Caude and R. Rosset, *J. Chromatog.*, 1991, **541**,
349. Copyright 1991, Elsevier Science Publishers BV.)

a series of selectors for optical resolution by liquid chromatography [201–203]. They
assumed that the hydrogen bond formation in the liquid stationary phase of the chiral
GC method used by Charles *et al.* [204] (cf. Section 6.1.1) could also be utilized in LC
when combined with a non-polar mobile phase. The suggested principle of optical
resolution by diastereomeric selector–solute complexes involving two hydrogen
bonds is shown in Fig. 7.22.

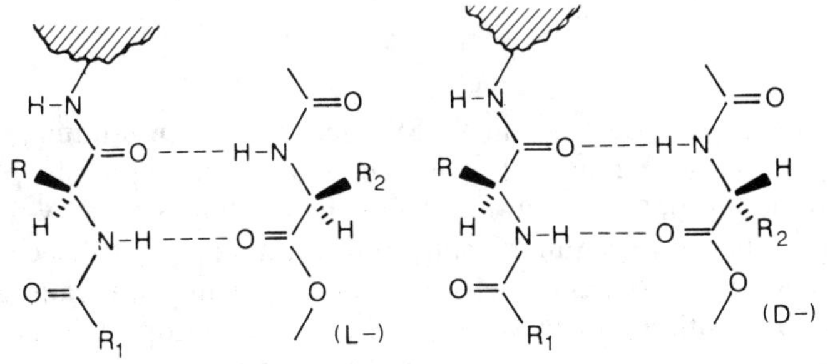

Fig. 7.22 — Diastereomeric sorption complexes formed by a two-point hydrogen-bonding
interaction.

The selectors were constructed from L-valine and D-tartaric acid as optically active starting materials. Various N-acyl derivatives of L-valine were prepared and covalently bound to 3-aminopropyl-silica. From tartaric acid an analogously amide-bound isopropyl amide was synthesized, as well as the free selector, di-isopropyl tartaric diamide [(R,R)-DIPTA], and used as a mobile phase additive. The various selectors are represented in Scheme 7.15.

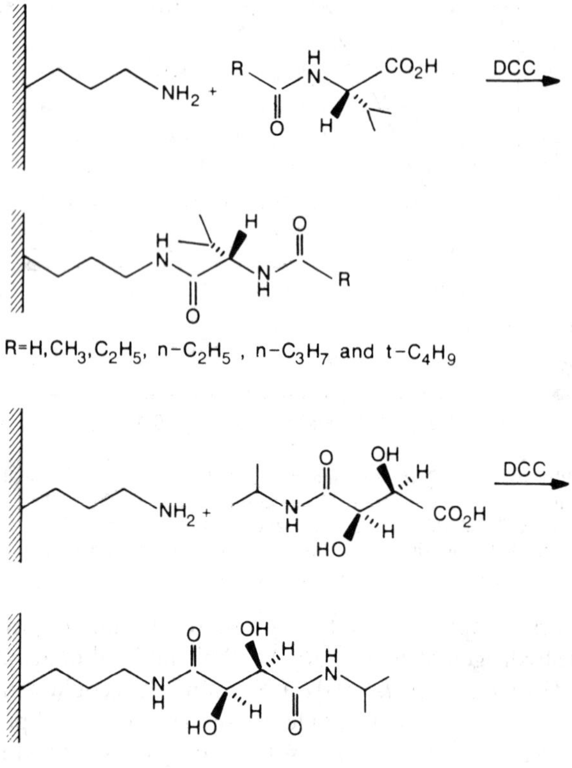

R=H,CH$_3$,C$_2$H$_5$, n-C$_2$H$_5$, n-C$_3$H$_7$ and t-C$_4$H$_9$

Scheme 7.15 — Routes to immobilized, hydrogen-bonding chiral selectors.

As expected, the N-acyl-L-valine CSPs were able to discriminate between the enantiomers of N-acylamino-acid esters in non-aqueous mobile phase systems (usually hexane/2-propanol mixtures) [205]. It was found that (N-formyl-L-valyl-amino)propyl-silica gave optimal results in tests on a series of N-acetylamino-acid methyl esters. Further, a change to a *tert*-butyl ester function caused increased α values. The D-enantiomers always eluted before the L-forms. This means that a solute–sorbent association, as depicted in Fig. 7.22 for the most strongly retained enantiomer, can be formulated. Interestingly, the same stereochemistry was found when N-acetyl-L-valine-*tert*-butylamide was added to the mobile phase as a selector.

An implication from this association stereochemistry is that the complex corresponding to the most strongly retained enantiomer will have both amino-acid α-substituents located on the same side and directed away from the surface of the silica matrix.

The mobile phase effects observed were completely consistent with the hydrogen bond association model. Capacity factors increased in a predictable way with decreasing 2-propanol content in hexane, meaning that there is then less competition by the solvent for hydrogen-bonding sites of the CSP. The same effect could be achieved by substituting 2-propanol for less competitive, aprotic, solvents such as chloroform, dichloromethane and diethyl ether.

The number of possibilities of hydrogen-bond formation is increased considerably in the selector based on tartaric acid. There is also a greater conformational freedom which would allow an increased adaptability for association with a wide range of compounds and therefore, in principle, also for broader enantioselectivity. This immobilized selector is quite analogous to (R,R)-DIPTA and was developed on the basis of the successful use of the latter as a mobile phase additive, as will be described in Section 7.3.

7.2.5 Other types of selectors

Though the bonded selectors described in Sections 7.2.1–7.2.4 are reasonably well understood in terms of their chiral recognition behaviour, a number of other selectors have been developed [206–215] which probably act by more complicated and less well understood mechanisms.

Very promising CSPs based on amide [211–213] or urea [214,215] derivatives have been prepared and studied by Oi and colleagues. The latter phases all contain an asymmetric carbon atom directly attached to a urea nitrogen atom (**12,14**), (Scheme 7.16). Thus, **12** contains L-valine as the chiral component [coupled to 3-aminopropyl-silica as the *N*-(*tert*-butylaminocarbonyl) derivative], **13** the chiral 1-(α-naphthyl)ethylamine substituent, and **14** contains both these chiral elements.

It seems most likely that phase **12** which is structurally similar to the diamide phases used in GC and by Hara in LC, operates by a combination of hydrogen bonding and steric effects. Results from the use of phase **13**, however, indicated that additional charge-transfer interactions could contribute, because resolution of compounds with π-accepting substituents was facilitated. Interestingly, Oi found that phase **14**, which embodies two chiral centres, was superior to the first two, owing to its broader applicability. It gave particularly good resolution of *N*-3,5-dinitrobenzoyl derivatives of amines, 3,5-dinitroanilide derivatives of carboxylic acids and 3,5-dinitrophenylurethane derivatives of alcohols. Some esters and alcohols were also well resolved without prior derivatization.

Owing to their long-standing and frequent use for classical resolution purposes, alkaloids are attractive chiral selectors, since they are readily available, naturally occurring, low molecular weight compounds, yet possess a structural complexity not easily obtained by synthetic means. Their application in asymmetric synthesis [216] adds further to their potential usefulness as chromatographic chiral auxiliaries.

By utilizing the olefinic bond in *Cinchona* alkaloids (cf. Section 11.3), Salvadori and collaborators were the first to immobilize quinine onto a silica matrix without

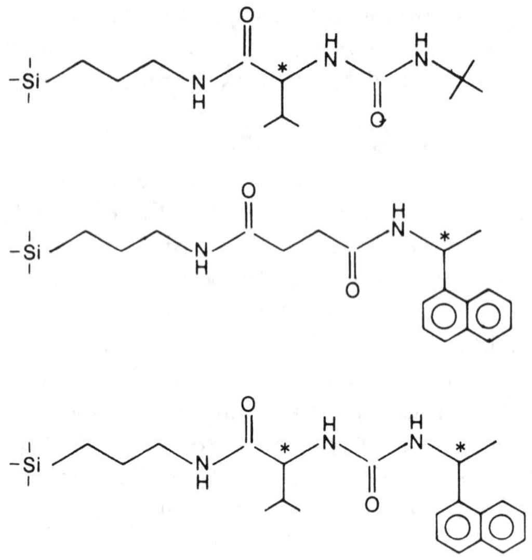

Scheme 7.16 — Amide and urea-linked chiral selectors developed by Oi.

blocking any stereogenic centre in the molecule [217,218]. The same principle was later used to immobilize quinidine and cinchonidine [219] as well as N-methylquini-nium iodide [220]. The latter selector was found to be particularly useful for the optical resolution of some pharmaceutically active benzodiazepinones (cf. Section 8.3.1.4). Figure 7.23 shows this, and also demonstrates the good column perfor-mance and the simultaneous determination of elution order by chiroptical (CD) detection.

7.3 TECHNIQUES BASED ON ADDITION OF CHIRAL CONSTITUENTS TO THE MOBILE PHASE

Modern reversed-phase HPLC sorbents ($C_2–C_{18}$ silicas) are excellent materials for achievement of high column efficiency. The principle of using mobile phase additives in reversed-phase LC in order to regulate the retention behaviour of an analyte is widely used today. Basically, the technique makes use of the well-known phenome-non of ion-pair formation in organic media, where the partition of a positively charged analyte such as a protonated amine will be greatly influenced by the nature of the counter-ion (cf. Section 5.2). By the use of an optically active counter-ion, diastereomeric ion-pairs will be formed, which may be well separated on an achiral ordinary reversed-phase column. As discussed previously, the technique of using amphiphilic additives can be regarded as a dynamic coating of the achiral sorbent, i.e. a physical immobilization of the amphiphile. Thus, if this additive is optically active, the achiral sorbent will be converted into a chiral one.

Consequently, there is no fundamental difference between the techniques given below and those described above. The ionically bound Pirkle-phases (Section 7.2.3)

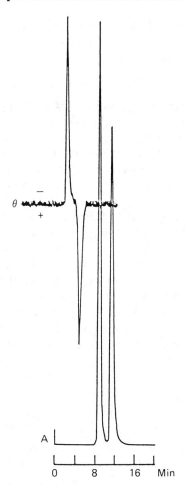

Fig. 7.23 — Optical resolution of the benzodiazepinone lormethazepam by the use of a column containing *N*-methylquininium iodide immobilized to silica. Mobile phase: Hexane/2-propanol/dichloromethane (100:28:24 by volume), flow rate 1.0 ml/min. UV and CD detectors in series, both operating at 254 nm. (−)-(R)-lormethazepam is the first eluted enantiomer. (Reprinted, with permission, from C. Rosini, C. Bertucci, D. Pini, P. Altemura and P. Salvadori, *Chromatographia*, 1987, **24**, 671. Copyright 1987, Friedr. Vieweg & Sohn Verlagsgesellschaft mbH.)

are excellent CSPs as long as they are used in non-polar solvents where the mobile phase has very little tendency to displace the selector from its adsorption sites. Under these conditions, therefore, the mobile phase can be used without any added selector. In cases where the selector is immobilized by strong hydrophobic interaction to an alkyl-silica or other hydrophobic matrix, the situation may be similar but normally a constant coverage of the matrix with the selector will require its presence in the mobile phase.

Many of the principles already described, which are based on covalently bound chiral phases, can also be applied to the technique of adding the chiral selector to the mobile phase. They can be divided into three categories, viz. the metal complexation used in CLEC, the use of various uncharged additives, and finally the ion-pairing techniques used for charged analytes.

7.3.1 Metal complexation

Karger was the first to use metal complexation for CLEC applications [221,222]. By the use of chiral triamines (L-2-ethyl- and L-2-isopropyl-4-octyldiethylenetriamine — each of which has a hydrophobic C_8-substituent) as mobile phase additives in the presence of Zn(II) and other transition metal ions, a series of dansyl amino-acids was well resolved on a C_8–reversed-phase column. A similar system utilized L-prolyl-N-octylamide and Ni(II) and was successfully applied to the same type of resolutions with a C_{18}-column [223].

The technique was studied in detail by Davankov *et al.* [152] with the intention of arriving at a useful method for analytical as well as preparative resolution of free amino-acids. N-Alkyl-L-hydroxyprolines (C_7, C_{10} and C_{16} chain lengths) were used to coat a C_{18}-silica column and Cu(II) acetate ($0.1\text{m}M$) in methanol/water (15/85 v/v) was used as the mobile phase. All additives were adsorbed by means of strong hydrophobic interaction with the C_{18} sorbent, giving essentially no column bleed. Therefore, it was possible in this case to use the mobile phase without the chiral additive, provided the water content was high enough. The postulated structures of the mixed-ligand sorption complexes formed are shown in Fig. 7.24.

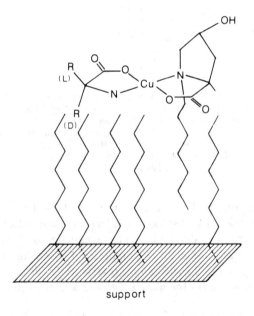

support

Fig. 7.24 — Structures proposed for the mixed-ligand sorption complexes.

In general, decreasing alkyl chain length was found to give larger separation factors. The reason for this is obscure as the amounts of chiral additive adsorbed could not be determined. The effects caused by a variation of the mobile phase composition are essentially as follows: (a) an increase in pH (above 5.5) causes increased retention and larger separation factors, (b) a decrease in Cu(II) concentration causes a small increase in retention but no significant effects on separation factors, (c) increasing the ammonium acetate concentration strongly decreases the retention but the effects on the separation factors are quite complex.

As pointed out previously (cf. Section 7.2.2), the relative stabilities of the mixed-ligand sorption complexes formed in CLEC are highly dependent on the method used for immobilization. In the present case, where the selector is physically immobilized by hydrophobic interaction, the elution order of all amino-acids is $k'(L) < k'(D)$. This experimental result, combined with the mobile phase effects, points to the enantioselective mechanism outlined in Fig. 7.24. The N-alkyl chains of the selector are assumed to be oriented parallel to the C_{18}-chains. By co-ordination with a Cu(II) ion, this fixed ligand will adopt a conformation such that the hydroxypyrrolidine ring and its N-alkyl group extend in a direction opposite from the main co-ordination plane of the Cu(II) chelate. Therefore, in a mixed-ligand sorption complex formed by the D-enantiomer of the analyte, the α-substituent of this enantiomer will be directed towards the hydrophobic (C_{18}) sorbent surface. This will cause an increased stabilization by means of hydrophobic interaction. An L-enantiomer, on the other hand, will lack this possibility and will therefore be eluted faster than the D-enantiomer.

The same principle was used in a technique based on coating alkyl-silicas with N-decyl-L-histidine [224]. The separation factors were generally lower than those found when using N-alkyl-L-hydroxyproline selectors, but were still useful for determination of enantiomer composition. As in the previous case, the highest enantioselectivity was found for the amino-acids with the largest α-substituents (alkyl and aryl groups).

If the alkyl chain of the chiral selector is omitted, conditions are obtained under which there should no longer be any strong hydrophobic interactions with the alkyl-silica and no actual physical immobilization of the selector. Consequently, the chromatographic process might then be best thought of as an *in situ* generation and separation of diastereomeric complexes in a reversed-phase mode. A variety of methods based on this principle of chiral metal complexation in the mobile phase have been developed during recent years and these are summarized in Table 7.12.

7.3.2 Uncharged chiral mobile phase additives

As described in Section 7.2.4, many diamide-type CSPs give rise to enantioselective hydrogen-bond interactions and these bonded selectors should therefore also be useful as additives in a non-polar mobile phase in normal-phase chromatography. Under these conditions they are adsorbed rather strongly on a silica surface, which can then be regarded as coated with a CSP. In particular, N-acetyl-L-valine *tert*-butylamide (**15**) and (R,R)-DIPTA (**16**; Section 7.2.4), have been found useful for the optical resolution of a variety of polar solutes [249–252]. From the point of view of chiral recognition, the behaviour shown by the *threo-* and *erythro-* forms of solutes

Table 7.12 — Chiral metal complexes used as mobile phase additives for optical resolution by CLEC

Selector	Metal ion	Stationary phase	Analyte	References
L-proline	Cu^{2+}	octyl-silica	amino-acids	[225–227]
L-proline	Cu^{2+}	octyl-silica	dansylamino-acids	[228]
L-proline	Cu^{2+}	silica	thyroid hormones	[229]
L-proline	Cu^{2+}	cation exchanger	amino-acids	[230]
L-histidine	Cu^{2+}	octyl-silica	amino-acids	[227,231]
L-histidine methyl ester	Cu^{2+}	octadecyl-silica	amino-acids	[232,233]
L-arginine	Cu^{2+}	octyl-silica	amino-acids	[227]
L-phenylalanine	Cu^{2+}	octadecyl-silica	aromatic amino-acids	[229]
L-phenylalanine	Cu^{2+}	octadecyl-silica	mandelic acids	[234]
L-aspartic acid monoalkylamides	Cu^{2+}	octadecyl-silica	amino-acids	[235–237]
L-aspartyl-L-phenylalanine methyl ester (Aspartame)	Cu^{2+}, Zn^{2+}	octadecyl-silica	amino-acids	[238,239]
N,N-dipropyl-L-alanine	Cu^{2+}	octadecyl-silica	dansylamino-acids methylamino-acids	[240–243]
N,N-dialkyl-L-amino-acids	Cu^{2+}	octadecyl-silica	amino-acids	[244]
N-(p-tosyl)-L- (and D-) phenylalanine	Cu^{2+}	octadecyl-silica	amino-acids	[245,246]
L-amino-acids	Cu^{2+}	octadecyl-silica	hydroxy-acids	[247]
(R,R)-tartaric acid monooctylamide	Cu^{2+}, Ni^{2+}	octadecyl-silica	amino-acids	[248]

containing a 1,2-diol structure (exemplified by **17**) is quite relevant. Whereas the *threo*-compounds (T) yield larger separation factors with increasing bulkiness of the substituent R, the reverse is found for the *erythro*-forms (E). This is quite understandable in view of the preferred conformations (Scheme 7.17) of the two forms,

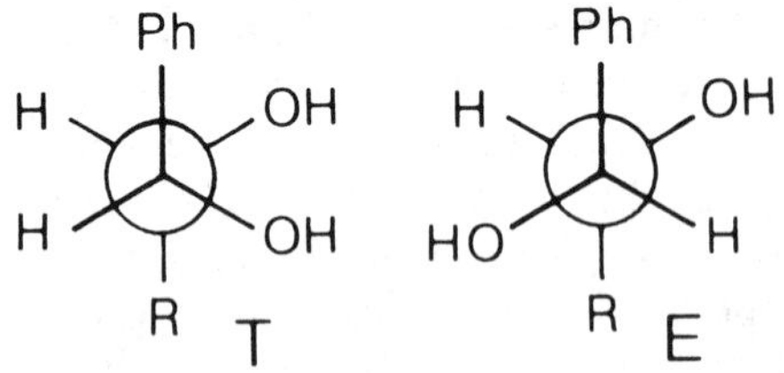

Scheme 7.17 — The role of conformational stability for the substituent effect in *erythro*- and *threo*-isomers on α.

which become more populated as the steric requirements of R increase. Because a two-point hydrogen-bond formation between the CSP and the solute will require a *gauche* conformation of the two hydroxyl groups in **17**, it is clear that the experimental results strongly support such a mechanism of formation.

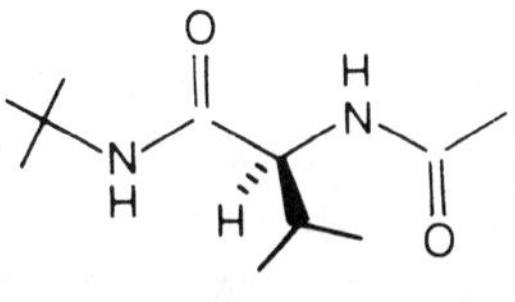

15 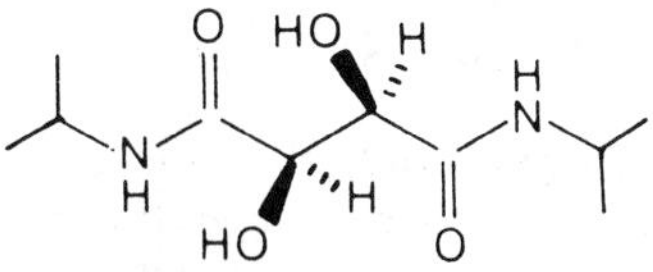   **16**

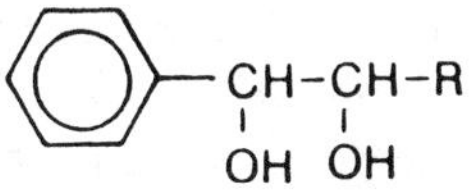

17

Another type of uncharged, chiral selector which has been used as a mobile phase additive is cyclodextrin (CD), mainly in the β-form. This is generally used in a reversed-phase system employing C_{18}-silica and an aqueous buffer system [253–255]. The first studies were performed on substituted mandelic acids, and it was shown that substituent effects were very large, and retention decreased with increasing pH or increasing β-CD concentration. Complete optical resolution ($\alpha = 1.8$) was achieved for o-chloromandelic acid at pH 2.1 and with 14.4mM β-CD in the buffer, otherwise separation factors were low and decreased with increasing pH.

Further insight into the mechanism of enantioseparation by this technique was obtained recently from a study of some barbiturates and related compounds [256]. It was assumed that the analyte (G) was present in an uncharged form (pH $<$ pK_a), that it formed a 1:1 inclusion complex with the cyclodextrin molecule, and that the properties of the sorbent were unaffected by the added β-CD. With these assumptions the following equilibria can be written:

$$G_m + \text{β-CD} \underset{}{\overset{K_G}{\rightleftharpoons}} (G\text{β-CD})_m \qquad \text{(mobile phase equilibrium)}$$

$$G_s \underset{}{\overset{k'_G}{\rightleftharpoons}} G_m$$

$$(G\text{β-CD})_s \underset{}{\overset{k'_{(G\text{β-CD})}}{\rightleftharpoons}} (G\text{β-CD})_m \qquad \text{(mobile phase/ stationary phase equilibria)}$$

A system of this kind will yield an observed capacity ratio k', which can be expressed by Eq. (7.3). Rearrangement gives Eq. (7.4).

$$k' = \frac{k'_G + k'_{(G\text{β-CD})}K_G[\text{β-CD}]}{1 + K_G[\text{β-CD}]} \qquad (7.3)$$

$$k' = \frac{k'_G - k'}{K_G[\beta\text{-CD}]} + k'_{(G\beta\text{-CD})} \qquad\qquad (7.4)$$

Plots of k' as a function of $(k'_G - k')/[\beta\text{-CD}]$ were linear, as expected from Eq. (7.4), and thus in accordance with 1:1 stoichiometry of the complex. The value of k'_G is, of course, easily obtained experimentally as k' in the absence of β-CD. From the slopes and intercepts of such plots, values of K_G and of $k'_{(G\beta\text{-CD})}$ are readily determined for both enantiomers. Interestingly, this led in the case cited [256] to the result that the barbiturates were optically resolved only by virtue of different stability constants (K_G) of the diastereomeric β-CD inclusion complexes, as the calculated $k'_{(G\beta\text{-CD})}$ values were found to be close to zero (in fact they were slightly negative). This means that the enantiomer corresponding to the least stable complex will be preferentially retained.

In contrast to this result it was found that a substituted hydantoin (mephenytoin) was optically resolved predominantly by a difference in the retention of the diastereomeric β-CD inclusion complexes formed, as the calculated $k'_{(G\beta\text{-CD})}$ values differed by a factor of almost 3. However, the slightly different K_G values obtained resulted in a decreased α value. This difference in behaviour was interpreted as due to incomplete immersion of the hydantoin in the β-CD cavity, a condition which should be different for the two enantiomers. Retention here is essentially caused by the part of the molecule which protrudes from the cavity.

Even if further experimental data are needed in order to fully verify the suggested mechanisms, the analysis given emphasizes the importance of the stability of the complex. Accordingly, the enantiomer differentiation observed can arise from either or both of the following effects:

(1) the differences in the complex stability constants, $K_{(-)\text{-}G}$ and $K_{(+)\text{-}G}$;
(2) the differences in retention of the complexes, i.e. their adsorption on the sorbent, $k'_{(-)\text{-complex}}$ and $k'_{(+)\text{-complex}}$.

It is also quite important to realize that these effects may work in opposite directions and in extreme cases yield $\alpha = 1$.

The technique described above has also been used in planar chromatography. By the use of reversed-phase (C_{18}) TLC plates and β-CD in the mobile phase, several racemates were separated into enantiomers, particularly a series of racemic amino acids [257]. Certain problems were found, however, arising from the low solubility of cyclodextrins. To overcome these, saturated solutions of urea were used to increase the β-CD solubility. Further, $0.6M$ sodium chloride had to be added to stabilize the binder of the TLC plates. It was found that optimum enantiomeric resolution occurs within a relatively narrow range of β-CD concentrations (ca. 0.1–$0.2M$), and also of organic modifier type and concentration. The modifiers were typically 20–30% of acetonitrile or 30–40% of methanol. In a mobile phase system composed of 30% of acetonitrile $+70$% of $0.1M$ β-CD, baseline optical resolution of 11 dansyl-DL-amino acids, applied on the same plate, was obtained.

7.3.3 Ion-pairing techniques
As a logical extension and elaboration of the ion-pair chromatographic technique [258,259], a chiral counter-ion, (+)-10-camphorsulphonic acid, was introduced as a

mobile phase additive for the optical resolution of some amino-alcohols [260]. The amino-alcohol in its protonated form combines with the camphorsulphonate by electrostatic interaction to form diastereomeric complexes. It is assumed that a second interaction, a hydrogen bond between the keto group and the hydroxy group, respectively, of the two partners, is present and gives rise to the different chromatographic retentions observed (Fig. 7.25). The separations were conducted in a normal-phase mode on silica or diol-silica [261] by using methylene chloride with a small amount of a polar retention modifier.

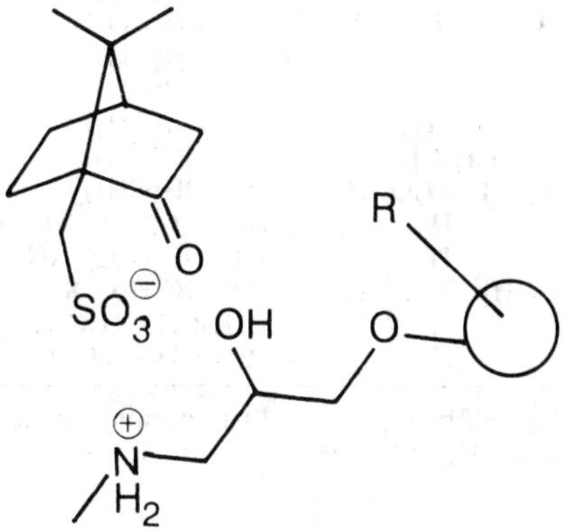

Fig. 7.25 — Proposed interaction between the camphorsulphonate ion and the amino-alcohol to account for the chromatographic enantioselectivity observed.

Methylene chloride is an excellent ion-pair solvent and it is most likely that the enantiomer separation should be regarded as a separation of labile diastereomeric ion-pair complexes, i.e. the chiral discrimination occurs in the mobile phase. As expected, the capacity factors decrease considerably with increasing concentration of the polar component of the mobile phase.

The requirement of a two-point interaction for enantioselection is supported by results illustrating the structural effects of the analyte on the separation factor (Table 7.13). No optical resolution takes place when the hydroxyl and amino groups are separated by more than two carbon atoms or if the hydroxyl group is absent. Further, oxprenolol (18a) is not resolved, probably due to an internal hydrogen bond between the hydroxyl group and the oxygen atom of the allyloxy group.

Improved results were obtained with the use of an N-protected dipeptide, N-carbobenzoxycarbonyl-glycine-L-proline, as the chiral counter-ion. In these cases α values as high as 1.4 were obtained for some of the amino-alcohols [262].

These results encouraged a reciprocal use of the ion-pair system for the enantioseparation of racemic sulphonic and carboxylic acids with optically active amino-alcohols as counter-ions. It was then found that whereas alprenolol (18b), with its binding groups located in a flexible alkyl chain, give a low degree of stereoselectivity

Table 7.13 — Separation factors obtained in chiral ion-pair chromatography of some amino-alcohols with camphorsulphonate as a counter-ion. (Reprinted, with permission, from C. Pettersson and G. Schill, *Chromatographia*, 1982, **16**, 192. Copyright 1982. Fr. Vieweg & Sohn Verlagsgesellschaft mbH).

Structure: R_2-benzene-R_4-CH-(CH$_2$)$_n$-NHR$_3$, OH, R_1

n	R₁	R₂	R₃	R₄	α
1	H	CH₂CH₂OCH₃	CH(CH₃)₂	OCH₂	1.09
2	H	CH₂CH₂OCH₃	CH(CH₃)₂	OCH₂	1.00
3	H	CH₂CH₂OCH₃	CH(CH₃)₂	OCH₂	1.00
1	H	CH₂CH₂OCH₃	CH(CH₃)₂	—	1.09
1	H	OCH₂CH=CH₂	CH(CH₃)₂	OCH₂	1.08
1	CH₂CH=CH₂	H	CH(CH₃)₂	OCH₂	1.08
1	OCH₂CH=CH₂	H	CH(CH₃)₂	OCH₂	1.00
1	H	OCH₃	CH(CH₃)₂	OCH₂	1.08
1	H	CH₂CH₃	CH(CH₃)₂	OCH₂	1.09
1	H	OCH₂CH₂OCH₃	CH(CH₃)₂	OCH₂	1.09
1	CH₂CH=CH₂	H	CH₂CH₂OC₆H₄CONH₂	OCH₂	1.11
1	Cl	H	CH₂OC₆H₄CONH₂	OCH₂	1.11
1	Br	CH₂CH₂OCH₃	CH₂OC₆H₄CONH₂	OCH₂	1.11
1	CH₃	H	CH₂CH₂CH₂C₆H₅	OCH₂	1.00
1	H	H	CH₂CH₂C₆H₄CH₃	OCH₂	1.00

Solid phase Lichrosorb Diol. Mobile phase: (+)-10-camphorsulphonate (2.2mM) in methylene chloride + 1-pentanol (199:1). $\alpha = k'(-)/k'(+)$

[263], the situation was considerably improved by the use of compounds with rigid ring systems such as quinine, quinidine and cinchonidine [264].

18

These and related techniques have recently been reviewed by Pettersson and Schill [265].

Although the use of chiral counter-ions for optical resolution by the technique described has proved to be successful for certain applications, the chromatographic system is rather complex and the effects of the many factors involved, on retention and resolution, are not always easy to interpret [264]. Some of these important factors are given below.

(1) The chromatographic sorbent. The surface properties of the strongly polar silica-sorbent are critical and it has been found [264] that diol-silica (which is a surface-modified silica with hydrophilic properties) is generally preferred because of its better performance.

(2) The water content of the mobile phase. This is critical owing to its pronounced influence on retention (and resolution). A water content of 80–90 ppm has been recommended [264]. Higher concentrations of water are quite deleterious. The small amount of water is probably essential in order to deactivate the silica surface, which otherwise might adsorb polar components too strongly.

(3) The capacity ratios of the enantiomers. These usually decrease as the concentration of counter-ion increases. This has been interpreted as due to competition of the counter-ion and the ion-pairs for the same adsorption sites on the sorbent.

(4) The polar components of the mobile phase. These, e.g. 1-pentanol, cause a drastic decrease in retention of the solute, which is usually accompanied by a decrease in optical resolution. The latter effect is thought to be caused by a competition for hydrogen-bonding groups in the ion-pair components, leading to reduced stereoselectivity.

(5) The optical purity of the mobile phase additive. This will, of course, influence the separation factor. It can easily be shown that

$$\alpha_{obs} = \frac{\alpha P + (100 - P)}{\alpha(100 - P) + P} \qquad (7.5)$$

where P is the fraction (%) of one enantiomer in the mobile phase additive, and α_{obs} is the observed separation factor [266]. Note that P is not equal to the optical purity, which is 0 for $P = 50$.

A quite different ion-pairing chromatographic technique, utilizing (+)-dibutyl tartrate (DBT) as chiral additive, has also been described [267]. The principle of the technique originates from studies by Prelog *et al.* [268] of the liquid–liquid distribution of enantiomeric amino-alcohols as ion-pairs in the presence of tartaric acid esters; an unequal distribution of the enantiomers was observed.

In the chromatographic method DBT is used as a physically immobilized CSP, applied by adsorption from an aqueous phase onto a hydrophobic matrix (alkyl-silica). This system has permitted partial optical resolution of a series of amino-alcohols as their ion-pairs with hexafluorophosphate in a reversed-phase system employing a completely aqueous buffer. The mechanism of this enantiodifferentiation is not quite clear.

BIBLIOGRAPHY

R. Audebert, Direct Resolution of Enantiomers in Column Liquid Chromatography, *J. Liquid Chromatog.*, 1979, **2**, 1063.

G. Blaschke, Chromatographische Racemattrennung, *Angew. Chem.*, 1980, **92**, 14 (*Int. Ed.*, 1980, **19**, 13).

V. A. Davankov, Resolution of Racemates by Ligand-Exchange Chromatography, *Adv. Chromatog.*, 1980, **18**, 139.

W. Lindner, Trennung von Enantiomeren mittels moderner Flüssigkeits-Chromatographie, *Chimia*, 1981, **35**, 294.

W. Lindner, Resolution of Optical Isomers by Gas and Liquid Chromatography, in *Chemical Derivatization in Analytical Chemistry*, Vol. 2, R. W. Frei and J. F. Lawrence (eds.), Plenum, New York 1982, p. 145.

V. A. Davankov, A. A. Kurganov and A. S. Bochkov, Resolution of Racemates by High-Performance Liquid Chromatography, *Adv. Chromatog.*, 1983, **22**, 71.

W. H. Pirkle and J. Finn, Separation of Enantiomers by Liquid Chromatographic Methods, in *Asymmetric Synthesis*, Vol. I, J. D. Morrison (ed.), Academic Press, New York, 1983.

S. Allenmark, Recent Advances in Methods of Direct Optical Resolution, *J. Biochem. Biophys. Methods*, 1984, **9**, 1.

D. Armstrong, Chiral Stationary Phases for High Performance Liquid Chromatographic Separation of Enantiomers: A Mini-review, *J. Liquid Chromatog.*, 1984, **7** (Suppl. 2), 353.

R. W. Souter, *Chromatographic Separation of Stereoisomers*, CRC Press, Boca Raton, 1985.

D. W. Armstrong, Optical Isomer Separation by Liquid Chromatography, *Anal. Chem.*, 1987, **59**, 84A.

W. L. Hinze and D. W. Armstrong (eds.), *Ordered Media in Chemical Separations*, ACS, Washington DC, 1987.

W. Lindner, Recent Developments in HPLC Enantioseparation — A Selected Review, *Chromatographia*, 1987, **24**, 97.

Y. Okamoto, Separate Optical Isomers by Chiral HPLC, *Chemtech.*, 1987, 176.

E. Gil-Av, Selectors for Chiral Recognition in Chromatography, *J. Chromatog. Libr.*, 1985, **32**, 111.

W. H. Pirkle and T. C. Pochapsky, Chiral Stationary Phases for the Direct LC Separation of Enantiomers, *Adv. Chromatog.*, 1987, **27**, 73.

M. Zief and L. J. Crane (eds.), *Chromatographic Chiral Separation*, Dekker, New York, 1988. (*Chromatog. Sci.*, 1988, **40**).

W. H. Pirkle and T. C. Pochapsky, Considerations of Chiral Recognition Relevant to the Liquid Chromatographic Separation of Enantiomers, *Chem. Rev.*, 1989, **89**, 347.

V. A. Davankov, Separation of Enantiomeric Compounds Using Chiral HPLC Systems. A Brief Review of General Principles, Advances, and Development Trends, *Chromatographia*, 1989, **27**, 475.

J. Hermansson and G. Schill, Separation of Chiral Compounds with α_1-Acid Glycoprotein as Selector, *Chem. Anal. (N.Y.)*, 1989, **98**, 337.

P. R. Brown and R. A. Hartwick (eds.), *High Performance Liquid Chromatography*, Wiley, New York, 1989.

W. H. Pirkle and T. C. Pochapsky, Theory and Design of Chiral Stationary Phases for the Direct Chromatographic Separation of Enantiomers, *Chromatog. Sci.*, 1990, **47**, 783.

J. Martens and R. Bhushan, Importance of Enantiomeric Purity and Its Control by Thin-Layer Chromatography, *J. Pharm. Biomed. Anal.*, 1990, **8**, 259.

F. Gasparrini, D. Misiti and C. Villani, Recent Progress in the Separation of Enantiomers by Chromatographic Technique, *Chim. Ind. (Milan)*, 1990, **72**, 341.

G. Gübitz, Separation of Drug Enantiomers by HPLC Using Chiral Stationary Phases — A Selective Review, *Chromatographia*, 1990, **30**, 555.

REFERENCES

[1] G. M. Henderson and H. G. Rule, *Nature* 1938, **141**, 917.
[2] V. Prelog and P. Wieland, *Helv. Chim. Acta*, 1944, **27**, 1127.
[3] C. E. Dent, *Biochem. J.*, 1948, **43**, 169.
[4] M. Kotake, T. Sakan, N. Nakamura and S. Senoh, *J. Am. Chem. Soc.*, 1951, **73**, 2973.
[5] C. E. Dalgliesh, *Biochem. J.*, 1952, **52**, 3.
[6] C. E. Dalgliesh, *J. Chem. Soc.*, 1952, 3940.
[7] R. Weichert, *Arkiv Kemi*, 1970, **31**, 517.
[8] S. F. Contractor and J. Wragg, *Nature*, 1965, **208**, 71.
[9] T. Alebic-Kolbah, S. Rendic, Z. Fuks, V. Sunjic and F. Kajfez, *Acta Pharm. Jugoslav.*, 1979, **29**, 53.
[10] S. Allenmark, *Chem. Scr.*, 1982, **20**, 5.
[11] O. A. Battista, *Microcrystal Polymer Science*, McGraw-Hill, New York, 1975.
[12] S. V. Rogizhin and V. A. Davankov, *Russ. Chem. Rev.*, 1968, **37**, 565.
[13] K. Bach and J. Haas, *J. Chromatog.*, 1977, **136**, 186.
[14] S. Yuasa and A. Shimada, *Sci. Rep. Coll. Gen. Educ. Osaka Univ.*, 1982, **13**, Nos. 1–2, 13.
[15] G. Gübitz, W. Jellenz and D. Schönleber, *J. High Resol. Chromatog., Chromatog. Commun.*, 1980, **3**, 31.
[16] S. Yuasa, A. Shimada, K. Kameyama, M. Yasui and K. Adzuma, *J. Chromatog. Sci.*, 1980, **18**, 311.
[17] S. Yuasa, M. Itoh and S. Shimada, *J. Chromatog. Sci.*, 1984, **22**, 288.
[18] A. Chimiak and J. Polonski, *J. Chromatog.*, 1975, **115**, 635.
[19] B. Frank and G. Schlingloff, *Liebigs Ann. Chem.*, 1962, **659**, 123.
[20] B. Frank and G. Blaschke, *Liebigs Ann. Chem.*, 1966, **695**, 144.

[21] W. Mayer and F. Merger, *Liebigs Ann. Chem.*, 1961, **644**, 651.
[22] B. Philipp, J. Kunze and H.-P. Fink, in *The Structures of Cellulose*, R. H. Attala (ed.), ACS, Washington DC, 1987, p. 178.
[23] W. Steckelberg, M. Bloch and H. Musso, *Chem. Ber.*, 1968, **101**, 1519.
[24] R. K. Haynes, H. Hess and H. Musso, *Chem. Ber.*, 1974, **107**, 3733.
[25] H. Hess, G. Burger and H. Musso, *Angew. Chem.*, 1978, **90**, 645.
[26] H. Krebs, J. A. Wagner and J. Diewald, *Chem. Ber.*, 1956, **89**, 1875.
[27] H. Krebs and W. Schumacher, *Chem. Ber.*, 1966, **99**, 1341.
[28] F. Schardinger, *Zentralbl. Bakteriol. Parasitenk.*, *Abt. II*, 1908, **22**, 98 (*Chem. Zentralbl.*, 1909, 68).
[29] M. L. Bender and M. Komiyama, *Cyclodextrin Chemistry*, Springer-Verlag, Berlin, 1978.
[30] W. L. Hinze, *Sep. Purif. Methods*, 1981, **10**, 159.
[31] D. W. Armstrong, *J. Liquid Chromatog.*, 1980, **3**, 895.
[32] W. L. Hinze and D. W. Armstrong, *Anal. Lett.*, 1980, **13**, 1093.
[33] N. Wiedenhof, *Stärke*, 1969, **21**, 119, 164.
[34] J. L. Hoffman, *Anal. Biochem.*, 1970, **33**, 209.
[35] A. Harada, M. Furue and S. Nozakura, *J. Polym. Sci.*, 1978, **16**, 187.
[36] B. Zsadon, L. Decsei, M. Szilasi and F. Tudös, *J. Chromatog.*, 1983, **270**, 127.
[37] K. Fujimura, T. Ueda and T. Ando, *Anal. Chem.*, 1983, **55**, 446.
[38] Y. Kawaguchi, M. Tanaka, M. Nakae, K. Funazo and T. Shono, *Anal. Chem.*, 1983, **55**, 1852.
[39] D. W. Armstrong and W. Demond, *J. Chromatog. Sci.*, 1984, **22**, 441.
[40] W. L. Hinze, T. E. Riehl, D. W. Armstrong, W. Demond, A. Alak and T. Ward, *Anal. Chem.*, 1985, **57**, 237.
[41] A. Lüttringhaus and K. C. Peters, *Angew. Chem.*, 1966, **78**, 603.
[42] A. Lüttringhaus, U. Hess and H. J. Rosenbaum, *Z. Naturforsch.*, 1967, **22b**, 1296.
[43] G. Hesse and R. Hagel, *Chromatographia*, 1973, **6**, 277.
[44] G. Hesse and R. Hagel, *Chromatographia*, 1976, **9**, 62.
[45] G. Hesse and R. Hagel, *Liebigs Ann. Chem.*, 1976, 996.
[46] H. Koller, K. H. Rimböck and A. Mannschreck, *J. Chromatog.*, 1983, **282**, 89.
[47] M. Mintas and A. Mannschreck, *Chem. Commun.*, 1979, 602.
[48] E. Francotte, R. M. Wolf, D. Lohmann and R. Mueller, *J. Chromatog.*, 1985, **347**, 25.
[49] T. Shibata, H. Nakamura, Y. Yuki and I. Okamoto, *Chem. Abstr.*, 1985, **102**, 97219p.
[50] Y. Yuki, I. Okamoto, T. Shibata and H. Nakamura, *Chem. Abstr.*, 1985, **102**, 26690z.
[51] T. Shibata, I. Okamoto and K. Ishii, *J. Liquid Chromatog.*, 1986, **9**, 313.
[52] H. Nakamura, H. Namikoshi, T. Shibata, I. Okamoto and K. Shimizu, 1985, *50th National Meeting Chemical Society of Japan*, Tokyo, April 1985, Abstr. 4Z08.
[53] A. Ichida, T. Shibata, I. Okamoto, Y. Yuki, H. Namikoshi and Y. Toga, *Chromatographia*, 1984, **19**, 280.
[54] I. Okamoto, M. Kawashima and K. Hatada, *Polymer Preprints*, Japan, 1984, **33**, 596, 597.
[55] R. Isaksson, J. Roschester, J. Sandström and L.-G. Wistrand, *J. Am. Chem. Soc.*, 1985, **107**, 4074.
[56] G. Blaschke, *Chem. Ber.*, 1974, **107**, 237, 2792.
[57] G. Blaschke and F. Donow, *Chem. Ber.*, 1975, **108**, 1188, 2792.
[58] G. Blaschke and A.-D. Schwanghart, *Chem. Ber.*, 1976, **109**, 1967.
[59] A.-D. Schwanghart, W. Backmann and G. Blaschke, *Chem. Ber.*, 1977, **110**, 778.
[60] G. Blaschke and H.-P. Kraft, *Makromol. Chem.*, *Rapid Commun.*, 1980, **1**, 85.
[61] G. Blaschke and H. Markgraf, *Chem. Ber.*, 1980, **113**, 2031.
[62] G. Blaschke, H.-P. Kraft and H. Markgraf, *Chem. Ber.*, 1980, **113**, 2318.
[63] G. Blaschke, *J. Liquid Chromatog.*, 1986, **9**, 341.
[64] Y. Okamoto, K. Suzuki, K. Ohta, K. Hatada and H. Yuki, *J. Am. Chem. Soc.*, 1979, **101**, 4763.
[65] Y. Okamoto, K. Suzuki and H. Yuki, *J. Polym. Sci. Polym. Chem. Ed.*, 1980, **18**, 3043.
[66] Y. Okamoto, H. Shohi and H. Yuki, *J. Polym. Sci. Polym. Chem. Ed.*, 1983, **21**, 601.
[67] Y. Okamoto, I. Okamoto and H. Yuki, *Chem. Lett.*, 1981, 835.
[68] Y. Okamoto, S. Honda, I. Okamoto, H. Yuki, S. Murata, R. Noyori and H. Takaya, *J. Am. Chem. Soc.*, 1981, **103**, 6971.
[69] H. Yuki, Y. Okamoto and I. Okamoto, *J. Am. Chem. Soc.*, 1980, **102**, 6356.
[70] Y. Okamoto, I. Okamoto and H. Yuki, *J. Polym. Sci. Polym. Chem. Ed.*, 1981, **19**, 451.
[71] Y. Okamoto and K. Hatada, *J. Liquid Chromatog.*, 1986, **9**, 369.
[72] Y. Okamoto, M. Ishikura, K. Hatada and H. Yuki, *Polym. J.*, 1983, **15**, 851.
[73] Y. Okamoto, S. Honda, E. Yashima and H. Yuki, *Chem. Lett.*, 1981, 1221.
[74] Y. Okamoto, E. Yashima and K. Hatada, *Chem. Commun.*, 1984, 1051.

[75] L. Pauling, *Chem. Eng. News.*, 1949, **27**, 313.
[76] G. Wulff, W. Wesper, R. Grobe-Einsler and A. Sarhan, *Makromol. Chem.*, 1977, **178**, 2799.
[77] G. Wulff, R. Grobe-Einsler, W. Wesper and A. Sarhan, *Makromol. Chem.*, 1977, **178**, 2817.
[78] G. Wulff and W. Wesper, *J. Chromatog.*, 1978, **167**, 171.
[79] G. Wulff, R. Kemmerer, J. Vietmeir and H.-G. Poll, *Nouv. J. Chim.*, 1982, **6**, 681.
[80] G. Wulff, H.-G. Poll and M. Minarik, *J. Liquid Chromatog.*, 1986, **9**, 385.
[81] G. Wulff, W. Dederichs, R. Grotstollen and C. Jupe in *Affinity Chromatography and Related Techniques*, T. C. J. Gribnau, J. Vesser and R. J. F. Nivard (eds.), Elsevier, Amsterdam, 1982, p. 207.
[82] R. Arshady and K. Mosbach, *Makromol. Chem.*, 1981, **182**, 687.
[83] L. Andersson, B. Sellergren and K. Mosbach, *Tetrahedron Lett.*, 1984, **25**, 5211.
[84] B. Sellergren, B. Ekberg and K. Mosbach, *J. Chromatog.*, 1985, **347**, 1.
[85] B. Sellergren, M. Lepistö and K. Mosbach, *J. Am. Chem. Soc.*, 1988, **110**, 5853.
[86] B. Sellergren, *Chirality*, 1989, **1**, 63.
[87] B. Sellergren, *Angew. Chem.*, in the press.
[88] N. A. Brown, E. Jähnchen, W. E. Müller and U. Wollert, *Mol. Pharmacol.*, 1977, **13**, 70.
[89] W. E. Müller and U. Wollert, *Mol. Pharmacol.*, 1975, **11**, 52.
[90] K. K. Stewart and R. F. Doherty, *Proc. Natl. Acad. Sci. U.S.A.*, 1973, **70**, 2850.
[91] C. Lagercrantz, T. Larsson and H. Karlsson, *Anal. Biochem.*, 1979, **99**, 352.
[92] C. Lagercrantz, T. Larsson and I. Denfors, *Comp. Biochem. Physiol.*, 1981, **69C**, 375.
[93] S. Allenmark, B. Bomgren and H. Borén, *J. Chromatog.*, 1982, **237**, 473.
[94] S. Allenmark and B. Bomgren, *J. Chromatog.*, 1982, **252**, 297.
[95] S. Allenmark, B. Bomgren and H. Borén, *J. Chromatog.*, 1983, **264**, 63.
[96] S. Allenmark, B. Bomgren and H. Borén, *J. Chromatog.*, 1984, **316**, 617.
[97] S. Allenmark, *LC, Liquid Chromatog., HPLC Mag.*, 1985, **3**, 348.
[98] S. Allenmark, *J. Liquid Chromatog.*, 1986, **9**, 425.
[99] R. A. Thompson, S. Andersson and S. Allenmark, *J. Chromatog.*, 1989, **465**, 263.
[100] S. Andersson, S. Allenmark, P. Erlandsson and S. Nilsson, *J. Chromatog.*, 1990, **498**, 81.
[101] S. Andersson, R. A. Thompson and S. Allenmark, *J. Chromatog.*, in the press.
[102] S. Allenmark, in *Chiral Separations*, S. Ahuja (ed.), ACS Symp. Ser., 1991, in the press.
[103] K. Schmid in *The Plasma Proteins*, F. W. Putnam (ed.), Academic Press, New York, 1975, p. 184.
[104] J. Hermansson, Eur. Pat. Appl. No. 84850169.8, Publ. No. EP0128886A2.
[105] E. Pike, B. Skuterud, D. Kierulf, S. M. Abdel Sayed and P. K. M. Lunde, *Clin. Pharmacokinet.*, 1981, **6**, 367.
[106] J. Hermansson, *J. Chromatog.*, 1984, **298**, 67.
[107] G. Schill, I. W. Wainer and S. A. Barkan, *J. Liquid Chromatog.*, 1986, **9**, 641.
[108] G. Schill, R. Modin, K.-O. Borg and B.-A. Persson, in *Handbook of Derivatives for Chromatography*, K. Blau and G. S. King (eds.), Heyden, London 1978, p. 550.
[109] A. Tilly-Melin, Y. Askemark, K. G. Wahlund and G. Schill, *Anal. Chem.*, 1979, **51**, 976.
[110] G. Schill, I. W. Wainer and S. A. Barkan, *J. Chromatog.*, 1986, **365**, 73.
[111] J. Hermansson and M. Eriksson, *J. Liquid Chromatog.*, 1986, **9**, 621.
[112] J. Hermansson, *Tr. Anal. Chem.*, 1989, **8**, 251.
[113] T. Miwa, M. Ichikawa, M. Tsuno, T. Hattori, T. Miyakawa, M. Kayano and Y. Miyake, *Chem. Pharm. Bull.*, 1987, **35**, 682.
[114] T. Miwa, T. Miyakawa, M. Kayano and Y. Miyake, *J. Chromatog.*, 1987, **408**, 316.
[115] T. Miwa, H. Kuroda, S. Sakashita, N. Asakawa and Y. Miyake, *J. Chromatog.*, 1990, **511**, 89.
[116] J. Haginaka, J. Wakai, K. Takahashi, H. Yasuda and T. Katagi, *Chromatographia*, 1990, **29**, 587.
[117] T. Miwa, T. Miyakawa and Y. Miyake, *J. Chromatog.*, 1988, **457**, 227.
[118] P. Erlandsson, I. Marle, L. Hansson, R. Isaksson, C. Pettersson and G. Pettersson, *J. Am. Chem. Soc.*, 1990, **112**, 4573.
[119] I. Marle, P. Erlandsson, L. Hansson, R. Isaksson, C. Pettersson and G. Pettersson, *J. Chromatog.*, in the press.
[120] E. Domenici, C. Bertucci, P. Salvadori, G. Felix, I. Cahagne, S. Motellier and I. W. Wainer, *Chromatographia*, 1990, **29**, 170.
[121] E. Domenici, C. Bertucci, P. Salvadori, S. Motellier and I. W. Wainer, *Chirality*, 1990, **2**, 263.
[122] E. Domenici, C. Bertucci, P. Salvadori and I. W. Wainer, *J. Pharm. Sci.*, 1991, **80**, 164.
[123] S. Jacobson, S. Golshan-Shirazi and G. Guiochon, *J. Am. Chem. Soc.*, 1990, **112**, 6492.
[124] F. Vögtle and E. Weber (eds.), *Top. Curr. Chem.*, 1984, **121**, 1 and earlier volumes in the series.
[125] G. W. Gokel and H. D. Durst, *Aldrichim. Acta*, 1976, **9**, 3.
[126] W. D. Curtis, D. A. Laidler, J. F. Stoddart and G. H. Jones, *Chem. Commun.*, 1975, 833, 835.

[127] E. P. Kyba, G. W. Gokel, F. de Jong, K. Koga, L. R. Sousa, M. G. Siegel, L. Kaplan, G. D. Y. Sogah and D. J. Cram, *J. Org. Chem.*, 1977, **42**, 4173.
[128] E. P. Kyba, J. M. Timko, L. Kaplan, F. de Jong, G. W. Gokel and D. J. Cram, *J. Am. Chem. Soc.*, 1978, **100**, 4555.
[129] L. R. Sousa, G. D. Y. Sogah, D. H. Hoffman and D. J. Cram, *J. Am. Chem. Soc.*, 1978, **100**, 4569.
[130] S. V. Rogozhin and V. A. Davankov, *Dokl. Akad. Nauk SSSR*, 1970, **192**, 1288.
[131] S. V. Rogozhin and V. A. Davankov, *Chem. Commun.*, 1971, 490.
[132] V. A. Davankov, A. A. Kurganov and A. S. Bochov, *Adv. Chromatog.*, 1983, **22**, 71.
[133] V. A. Davankov and Y. A. Zolotarev, *J. Chromatog.*, 1978, **155**, 285.
[134] B. Lefebvre, R. Audebert and C. Quivoron, *J. Liquid Chromatog.*, 1978, **1**, 761.
[135] D. Müller, J. Jozefonvicz and M. A. Petit, *J. Inorg. Nucl. Chem.*, 1980, **42**, 1083.
[136] J. Boué, R. Audebert and C. Quivoron, *J. Chromatog.*, 1981, **204**, 185.
[137] Y. Zolotarev, N. Myasoedov, V. Penkina, O. Petrenik and V. Davankov, *J. Chromatog.*, 1981, **207**, 63.
[138] A. Foucault, M. Caude and L. Oliveros, *J. Chromatog.*, 1979, **185**, 345.
[139] G. Gübitz, W. Jellenz, G. Löffler and W. Santi, *J. High Resol. Chromatog., Chromatog. Commun.*, 1979, **2**, 145.
[140] H. Engelhardt and S. Kromidas, *Naturwiss.*, 1980, **67**, 353.
[141] W. Lindner, *Naturwiss.*, 1980, **67**, 354.
[142] K. Sugden, C. Hunter and G. Lloyd-Jones, *J. Chromatog.*, 1980, **192**, 228.
[143] G. Gübitz, W. Jellenz and W. Santi, *J. Liquid Chromatog.*, 1981, **4**, 701.
[144] G. Gübitz, W. Jellenz and W. Santi, *J. Chromatog.*, 1981, **203**, 377.
[145] G. Gübitz, F. Juffman and W. Jellenz, *Chromatographia*, 1982, **16**, 203.
[146] P. Roumeliotis, K. K. Unger, A. A. Kurganov and V. A. Davankov, *J. Chromatog.*, 1983, **255**, 51.
[147] P. Roumeliotis, A. A. Kurganov and V. A. Davankov, *J. Chromatog.*, 1983, **266**, 439.
[148] B. Feibush, M. J. Cohen and B. L. Karger, *J. Chromatog.*, 1983, **282**, 3.
[149] A. A. Kurganov, A. B. Telvin and V. A. Davankov, *J. Chromatog,*, 1983, **261**, 233.
[150] H. G. Kicinski and A. Kettrup, *Z. Anal. Chem.*, 1983, **316**, 39.
[151] G. Gübitz, *J. Liquid Chromatog.*, 1986, **9**, 519.
[152] V. A. Davankov, A. S. Bochkov, A. A. Kurganov, P. Roumeliotis and K. K. Unger, *Chromatographia*, 1980, **13**, 677.
[153] L. H. Klemm and D. Reed, *J. Chromatog.*, 1960, **3**, 364.
[154] L. H. Klemm, K. B. Desai and J. R. Spooner, *J. Chromatog.*, 1964, **14**, 300.
[155] M. S. Newman, W. B. Lutz and D. Lednicer, *J. Am. Chem. Soc.*, 1955, **77**, 3420.
[156] M. S. Newman and W. B. Lutz, *J. Am. Chem. Soc.*, 1956, **78**, 2469.
[157] P. Block, Jr. and M. S. Newman, *Org. Synth.*, 1968, **48**, 120.
[158] F. Mikeš, G. Boshart and E. Gil-Av, *Chem. Commun.*, 1976, 99.
[159] F. Mikeš, G. Boshart and E. Gil-Av, *J. Chromatog.*, 1976, **122**, 205.
[160] H. Numan, R. Helder and H. Wynberg, *Rec. Trav. Chim. Pays-Bas*, 1976, **95**, 211.
[161] B. Feringa and H. Wynberg, *Rec. Trav. Chim. Pays-Bas*, 1978, **97**, 249.
[162] Y. H. Kim, A. Tishbee and E. Gil-Av, *Chem. Commun.*, 1981, 75.
[163] H. Nakagawa, S. Ogashiwa, H. Tanaka, K. Yamada and H. Kawazura, *Bull. Chem. Soc. Japan*, 1981, **54**, 1903.
[164] W. J. C. Prinsen and W. H. Laarhoven, *J. Chromatog.*, 1987, **393**, 377.
[165] C. H. Lochmüller and R. Ryall, *J. Chromatog.*, 1978, **150**, 511.
[166] F. Mikeš and G. Boshart, *J. Chromatog.*, 1978, **149**, 455.
[167] F. Mikeš and G. Boshart, *Chem. Commun.*, 1978, 173.
[168] W. H. Pirkle and J. Finn, *J. Org. Chem.*, 1981, **46**, 2935.
[169] W. H. Pirkle, J. Finn, J. Schreiner and B. Hamper, *J. Am. Chem. Soc.*, 1981, **103**, 3964.
[170] W. H. Pirkle, J. M. Finn, B. C. Hamper, J. Schreiner and J. R. Pribish, in *Asymmetric Reactions and Processes in Chemistry*, E. L. Eliel and S. Otsuka (eds.), ACS, Washington DC, 1982, p. 245.
[171] W. H. Pirkle and D. L. Sikkenga, *J. Chromatog.*, 1976, **123**, 400.
[172] W. H. Pirkle and D. W. House, *J. Org. Chem.*, 1979, **44**, 1957.
[173] W. H. Pirkle, D. W. House and J. M. Finn, *J. Chromatog.*, 1980, **192**, 143.
[174] W. H. Pirkle and C. J. Welch, *J. Org. Chem.*, 1984, **49**, 148.
[175] W. H. Pirkle, T. C. Pochapsky, G. S. Mahler and R. E. Field, *J. Chromatog.*, 1985, **348**, 89.
[176] W. H. Pirkle and T. C. Pochapsky, *J. Am. Chem. Soc.*, 1986, **108**, 352.
[177] W. H. Pirkle, T. C. Pochapsky, G. S. Mahler, D. E. Corey, D. S. Reno and D. M. Alessi, *J. Org. Chem.*, 1986, **51**, 4991.
[178] W. H. Pirkle and M. H. Hyun, *J. Chromatog.*, 1985, **322**, 309.
[179] W. H. Pirkle and M. H. Hyun, *J. Org. Chem.*, 1984, **49**, 3043.

[180] W. H. Pirkle, M. H. Hyun and B. Bank, *J. Chromatog.*, 1984, **316**, 585.
[181] W. H. Pirkle and M. H. Hyun, *J. Chromatog.*, 1985, **322**, 287.
[182] W. H. Pirkle and M. H. Hyun, *J. Chromatog.*, 1985, **322**, 295.
[183] W. H. Pirkle, G. S. Mahler, T. C. Pochapsky and M. H. Hyun, *J. Chromatog.*, 1987, **388**, 307.
[184] W. Pirkle, D. M. Alessi, M. H. Hyun and T. C. Pochapsky, *J. Chromatog.*, 1987, **398**, 203.
[185] W. H. Pirkle and T. J. Sowin, *J. Chromatog.*, 1987, **396**, 83.
[186] W. H. Pirkle and R. Däppen, *J. Chromatog.*, 1987, **404**, 107.
[187] W. H. Pirkle and J. A. Burke, III, *Chirality*, 1989, **1**, 57.
[188] A. Tambuté, A. Bégos, M. Lienne, P. Macaudière, M. Caude and R. Rosset, *New J. Chem.*, 1989, **13**, 625.
[189] L. Siret, A. Tambuté, M. Caude and R. Rosset, *J. Chromatog.*, 1990, **498**, 67.
[190] M. Lienne, P. Macaudière, M. Caude, R. Rosset and A. Tambuté, *Chirality*, 1989, **1**, 45.
[191] A. Tambuté, L. Siret, M. Caude, A. Begos and R. Rosset, *Chirality*, 1990, **2**, 106.
[192] H. Hyun and W. H. Pirkle, *J. Chromatog.*, 1987, **393**, 357.
[193] N. Oi, H. Kitahara, Y. Matsumoto, H. Nakajima and Y. Horikawa, *J. Chromatog.*, 1989, **462**, 382.
[194] F. Gasparrini, D. Misiti and C. Villani, *Abstr. Papers Am. Chem. Soc.*, 1989, **198**, 40.
[195] G. Gargaro, F. Gasparrini, D. Misiti, G. Palmieri, M. Pierini and C. Villani, *Chromatographia*, 1987, **24**, 505.
[196] F. Gasparrini, D. Misiti, C. Villani, F. La Torre and M. Sinibaldi, *J. Chromatog.*, 1988, **457**, 235.
[197] G. Uray and W. Lindner, *Chromatographia*, 1990, **30**, 323.
[198] E. J. Corey, R. Imwinkelried, S. Pikul and Y. B. Xiang, *J. Am. Chem. Soc.*, 1989, **111**, 5493.
[199] A. Tambuté, L. Siret, M. Caude and R. Rosset, *J. Chromatog.*, 1991, **541**, 349.
[200] P. Pescher, M. Caude, R. Rosset and A. Tambuté, *J. Chromatog.*, 1986, **371**, 159.
[201] S. Hara and A. Dobashi, *J. High Resol. Chromatog.*, *Chromatog. Commun.*, 1979, **2**, 531.
[202] S. Hara and A. Dobashi, *J. Liquid Chromatog.*, 1979, **2**, 883.
[203] S. Hara and A. Dobashi, *J. Chromatog.*, 1979, **186**, 543.
[204] R. Charles, U. Beitler, B. Feibush and E. Gil-Av, *J. Chromatog.*, 1975, **112**, 121.
[205] A. Dobashi, K. Oka and S. Hara, *J. Am. Chem. Soc.*, 1980, **102**, 7122.
[206] Y. H. Kim, A. Tishbee and E. Gil-Av, *J. Am. Chem. Soc.*, 1980, **102**, 5915.
[207] Y. H. Kim, A. Tishbee and E. Gil-Av., *Science*, 1981, **213**, 1379.
[208] A. Tambuté, A. Begos, M. Lienne, M. Caude and R. Rosset, *J. Chromatog.*, 1987, **396**, 65.
[209] J. Schultze and W. A. König, *J. Chromatog.*, 1986, **355**, 165.
[210] N. Oi and H. Kitahara, *J. Chromatog.*, 1983, **265**, 117.
[211] N. Oi, M. Nagase and T. Doi, *J. Chromatog.*, 1983, **257**, 111.
[212] N. Oi, M. Nagase, Y. Inda and T. Doi, *J. Chromatog.*, 1983, **259**, 487.
[213] N. Oi, M. Nagase, Y. Inda and T. Doi, *J. Chromatog.*, 1983, **265**, 111.
[214] N. Oi and H. Kitahara, *J. Chromatog.*, 1984, **285**, 198.
[215] N. Oi and H. Kitahara, *J. Liquid Chromatog.*, 1986, **9**, 511.
[216] H. Wynberg, *Topics Stereochem.*, 1986, **16**, 87.
[217] C. Rosini, C. Bertucci, D. Pini, P. Altemura and P. Salvadori, *Tetrahedron Lett.*, 1985, **26**, 3361.
[218] C. Rosini, P. Altemura, D. Pini, C. Bertucci, G. Zullino and P. Salvadori, *J. Chromatog.*, 1985, **348**, 79.
[219] P. Salvadori, C. Rosini, D. Pini, C. Bertucci, P. Altemura, G. Uccello-Baretta and A. Raffaelli, *Tetrahedron*, 1987, **43**, 4969.
[220] C. Rosini, C. Bertucci, D. Pini, P. Altemura and P. Salvadori, *Chromatographia*, 1987, **24**, 671.
[221] J. LePage, W. Lindner, G. Davies, D. Seitz and B. L. Karger, *Anal. Chem.*, 1979, **51**, 433.
[222] W. Lindner, J. LePage, G. Davies, D. Seitz and B. L. Karger, *J. Chromatog.*, 1979, **185**, 323.
[223] Y. Taphui, N. Miller and B. L. Karger, *J. Chromatog.*, 1981, **205**, 325.
[224] V. A. Davankov, A. S. Bochkov and Y. P. Belov, *J. Chromatog.*, 1981, **218**, 547.
[225] E. Gil-Av, A. Tishbee and P. E. Hare, *J. Am. Chem. Soc.*, 1980, **102**, 5115.
[226] S. Lam and F. Chow, *J. Liquid Chromatog.*, 1980, **3**, 1579.
[227] S. Lam, F. Chow and A. Karmen, *J. Chromatog.*, 1980, **199**, 295.
[228] S. Lam, *J. Chromatog.*, 1982, **234**, 485.
[229] E. Oelrich, H. Preusch and E. Wilhelm, *J. High Resol. Chromatog.*, *Chromatog. Commun.*, 1980, **3**, 269.
[230] P. E. Hare and E. Gil-Av, *Science*, 1979, **204**, 1226.
[231] S. Lam and A. Karmen, *J. Chromatog.*, 1982, **239**, 451.
[232] S. Lam and A. Karmen, *J. Chromatog.*, 1984, **289**, 339.
[233] S. Lam, *J. Chromatog. Sci.*, 1984, **22**, 416.
[234] W. Klemisch, A. von Hodenberg and K. Vollmer, *J. High Resol. Chromatog.*, *Chromatog. Commun.*, 1981, **4**, 535.

[235] C. Gilon, R. Leshem and E. Grushka, *J. Chromatog.*, 1981, **203**, 365.
[236] C. Gilon, R. Leshem and E. Grushka, *Anal. Chem.*, 1980, **52**, 1206.
[237] E. Grushka, R. Leshem and C. Gilon, *J. Chromatog.*, 1983, **255**, 41.
[238] C. Gilon, R. Leshem, Y. Taphui and E. Grushka, *J. Am. Chem. Soc.*, 1979, **101**, 7612.
[239] G. Gundlach, E. Sattler and U. Wagenbach, *Z. Anal. Chem.*, 1982, **311**, 684.
[240] S. Weinstein, M. Engel and P. E. Hare, *Anal. Biochem.*, 1982, **121**, 370.
[241] S. Weinstein and S. Weiner, *J. Chromatog.*, 1984, **303**, 244.
[242] N. Grinberg and S. Weinstein, *J. Chromatog.*, 1984, **303**, 251.
[243] S. Weinstein, *Trends Anal. Chem.*, 1984, **3**, 16.
[244] S. Weinstein, *Angew. Chem.*, Suppl., 1982, 425.
[245] N. Nimura, A. Toyama and T. Kinoshita, *J. Chromatog.*, 1982, **234**, 482.
[246] N. Nimura, A. Toyama, Y. Kasahara and T. Kinoshita, *J. Chromatog.*, 1982, **239**, 671.
[247] R. Horikawa, H. Sakamoto and T. Tanimura, *J. Liquid Chromatog.*, 1986, **9**, 537.
[248] W. F. Lindner and I. Hirschböck, *J. Liquid Chromatog.*, 1986, **9**, 551.
[249] A. Dobashi, Y. Dobashi and S. Hara, *J. Liquid Chromatog.*, 1986, **9**, 243.
[250] A. Dobashi and S. Hara, *J. Chromatog.*, 1983, **267**, 11.
[251] Y. Dobashi, A. Dobashi and S. Hara, *Tetrahedron Lett.*, 1984, **25**, 329.
[252] Y. Dobashi and S. Hara, *J. Am. Chem. Soc.*, 1985, **107**, 3406.
[253] J. Debowski, D. Sybilska and J. Jurczak, *J. Chromatog.*, 1982, **237**, 303.
[254] J. Debowski, D. Sybilska and J. Jurczak, *Chromatographia*, 1982, **16**, 198.
[255] J. Debowski, J. Jurczak and D. Sybilska, *J. Chromatog.*, 1983, **282**, 83.
[256] D. Sybilska, J. Zukowski and J. Bojarski, *J. Liquid Chromatog.*, 1986, **9**, 591.
[257] D. W. Armstrong, F.-Y. He and S. M. Han, *J. Chromatog.*, 1988, **448**, 345.
[258] J. H. Knox and J. Jurand, *J. Chromatog.*, 1981, **218**, 341.
[259] J. H. Knox and J. Jurand, *J. Chromatog.*, 1982, **234**, 222.
[260] C. Pettersson and G. Schill, *J. Chromatog.*, 1981, **204**, 179.
[261] C. Pettersson and G. Schill, *Chromatographia*, 1982, **16**, 192.
[262] C. Pettersson and M. Josefsson, *Chromatographia*, 1986, **21**, 321.
[263] C. Pettersson and K. No, *J. Chromatog.*, 1983, **282**, 671.
[264] C. Pettersson, *J. Chromatog.*, 1984, **316**, 553.
[265] C. Pettersson and G. Schill, *J. Liquid Chromatog.*, 1986, **9**, 269.
[266] C. Pettersson, A. Karlsson and C. Gioeli, *J. Chromatog.*, 1987, **407**, 217.
[267] C. Pettersson and H. W. Stuurman, *J. Chromatog. Sci.*, 1984, **22**, 441.
[268] V. Prelog, Ž. Stojanac and K. Konačević, *Helv. Chim. Acta*, 1982, **65**, 377.

8

Analytical applications in academic research and industry

Determination of enantiomeric composition or enantiomeric purity on a very small scale is only possible by chromatographic techniques. The most reliable way to carry out such determinations is by direct observation of the separated enantiomers as obtained by chiral chromatography, i.e. without any chiral derivatization prior to separation (cf. Section 4.3). Therefore, this chapter will be mainly devoted to analytical applications of the chromatographic methods described in Chapters 6 and 7.

Applications develop rapidly as new chiral sorbents are developed and investigated. Owing to the complex nature of chiral recognition phenomena, there exists no 'universal' sorbent which solves all optical resolution problems, but rather a spectrum of different sorbents, each with its own merits, areas of application and limitations.

8.1 AMINO-ACIDS

The determination of amino-acids has been of tremendous importance in biochemistry because of the role of amino-acids as building blocks of peptides and proteins. The widely used, now classical technique of Moore and Stein [1], based on ion-exchange chromatography, permits no distinction between enantiomers. However, there is a definite demand for *chiral* amino-acid analysis such as in peptide synthesis where the optical purity of a starting material may be critical, or racemization reactions may impair the stereochemical result. Other areas include structure elucidation of many microbial products, such as polypeptide antibiotics, where the presence of D-amino-acids, not found in mammalian systems, are common [2].

Chiral amino-acid analysis can, in principle, be used for two fundamentally different purposes. The first is determination of L-amino-acids, with use of the 'unnatural' D-forms as internal standards, a technique called 'enantiomer labelling'.

The second is to obtain the quantitative D-/L-amino-acid ratio in the analyte. These application modes will be treated separately in the following.

8.1.1   The enantiomer labelling technique

A well-known problem, particularly in bioanalytical chemistry, is how to determine recovery, i.e. the percentage of a compound present after isolation, from (say) a biological matrix. The recovery is often estimated by an internal standard technique and the general guideline has been that these standards should be as similar to the analytes as possible. Quite often this is very difficult to achieve, which, of course, will have an impact on the reliability of the results. Almost ideal internal standards are found in the isotopically labelled analogues, which has made mass spectrometric detection so important for quantitative use in gas chromatography. In that case, stable isotopes are used for labelling (deuterium analogues of the analytes are common) and the ratio between labelled internal standard and unlabelled analyte is obtained exactly because of the mass-resolution ability of the detector system. The chemical difference caused by the isotope substitution is usually negligible in the procedures for work-up and isolation commonly used. The isotope effect on retention in a chromatographic column is often not observable, although slightly different retention times of ^1H and ^2H isotopic isomers may be obtained in capillary GC, owing to small differences in the strength of hydrogen bonds formed with the stationary phase. As with radioactively labelled standards, the determination of labelled and unlabelled compounds is entirely based on the specific mode of *detection* of both species simultaneously.

A certain disadvantage with these methods is the relatively high cost of many isotopically labelled compounds. Also, for many applications standards of this type are not easily available.

Another elegant technique, viz. the use of internal standards which are optical antipodes of the compounds to be determined, called 'enantiomer labelling', makes use of chiral chromatographic *separation* to resolve standard from analyte for simultaneous quantitative evaluation of both [3]. Thus the only, but very significant, difference in comparison with the usual chromatographic internal standard procedure is that since the standard and analyte show identical chemical behaviour under achiral conditions, the separation must be done on a chiral sorbent. The different modes of using internal standards are outlined in Fig. 8.1.

Although of general applicability, the enantiomer labelling technique has been mainly designed and used for amino-acid analysis by GC. All D-amino-acids used as internal standards are commercially available. The method makes use of the fact that a mixture of all naturally occurring protein L-amino-acids is separable from the corresponding D-enantiomers as N,O,S-PFP isopropyl ester derivatives on "Chirasil-Val" columns by temperature programming (Fig. 8.2). The enantiomer labelling technique thus compensates for losses during work-up and derivatization, but does not account for possible racemization losses of L-amino-acids [4]. The degree of racemization is, however, readily obtained in a separate run under identical conditions with a mixture of pure standards. A valuable feature of the technique is that it compensates completely for losses of certain acid-labile amino-acids such as tryptophan, cysteine, threonine and serine during protein hydrolysis.

| Analyte | Internal standard | Separation mode | Detection mode |

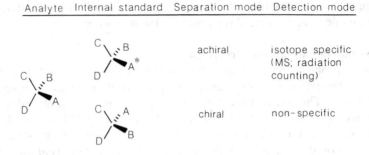

achiral — isotope specific (MS; radiation counting)

chiral — non-specific

Chromatographic appearance:

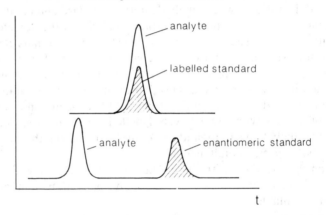

Fig. 8.1 — The different principles applied during use of isotopic isomers and enantiomers as 'ideal' internal standards. (For illustrative reasons an optically active analyte is assumed in both cases).

8.1.2   Other applications of chromatographic amino-acid resolution

When applied to amino-acids, GC has the disadvantage of requiring two derivatization reactions prior to separation. As we have seen in the previous chapter, some LC techniques have been designed for free amino-acids while others incorporate a derivatization of the amino group which will also simplify detection and improve sensitivity.

The moderate column efficiency obtained in LC, however, does not permit all common amino-acids to be resolved in a single chromatogram and therefore the use of column-switching techniques is of great interest.

Separation of amino-acid enantiomers is very important as it is the only way to determine low concentrations of the enantiomers in a given sample with high precision. Such determinations are often required and some more specific examples will be discussed in Sections 8.2 and 8.3. General applications include its use for enantiomeric purity determination and for determination of configuration (a common problem in structure elucidation of natural products).

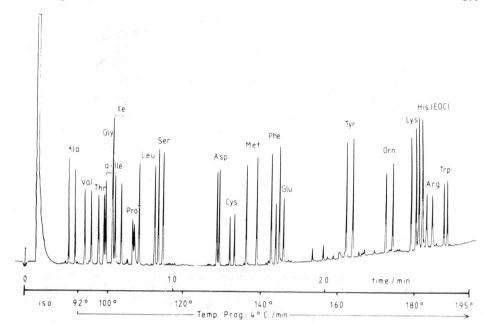

Fig. 8.2 — Optical resolution of nineteen enantiomeric amino-acid pairs (as PFP isopropyl esters) in one chromatogram by GC on a glass capillary column (0.25 mm×20 m) coated with "Chirasil-Val". A 4°C/min temperature program was used. (Reprinted from E. Bayer and H. Frank, in *Modification of Polymers*, C. E. Carraber, Jr. and M. Tsuda (eds.) with permission. Copyright 1980, American Chemical Society).

High optical purity of amino-acids or small peptides is required in peptide synthesis and conventional polarimetric determination of specific rotation is often not entirely adequate for such purposes. In such cases direct observation of the enantiomer ratio in a chromatogram obtained from a chiral separation is far more precise. An excellent example of the small quantities of contaminating enantiomer that can be determined in an amino-acid sample of high optical purity is given in Fig. 8.3, where less than 0.05% is determined with high precision in separate runs on stationary phases of opposite chirality.

LC-columns based on albumin have shown remarkable enantioselectivity towards N-acylated amino acids. Among the derivatives that have been shown to be resolvable (Table 8.1) are some which possess highly fluorescent properties, such as the dansyl- and FITC-derivatives. Recently, a new type of derivative, containing a carbazole moiety as the fluorescent part, has been reported [5]. These amino acid derivatives (here called CC-derivatives) were found to be easily prepared by the use of the reagent N-(chloroformyl)-carbazole, in an acetone/buffer mixture at pH 9 (Scheme 8.1). The derivatives of all protein amino acids were resolvable into enantiomers, often with high separation factors [6]. These derivatives give higher sensitivity than the corresponding FMOC-derivatives and are stable even in very dilute solutions. Because of the direct resolvability and low detection limit, the method could be used for chiral amino acid analysis. If a mixture of amino acids is

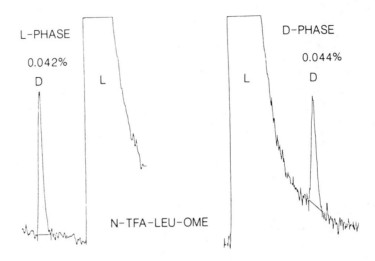

Fig. 8.3 — Determination of the optical purity of leucine enantiomers (as *N*-TFA methyl esters) by capillary GC on "Chirasil-Val" phases of opposite chirality (Courtesy of Dr. B. Koppenhöfer, University of Tübingen, FRG).

Table 8.1 — Some useful derivatives of amino acids for optical resolution by chiral LC on BSA-based columns

A. UV-detection Derivative	λ, nm	elution order of ala-derivative
N-(4-nitrobenzoyl)-	269	L<D
N-(3,5-dinitrobenzoyl)-		L<D
N-(2,4-dinitrophenyl)-	340	L<D
N-(phthalimido)-	225	D<L

B. Fluorescence detection Derivative	λ, nm	elution order of ala-derivative
N-(5-dimethylamino-1-napththalene-sulphonyl)-(DANSYL)	347/396	D<L
N-(9-fluorenylmethoxycarbonyl)-(FMOC)	260/310	L<D
N-(Fluoresceinthiocarbamoyl)-(FITC)	488/520	n.d.

derivatized and then subjected to fractionation on an analytical alkylsilica column, the separate fractions collected can be reinjected onto a BSA column and be analysed for enantiomer composition. Owing to the low detection limit, very small amounts of the second enantiomer can be detected without column overload, thus permitting samples of over 99.5% enantiomeric purity to be determined with precision [7]. Large differences in retention between the enantiomers are found for

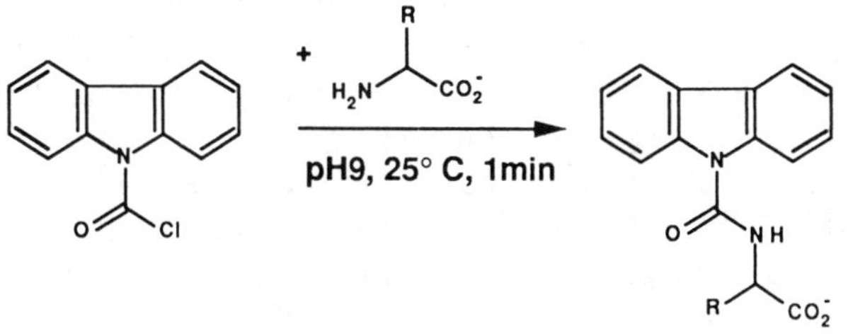

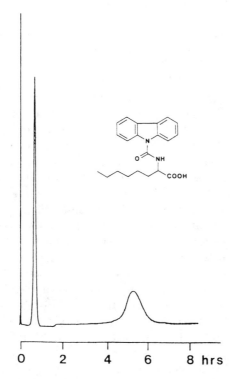

Scheme 8.1 — Fluorogenic derivatization of amino acids with *N*-(chloroformyl)carbazole.

the derivatives of synthetic amino acids having a long alkyl chain in the α-position, as illustrated in Fig. 8.4.

Fig. 8.4 — Optical resolution of the CC-derivative of rac. 2-amino-octanoic acid. (Resolvosil column (4.6×150 mm), mobile phase: 20m*M* phosphate buffer, pH 8.0, with 25% of acetonitrile, flow rate 1.5 ml/min). (S. Allenmark and S. Andersson, unpublished work.)

8.2 STEREOCHEMICAL PROBLEMS IN NATURAL PRODUCTS AND BIODEGRADATION CHEMISTRY

Often the final characterization of an organic compound isolated from natural sources involves a determination of its stereochemical integrity, i.e. optical purity and absolute configuration. In many cases the isolated sample is very small and will not allow investigations by chiroptical or NMR techniques. Here application of chiral chromatography is of enormous value. Provided optical resolution is achieved, the chromatographic technique will give direct information regarding the chemical and optical purity of the sample. Further, if synthetic reference samples are available, stereochemical correlations can easily be made.

Very interesting work in this field has been carried out on pheromones, for which it is known that olfactory receptors in insects can discriminate between enantiomers [8]. Sometimes the 'wrong' enantiomer can be totally inactive or even repellent [9]. The pheromones are, typically, relatively volatile compounds of low molecular weight and therefore lend themselves ideally to chiral GC.

Another field of great importance concerns the structure elucidation of polypeptide antibiotics and related compounds, where a key problem consists in determining the absolute configurations of the amino-acid components.

8.2.1 Pheromone stereochemistry

Many insect pheromones contain chiral, oxygen-containing structure elements and some examples are given in Table 8.2.

One class of widely distributed pheromone components is found in the spiroketals. These compounds have been isolated from various insects, notably wasps, bees and bark beetles [18–21] and identified. Their relatively non-polar nature and the absence of useful UV-absorbing groups make them attractive subjects for chiral gas chromatography. The first compound of this type to be optically resolved was chalcogran (2-ethyl-1,6-dioxaspiro[4.4]nonane), a main component of the aggregation pheromone of the beetle *Pityogenes chalcographus (L)*. This was done by complexation GC with the use of a capillary column coated with nickel(II)-bis(6-heptafluorobutyryl)-(R)-pulegonate (0.12M in squalane) [22]. Because of the 2-alkyl substituent, chalcogran can exist in four optically active forms, i.e. two pairs of enantiomers. The configurational relationships are given in Scheme 8.2.

For the parent compound (R=H) the Z/E-isomerism will vanish and only two enantiomers remain. Figure 8.5 shows the complete resolution of all optically active forms of a mixture of chalcogran with its two lower homologues (2-methyl and unsubstituted 1,6-dioxaspiro[4.4]nonane), respectively. It is evident that the CSP has consistently a certain small preference for the 5(R)-forms in each enantiometric pair. The order of elution is: (2S,5S); (2R,5R); (2R,5S); (2S,5R).

The method thus established was used in later work by Schurig to establish, very accurately, the enantiomer composition of various spiroketal pheromones from different species as well as from species from different locations [23,24]. For this purpose the combination with selected ion monitoring mass spectrometry (SIM) was particularly useful [25].

A symmetrically substituted spiroketal, 2,8-dimethyl-1,7-dioxaspiro[5.5]undecane, present in the mandibular glands of *Adrena* bees, was investigated recently by the use of LC on microcrystalline triacetyl-cellulose [26]. The compound exists as the *trans,trans-* (E,E) and *cis,trans-* (Z,E) geometric isomers, each composed of two enantiomeric forms (Scheme 8.3).

All four optically active compounds could be isolated in mg-amounts by the use of 180 g of the MCTA sorbent (60×2.5 cm glass column, particle size 30–45 μm, 96%

Table 8.2 — Examples of chirality present in various insect pheromones

Compound	Structure	Absolute configuration	Insect	References
2-Heptanol		(R)	different ant species	[10]
(−)-3-Octanol		(R)		[11]
(−)-4-Methyl-3-heptanol		(3S,4S)	*Scolytus* bark beetle (aggregation pheromone)	[11]
4-Methyl-3-hexanol		(3S,4S)	*Myrmicinae* ant	[12]
4-Methyl-3-heptanone		(4S)	*Atta texana, Atta cephalotes*	[13]
(+)-Frontalin			*Dendroctonus* beetles	[14]
exo-(+)-Brevicomin			*Dendroctonus brevicomis* (Western Pine beetle; aggregation pheromone)	[15]
(−)-Serricornin		(4S,6S,7S)	*Lasioderma serricorne* (female cigarette beetle; sex pheromone)	[16]
4,6-Dimethyl-4-octen-3-one (Manicon)	$R = C_2H_5$ $R = n\text{-}C_4H_9$		*Manica mutica* *Manica bradleyi* (ants)	[17]

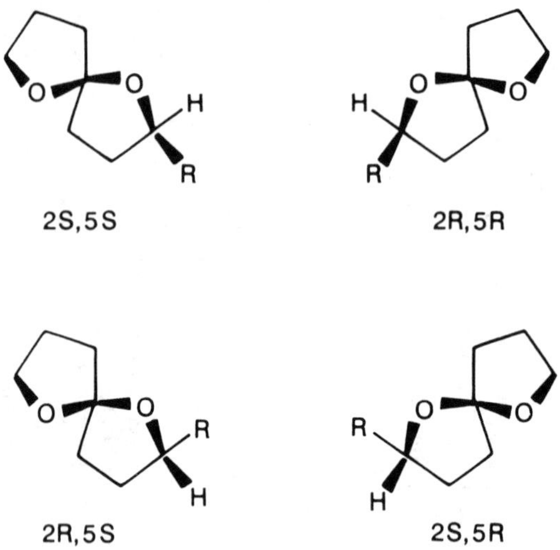

2S,5S   2R,5R

2R,5S   2S,5R

Scheme 8.2 — Stereoisomeric forms of chalcogran.

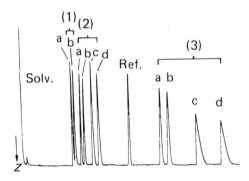

Fig. 8.5 — Resolution of all stereoisomers of chalcogran (3), its 2-methyl analogue (2) and the parent compound (1) by complexation gas chromatography. (Reprinted, with permission, from B. Koppenhöfer, K. Hintzer, R. Weber and V. Schurig, *Angew. Chem.*, 1980, **92**, 473. Copyright 1980, Verlag Chemie GmbH).

ethanol, flow-rate 100 ml/hr, pressure 3.6 bar): 140 mg of E,E and 80 mg of Z,E gave ca. 50 mg and 30 mg respectively of each enantiomer. The components were isolated by dilution of the ethanol fractions with water and extraction with pentane. Figure 8.6 shows a polarimetric recording of the chromatograms obtained for compound 2.

8.2.2   Structure elucidation of polypeptide antibiotics and related natural products
Another fascinating topic in the chemistry of natural products concerns the polypeptide antibiotics produced by bacteria and fungi. Such polypeptides are often found to

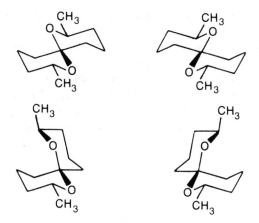

Scheme 8.3 — Stereoisomeric forms of 2,8-dimethyl-1,7-dioxaspiro[5.5]undecane.

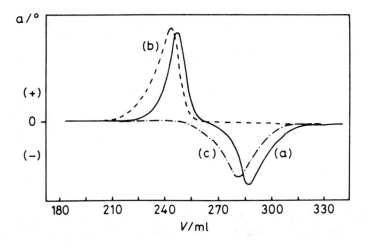

Fig. 8.6 — Polarimetric recording of a chromatogram obtained from an MCTA column, showing the optical resolution of *cis-*,*trans*-2,8-dimethyl-1,7-dioxaspiro[5.5]undecane. (Reproduced, with permission, from R. Isaksson, T. Liljefors and P. Reinholsson, *Chem. Commun.*, 1984, 137. Copyright 1984, Royal Society of Chemistry).

contain structure elements composed of 'unnatural' amino-acids, i.e. of D-configuration or possessing structures not found in proteins. Purification and structure determination of such complex compounds, often obtained in very small quantities, require qualified separation and analytical techniques. In this respect the direct determination of amino-acid configuration by chiral chromatographic techniques is important. Particularly, use of chiral GC as a tool for chiral amino-acid analysis and mapping of hydrolysates is of great value. The examples [27] given below should clearly illustrate its importance.

Epidermin is a polypeptide antibiotic isolated from cultures of a strain of *Staphylococcus epidermidis*. By total acid hydrolysis it was found to contain thirteen protein amino-acids, two lanthionines and one 3-methyllanthionine. Two unsaturated amino-acids, detectable by NMR, were totally degraded in the hydrolysis. By a combination of enzymatic partial hydrolysis, Raney-nickel desulphurization, Edman degradation, fast atom bombardment mass spectrometry (FAB–MS) and chiral GC, the total structure of epidermin, represented in Fig. 8.7, could be established. The

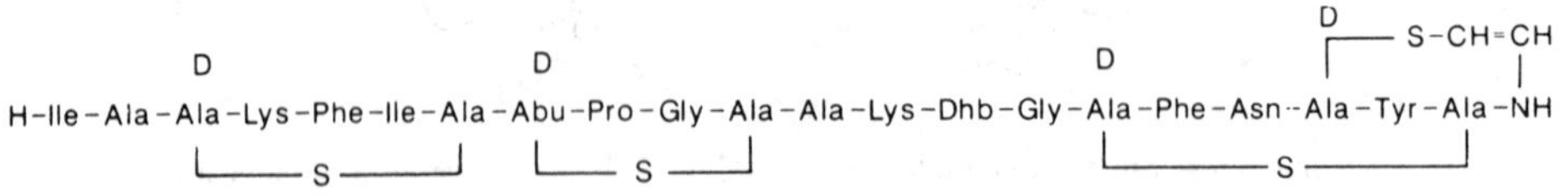

Fig. 8.7 — Structure of epidermin, showing the four ring structures formed by sulphide bonds.

(2R,6S)- (i.e. meso-) and (2S,3S,6R)-configuration of lanthionine (**1**) and 3-methyllanthionine (**2**), respectively, were determined as *N*-PFP methyl esters with a "Chirasil-Val" glass capillary (0.3 mm × 20 m) column [28].

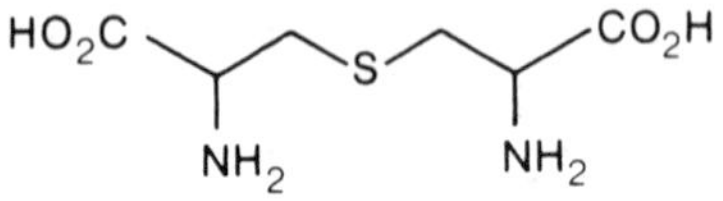

These rather unusual amino-acids which, when inserted into a polypeptide chain, give rise to far more stable cyclic partial structures than those formed by disulphide bridges, are also present in other microbial products. Examples are nisin [29], subtilin [30] and ancovenin [31], for which chiral GC has also been used to establish the amino-acid configurations. The extraordinary usefulness of the technique is illustrated by Fig. 8.8, which shows in the 3-methyllanthionine region of a chromatogram from a total hydrolysate of nisin, only one (2S,3S,6R) of the four possible stereoisomers, which are all well separated under these conditions.

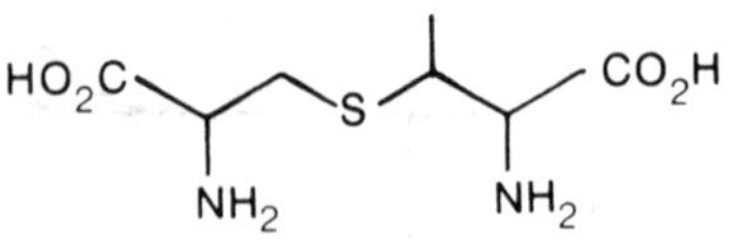

The reliability of modern chiral capillary GC has also encouraged its use as a tool for *sequential* configuration analysis of peptides [32]. This new procedure is worth more detailed consideration. The principle is outlined in Scheme 8.4. The peptide (0.5–1 mg) is first converted into a *tert*-butylcarbamoyl peptide by reaction [33] with *tert*-butyl isocyanate and pyridine (3:5, 30°C, 45 min) and is then heated with a solution of hydrogen chloride in 2-propanol (1.0*M*, 100°C, 30 min). This results in

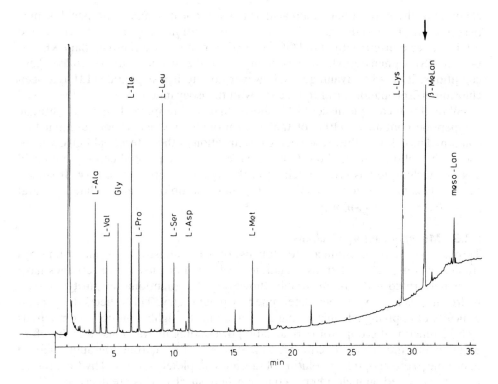

Fig. 8.8 — Gas chromatogram showing the single enantiomer of 3-methyllanthionine obtained on hydrolysis of nisin. (Reprinted, with permission, from E. Küsters, H. Allgaier, G. Jung and E. Bayer, *Chromatographia*, 1984, **18**, 287. Copyright 1984, F. Vieweg & Sohn Verlagsgesellschaft mbH, Braunschweig.).

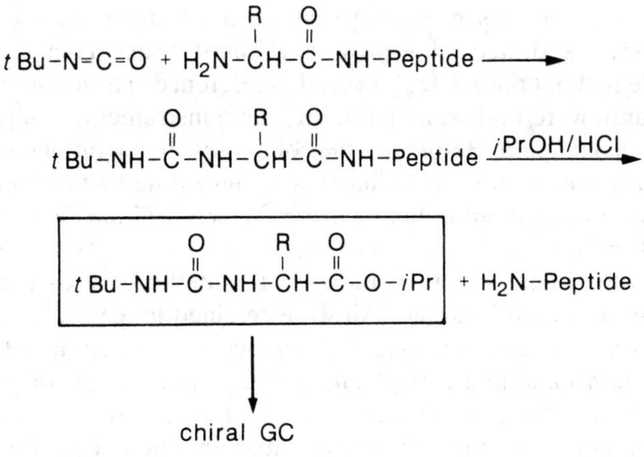

$$t\,\text{Bu-N=C=O} + \text{H}_2\text{N-CH-C-NH-Peptide} \longrightarrow$$

$$t\,\text{Bu-NH-C-NH-CH-C-NH-Peptide} \xrightarrow{\;i\text{PrOH/HCl}\;}$$

$$\boxed{t\,\text{Bu-NH-C-NH-CH-C-O-}i\text{Pr}} + \text{H}_2\text{N-Peptide}$$

chiral GC

Scheme 8.4 — The principle for sequential peptide degradation and chiral amino-acid analysis.

cleavage of the *N*-terminal amino-acid as a *tert*-butylcarbamoyl amino-acid isopropyl ester. After extraction with chloroform, the enantiomeric purity of this liberated ester can be determined by chiral GC ($\alpha = 1.07$–1.11 on a 25 m glass capillary XE-60–L-valine-(S)-α-phenylethylamide column). No racemization occurs under these conditions. It may be advantageous, however, to esterify the peptide (1*M* hydrogen chloride in 2-propanol) prior to reaction with the isocyanate.

With this technique it could be shown that a *C*-terminal Leu in a synthetic octapeptide contained >10% of the D-enantiomer, whereas a non-terminal Leu contained only 1.1%, the reason being racemization of the *C*-terminal Leu owing to use of the solid-phase method. Quite obviously, this sequential chiral amino-acid analysis technique has great potential in the field of peptide chemistry. It should further facilitate structure analysis of peptide antibiotics and related natural products.

8.2.3 Miscellaneous applications

Terpenes are natural products from plants, of wide industrial use, and numerous studies of stereoselective transformations involving optically active terpenes have been performed. Almost invariably, however, the enantiomeric purity of such hydrocarbons has so far been determined by polarimetry. This reflects the general difficulty of applying adequate separation methods to highly non-polar compounds lacking functional groups for derivatization or bonding interactions. It is therefore quite understandable (cf. Section 5.2.3) that in order for a chromatographic method to be able to effect optical resolution of racemic completely saturated hydrocarbons, it has to be based on inclusion phenomena where steric effects are decisive.

It is therefore worthwhile considering a particular chiral GC separation in some detail, as it describes a very elegant and efficient technique for solving a difficult problem, viz. the complete separation of the enantiomers of terpene hydrocarbons.

In these studies [34,35] a glass column (4 mm×2 m) packed with Celite coated with a formamide solution of α-cyclodextrin (4.5 g of formamide/20 g of Celite; 0–1.2 mole% α-CD in formamide) was used. The Celite was slurried with an aqueous solution of the α-CD and formamide by shaking for about 10 min, then the excess of water was slowly evaporated at low pressure (20 mmHg) and a temperature of 50°C. Columns made without the α-CD were used as reference. The chiral column and the reference column were both mounted in the same instrument, equipped with dual flame ionization detection. Before use the columns were conditioned by heating for 2 hr at 70°C. Injections onto the columns were performed almost simultaneously. Separations were carried out at temperatures between 35 and 50°C, with a helium flow-rate of 40 ml/min.

Five racemic hydrocarbons were studied by this method, viz. α- and β-pinene, *cis*-pinane, *trans*-pinane and 2-carene. All were retained longer on an α-CD column than on a reference column, showing that interaction between the CD-selector and both hydrocarbon enantiomers had taken place. The degree of interaction (by inclusion) with the CSP is highly dependent on the absolute configuration of the hydrocarbon, however, causing large differential retention. The effect of 0.65% α-CD on the chromatogram of racemic *cis*-pinane is shown in Fig. 8.9.

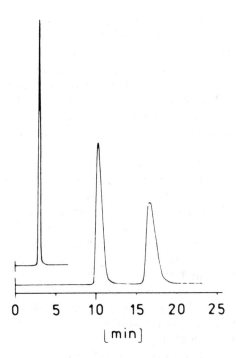

0   5   10   15   20   25

⌊min⌋

Fig. 8.9 — Chromatograms of racemic *cis*-pinane at 35°C; (a) reference column (upper curve), (b) 0.65% α-CD in formamide (lower curve). (Reprinted, with permission, from T. Koscielski, D. Sybilska and J. Jurczak, *J. Chromatog.*, 1986, **364**, 299. Copyright 1984, Elsevier Science Publishers B.V.).

In all cases the (+)-form was found to be the least strongly retained. The separation factors obtained were generally very large for a GC separation, even though the operating temperature was quite low. As seen in Table 8.3, the *trans-*

Table 8.3 — Data obtained from the chiral GC separation of terpene enantiomers. (Reprinted, with permission, from T. Koscielski, D. Sybilska and J. Jurczak, *J.Chromatog.*, 1986, **364**, 299. Copyright 1984, Elsevier Science Publishers B.V.).

Compound	Temperature, °C	α
α-Pinene	35	2.00
β-Pinene	35	1.40
cis-Pinane	35	1.60
trans-Pinane	35	1.20
2-Carene	50	2.17

Column: 4 mm×2 m packed with Celite coated with 0.65 mole% α-CD in formamide
$\alpha = k'(-)/k'(+)$

isomer of pinane was more difficult to resolve than the *cis*-isomer under these conditions. Both enantiomers of the *trans*-isomer are less strongly retained than the first-eluted enantiomer of the *cis*-isomer, indicating that the *trans*-configuration of pinane is sterically more unfavourable in the inclusion process.

Interesting investigations concerning chiral flavour compounds in fruit extracts (e.g. from apricots and strawberries) have been made with the use of multidimensional gas chromatography incorporating a Lipodex chiral column [36,37]. A modification of the technique to incorporate a liquid chromatographic step prior to the chiral GC has also been described [38]. The first method was used for studies of apricot aroma and flavour compounds, and the second was applied to strawberry extracts. Of particular importance here are the various γ-lactones that are different by virtue of the length of the 4-alkyl substituent (C_2–C_8) at the stereogenic centre. It was found that on both chiral columns used (Lipodex B and D), the elution order of the enantiomers was constantly (R) before (S). Also, from the analysis of both apricot and strawberry extracts, the 4(R)-enantiomers were always found to predominate. In apricots, the content of the 4(R)-form in γ-hepta, γ-octa, γ-deca and γ-dodecalactone was more than 80, 86, 91 and 98%, respectively. The odour quality differs between the different members of the homologous series and, perhaps even more importantly, differs also between the members of an enantiomeric pair [39]. The technique is very useful in distinguishing between flavour and aroma compounds of natural and synthetic origin [40].

Chiral GC has also found use as an analytical tool for biodegradation studies. In order to investigate the fate of racemic α-1,2,3,4,5,6-hexachlorocyclohexane (α-HCH) in the marine environment, various organs of the eider duck (*Somateria mollissima L.*) were analysed [41]. The sample was homogenized, extracted with hexane, the extract purified by passage through a deactivated alumina column, and finally the α-HCH fraction was isolated by LC on a silica column. This fraction was then analysed by use of a Lipodex-type of chiral GC column [60-m glass capillary coated with heptakis-(3-*O*-butyryl-2,6-di-*O*-pentyl)-β-cyclodextrin] at an oven temperature of 150°C. All samples showed an excess of the (+)-enantiomer of HCH. The liver extract was of particular interest, since it proved to contain exclusively the (+)-form. These results indicate that the (−)-enantiomer of HCH is enzymatically preferentially degraded. The absolute configuration of this biodegradable form has not yet been determined, however.

8.3   PHARMACEUTICAL APPLICATIONS

The different behaviour of enantiomers in biological systems [42,43] has created a demand for analytical methods to determine the enantiomeric composition or purity of drugs or drug metabolites. If a racemic drug can be optically resolved, there is the possibility of studying the different pharmacokinetic behaviour of the drug enantiomers. Chiral chromatography is also potentially useful for the determination of the stereochemistry of metabolic transformations. So far, applications have mostly been concerned with the screening of new racemic drugs for optical resolution on various chiral columns. In the following sections some important classes of pharmaceuticals which have been recently investigated are treated.

8.3.1 Neutral or weakly acidic or basic pharmaceuticals

Many chiral compounds belonging to this category contain amide bonds which are of importance in the chiral recognition by the chromatographic stationary phase. The polarity of many of these compounds makes them generally less suitable for GC. A variety of LC methods can be used, however.

8.3.1.1 Barbiturates

Derivatives of barbituric acid (**3**) have long been used as sedatives and hypnotics. Variation of the carbon and nitrogen substituents has generated a multitude of racemic drugs for medical use. These compounds ($R_1 \neq R_2$; $R, R_1, R_2 \neq H$) have a pK_a of ca. 8.5 and are stable towards racemization. The potentially different pharmacological behaviour of the enantiomers has not yet been fully elucidated.

3

It has been demonstrated, however, that the (S)-(+)-enantiomer of hexobarbital ($R = R_1 = CH_3$, $R_2 = 1$-cyclohexenyl) is more potent as a hypnotic than the (R)-(−)-form [44]. Such differences have also been observed for other enantiomeric pairs of barbiturates [45,46].

Hexobarbital was first resolved by Blaschke [47] on microcrystalline cellulose triacetate (MCTA). Later it was resolved on poly(triphenylmethylmethacrylate) together with the analogue mephobarbital ($R = R_1 = CH_3$, $R_2 = $phenyl) [48] which has also been resolved on a cyclodextrin phase [49]. Resolution of a series of barbiturates on MCTA was recently reported by Shibata et al. [50]. A more extensive investigation was undertaken by Yang et al. [51] who used the Pirkle-type N-(3,5-dinitrobenzoyl)amino-acid chiral phases. Although the α values obtained were generally low (≤1.12), analytically useful separations were obtained. Two of the compounds contained only an exocyclic asymmetric carbon atom, i.e. R=H.

8.3.1.2 Hydantoins

These drugs have the general structure **4**, and are, like some of the barbiturates, used as antiepileptics.

4

Several compounds of this type have been resolved on Pirkle-type columns [52]. Mephenytoin (R=CH$_3$, R$_1$=C$_2$H$_5$, R$_2$=C$_6$H$_5$) can be resolved with α=1.06 by using a covalent DNB-leu sorbent and hexane/2-propanol/acetonitrile (89:10:1) [51]. Mephenytoin can also be directly resolved by GC on a "Chirasil-Val" column and this technique has been elegantly used for pharmacokinetic purposes [53]. The metabolic fate of ($\pm$)-mephenytoin presents a good example of biological enantiose-lectivity, and its monitoring by chiral GC is worth a further comment.

After oral administration of 300 mg of ($\pm$)-mephenytoin, blood sampling is performed over a two-week period and the plasma (1 ml samples) extracted with 6 ml of dichloromethane after addition of 100 μl of internal standard and acidification (1 ml of 10M acetic acid). The organic phase is evaporated and the residue dissolved in 1 ml of a 4:1 v/v solution of 0.1M sodium hydroxide in methanol/iodopropane and then incubated in a water-bath at 50°C for 18 hr. (This process converts the demethylated metabolite, not resolvable on the column, into the corresponding resolvable propyl-derivative). Derivatization is stopped by addition of 1 ml of 0.1M hydrochloric acid and 3 ml of dichloromethane. Extraction of the derivative and transfer to a clean, small conical vial and evaporation to dryness under nitrogen is followed by addition of 10 μl of ethyl acetate to dissolve the residue. A 0.5–1.0 μl portion is injected into the gas chromatograph, which is equipped with a 25 m $\times$ 0.25 mm "Chirasil-Val" glass capillary column and a thermionic nitrogen/phos-phorus-specific detector. The following conditions are suitable: injector temperature 225°C, column temperature 165°C, detector temperature 300°C, helium flow 1.2 ml/min with a 20:1 split ratio, make-up gas flow 2.5 ml/min, hydrogen flow 3.5 ml/min and air flow 175 ml/min. 3-Methyl-5-phenyl-5-isopropylhydantoin is preferably used as internal standard.

This procedure permits direct observation of the enantiomers of mephenytoin, its desmethyl metabolite (converted into the 3-propyl homologue) and the internal standard in one chromatogram. It was found from samples taken at different times that the elimination half-life of the (S)-enantiomer of mephenytoin is less than 3 hr, whereas that for the (R)-form is over 70 hr. Accordingly, the peak plasma levels showed an (R):(S) ratio of 5. Similarly, the (S)-metabolite level was barely detec-table, whereas the concentration of the (R)-metabolite (R-PEH) increased over 4–6 days before declining. The situation is illustrated by Fig. 8.10. Scheme 8.5 shows the metabolic conversions [45].

8.3.1.3 *Benzothiadiazines and structurally related compounds*
The title compounds (**5**) belong to a group of diuretics which are pharmacologically interesting from a stereochemical point of view, since closely related diuretics, such

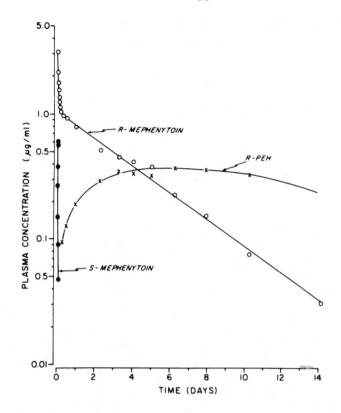

Fig. 8.10 — Elimination kinetics of racemic mephenytoin studied by chiral gas chromato-
graphy. (Reprinted, with permission, from P. J. Wedlund, B. J. Sweetman, C. B. McAllister,
R. A. Branch and G. R. Wilkinson, *J. Chromatog.*, 1984, **307,** 121. Copyright 1984, Elsevier
Science Publishers B.V.).

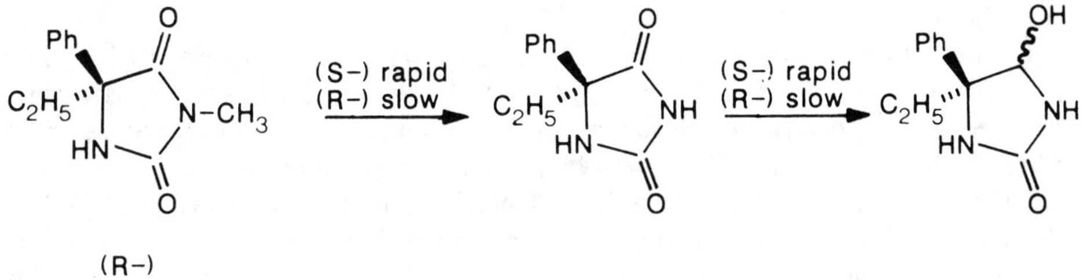

Scheme 8.5 — Metabolic conversion of mephenytoin.

as mefruside [4-chloro-N'-methyl-N'-(2-methyltetrahydrofurfuryl)-m-benzene-disulphonamide] have shown significant enantiospecificity in their renal effects [55,56].

X=SO$_2$ (benzothiadiazines)
X=CO   (tetrahydroquinazolinones)

5

Optical resolution of a series of racemic diuretics with X=SO$_2$ has been achieved by the use of LC on chiral polyacrylamide phases [57]. Although the separations were incomplete, high enantiomeric enrichment was often obtained. Further, the use of semipreparative scale separations permitted repetitive chromatography and isolation of some compounds in almost optically pure form.

The sorbents used were the poly-[(S)-N-acryloylphenylalanine ethyl and benzyl esters], Ia and Ib, (see Sections 7.1.2.1 and 11.2). Analytical scale separations were carried out with 5 g of sorbent, packed in a glass column (ca. 10×300 mm), at a flow-rate of 10–15 ml/hr (2.5–3.0 bar). The column was connected in series to a UV-detector operating at 285 nm and a polarimeter equipped with an 80 μl flow-cell. Volume fractions of the eluate were obtained with an automatic fraction collector. For semipreparative work, a glass column (ca. 38×800 mm) packed with 235 g of Ia was used at a flow-rate of 50 ml/hr (3.0 bar). A toluene/dioxan (1:1 v/v) mixture was used throughout as the mobile phase. Column loads were ca. 5 mg (dissolved in 0.5–2.0 ml of the eluent) and 200–250 mg for the analytical and semipreparative runs, respectively. Some relevant data are given in Table 8.4.

The role of hydrogen bonding in the retention and optical resolution process is evident from the fact that addition of 6% of methanol to the mobile phase was found to be detrimental to k' and α, and also that substitution of the sulphonamide hydrogen atom by a methyl group causes a similar effect. The k' values were quite consistently lower on Ib than on Ia. However, identical elution orders were obtained throughout.

Figure 8.11 shows the partial resolution of 5.0 mg of racemic penflutizide. The enantiomeric yields given in Table 8.4 are defined as the combined amount of the two enantiomers in optically pure form that can be isolated in a single run (hatched areas in Fig. 8.11), calculated in % of the total amount run (total area).

Both enantiomers of penflutizide and bendroflumethiazide were obtained in >97% optical purity and could be used for polarimetric studies of the racemization kinetics, which showed a perfect pseudo first-order reaction rate constant in aqueous ethanol under slightly alkaline conditions. The racemization studies showed that the half-life decreased with pH, being only a few minutes at pH >9. This base-catalysed racemization therefore proceeds simultaneously with the hydrolysis of the thiadiazine observed previously.

Table 8.4 — Experimental data for the separation of the drug enantiomers **5** (X=SO$_2$). (From G. Blaschke and J. Maibaum, *J. Pharm. Sci.*, 1985, **74**, 438, reproduced with permission of the copyright owner, the American Pharmaceutical Association).

Substituents			Adsorbent			
			Ia		Ib	
R_1	R_2	R_3	k_1'	k_2'	k_1'	k_2'
H	CH$_2$(CH$_2$)$_4$CH$_3$	CF$_3$	3.41(+)	3.98(−)	1.23(+)	1.60(−)
H	CH$_2$C$_6$H$_5$	CF$_3$	5.54(+)	6.31(−)	1.90(+)	2.11(−)
H	CH$_2$C$_6$H$_5$	Cl	6.60(+)	7.07(−)	2.86	
H	CH$_2$C$_6$H$_4$F-*p*	Cl	12.10(+)	12.90(−)	4.64(+)	5.03(−)
H	CH$_2$C$_6$H$_4$F-*m*	Cl	10.20(+)	11.30(−)	——	
H	CH$_2$C$_6$H$_4$F-*o*	Cl	5.89(+)	6.23(−)	2.27(+)	2.47(−)
H	CH$_2$CH(CH$_3$)$_2$	Cl	5.45(+)	5.83(−)	2.22(+)	2.56(−)
H	CH$_2$SCH$_2$CF$_3$	Cl	9.14(+)	10.20(−)	3.48(+)	4.23(−)
CH$_3$	CH$_2$SCH$_2$CF$_3$	Cl	4.70		1.85	
H	CHCl$_2$	Cl	6.29		2.38	
CH$_3$	CH$_2$Cl	Cl	5.22		2.20	

Toluene/dioxan (1:1) was used as mobile phase throughout

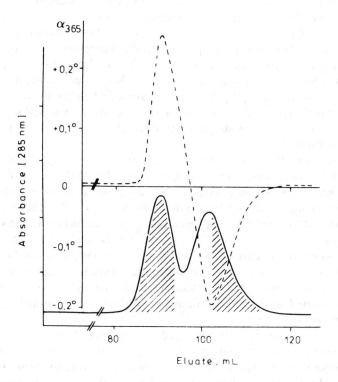

Fig. 8.11 — Semipreparative separation of the enantiomers of penflutizide. (From G. Blaschke and J. Maibaum, *J. Pharm. Sci.*, 1985, **74**, 438, reproduced with permission of the copyright owner, the American Pharmaceutical Association).

8.3.1.4   *Benzodiazepinones*
Compounds belonging to this group have long been used as sedatives and hypnotics. Although many of them are not chiral, a number of analogues produced are racemates of the general formula (**6**).

6

The parent compound (R=R$_1$=X=H), diazepam, which is achiral, is known as "Valium".

These compounds are known to exhibit enantiomer-dependent physiological effects [58]. They are too polar for GC separation but appear to be well suited for chiral LC. Optical resolutions have been carried out on an analytical as well as preparative scale and the main results are summarized below.

The first successful results were reported by Blaschke *et al.* [59] who used a cross-linked chiral polyacrylamide sorbent [ethyl (S)-phenylalaninate substituent] for preparative resolution of oxazepam (R$_1$=OH, X=R=H). Later the technique was improved by use of the polymer deposited on silica, HPLC conditions, and hexane/dioxan (65:35) as mobile phase [60]. Another silica-bound CSP, based on N-formyl-(S)-phenylalanine (cf. Section 7.2.5) was also found useful for analytical separations of enantiomers of this class (temazepam: R=CH$_3$, R$_1$=OH, X=H; camazepam: R=CH$_3$, R$_1$=OCONMe$_2$, X=H).

An exhaustive investigation by Pirkle *et al.* showed that a variety of benzo-diazepinones (42 in all) was readily resolvable on the N-(3,5-dinitrobenzoyl)-phenylglycine and -leucine sorbents [61]. The elution orders were found to be highly consistent with the suggested chiral recognition model. Thus, in all cases the (−)-form was eluted first on the (R)-phenylglycine sorbent but last on the (S)-leucine sorbent. The α values varied between 1.07 and 4.33. On the whole, the latter sorbent gave the largest separation factors. The most relevant data are given in Table 8.5.

In all compounds of known absolute configuration, the (R)-forms have negative rotations and (S)-forms positive. It is assumed that the most strongly retained enantiomer is adsorbed on the CSP by simultaneous three-site bonding interactions as illustrated by Fig. 8.12.

Note that the most stable conformation of a 3-substituted diazepam is the one shown, with the substituent occupying a pseudo-equatorial position. This conformation tends to direct the amide carbonyl group of the folded ring towards the amide hydrogen atom of the bound selector. The bonding interactions suggested are: (a) charge-transfer interaction between the 3,5-dinitrobenzoyl group and the benzene ring of the analyte, (b) hydrogen bonding between the 3,5-dinitrobenzamide

Table 8.5 — Data obtained from optical resolutions of a series of benzodiazepinones on N-(3,5-dinitrobenzoyl)amino-acid sorbents. (Reprinted, with permission, from W. H. Pirkle and A. Tsipouras, *J. Chromatog.*, 1984, **291**, 291. Copyright 1984, Elsevier Science Publishers B.V.).

Compound	Adsorbent			
$(X=R=H)$	(R)-DNB-phenylglycine		(S)-DNB-leucine	
R_1	k'_1	α	k'_1	α
CH_3	3.2(−)-(R)	1.6	1.8(+)-(S)	4.11
CH_2CH_3	2.3(−)	1.46	1.9(+)	2.22
$CH_2CH_2CH_3$	2.2(−)-(R)	1.42	2.7(+)-(S)	2.35
$CH(CH_3)_2$	2.7(−)-(R)	1.92	1.8(+)-(S)	3.53
$CH_2(CH_2)_2CH_3$	2.7(−)	1.92	1.7(+)	4.0
$CH_2CH(CH_3)_2$	3.2(−)-(R)	1.89	1.9(+)-(S)	4.20
$CH_2CH_2SCH_3$	4.0(−)	1.57	2.2(+)	3.0
$CH_2C_6H_5$	4.0(−)-(R)	1.93	2.2(+)-(S)	4.33
$CH_2C_6H_4OH$-p	12.3(−)-(R)	1.48	6.3(+)-(S)	2.18
$CO_2CH_2CH_3$	4.4(−)	1.36	4.3(+)	1.42
OH	16.9(−)-(R)	1.20	11.7(+)-(S)	1.13
OCH_3	10.0(−)	1.33	11.0(+)	1.19
$OCOCH_3$	7.4(−)	1.27	3.8(+)	1.62
$OCOC_6H_5$	3.1(−)	1.87	2.3(+)	2.84
$OCOC(CH_3)_3$	1.0(−)	2.05	0.7(+)	3.69

In each case the mobile phase was 10% 2-propanol in hexane

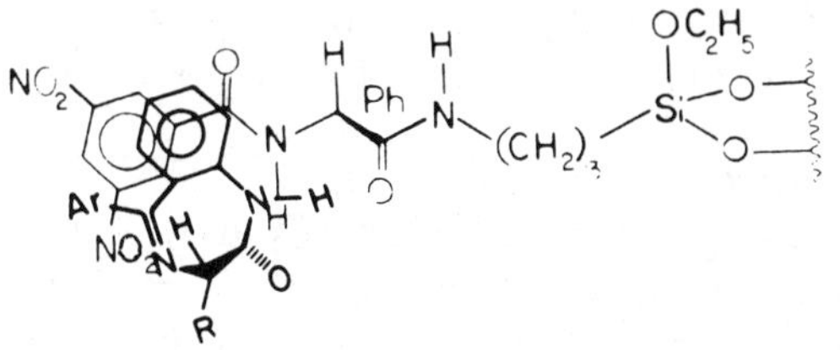

Fig. 8.12 — Chiral recognition model for retention by the stationary phase of the final benzodiazepinone enantiomer eluted. (Reprinted, with permission, from W. H. Pirkle and A. Tsipouras, *J. Chromatog.*, 1984, **291**, 291. Copyright 1984, Elsevier Science Publishers B.V.).

hydrogen atom and the amide carbonyl oxygen atom of the analyte and (c) hydrogen bonding between the amino-acid carbonyl oxygen atom of the CSP and the amide hydrogen atom of the analyte.

This model is also consistent with the findings that the length of the 3-substituent does not impair the optical resolution, which indicates that it does not contribute to steric interaction.

The large separation factors obtained for many of the compounds on these CSPs have encouraged preparative-scale resolutions, which will be treated in Chapter 9.

This type of pharmaceutical is also well resolvable by reversed-phase LC on albumin-silica sorbents. In investigations by Allenmark and Andersson [62,63], a series of compounds was shown to give separation factors up to about 7. The retention is highly dependent on the hydrophobic character of the 3-substituent, and the mobile phase has to be selected accordingly to give reasonable k' values. Table 8.6 gives an idea of the effects caused by the substituents and of the mobile phase composition.

Table 8.6 — Data obtained from the separation of benzodiazepinone enantiomers on a BSA-silica ("Resolvosil") column. (Reprinted from S. Allenmark, *J. Liquid Chromatog.*, 1986, **9**, 425, by courtesy of Marcel Dekker, Inc.).

Compound	Mobile phase composition					
(X=R=H) R_1	Molarity (mM)	pH	1-propanol (%)	k'_1	k'_2	α
CH_3	50	7.8	6	5.1	5.1	1.00
CH_3	20	7.5	1	7.25	14.7	2.03
$CH_2(CH_2)_2CH_3$	50	7.8	6	11.1	17.4	1.57
$CH_2C_6H_5$	50	7.8	6	41.0	110.0	2.68
$CH_2CH_2CH_2OH$	50	7.8	6	1.5	1.65	1.09
$CH_2CH_2CH_2OH$	20	7.5	1	3.0	5.0	1.67
$O_2CCH_2CH_3$	50	7.8	6	3.2	4.7	1.47
$O_2CCH_2CH_3$	20	7.5	1	6.5	10.75	1.65
$OCH_2CH_2CH(CH_3)_2$	50	7.8	6	14.0	28.2	2.01
OH	10	6.6	2	2.1	8.6	4.10

In conclusion, many chiral LC methods are well suited to a direct monitoring of enantiomer composition or enantiomeric purity of this type of pharmaceuticals. This should be of importance in connection with future development of optically pure products and pharmacokinetic studies of the antipodes.

8.3.1.5 Miscellaneous compounds

Various other racemic drugs have been resolved by direct chiral chromatographic methods thanks mainly to the work of Blaschke and co-workers [60]. Such resolutions include the anti-cancer drug "Ifosfamide" ("Holoxan"), **7**, and related compounds, which all contain an asymmetric phosphorus atom in a heterocyclic ring system [64]. The resolution of **7** was done on a semipreparative scale on poly[(S)-*N*-acryloylphenylalanine ethyl ester] with toluene/dioxan (1:1) as mobile phase [60].

X=CH₂,
R=R₁=CH₂CH₂Cl
R₂=H

7

8

The drug methaqualone, **8**, has been used for a long time as a hypnotic and anticonvulsive [65]. It was not until 1975, however, when it was shown by NMR that some methaqualone analogues had high barriers to rotation around the nitrogen–

aryl bond [66], that the chiral structure and resolvability of **8** was realized. It has been shown by Mannschreck *et al.* [67] that the barrier to enantiomer interconversion ($\Delta\Delta G^{\pm}$) in methaqualone is as high as 131.6 kJ/mole (at 135°C). This is sufficient to prevent any racemization of an enantiomer, which was also proved experimentally by semipreparative partial resolution of the racemate on an MCTA column [67]. Typically, 300 mg of racemic **8** gave ca. 100 mg of (+)-**8** (optical purity 70%, first eluted) and ca. 120 mg of (−)-**8** (optical purity 60%). Both enantiomers showed anticonvulsive activity, with (−)-**8** the more active form.

Compounds possessing a 1,4-dihydropyridine structure have come into focus as drugs influencing ion-channel transport of calcium [68]. Some of these compounds have been resolved into enantiomers, which have opposite effects on calcium channels [69,70]. Optical resolutions have been achieved by chiral LC on MCTA columns with 95% ethanol as mobile phase [71–73]. Recycling techniques in a two-column configuration have also been used to achieve baseline resolution [73].

Two drugs with the common structure **9** have been optically resolved by LC.

9

Baclophen-lactam (Ar=*p*-Cl-C$_6$H$_4$-; **9a**) shows a large α value on a "Resolvosil" column (50mM phosphate buffer, pH 7, 2% 1-propanol) [63]. Rolipram (**9b**) is completely separated into enantiomers on an MCTA column (96% ethanol) [60]. Other racemic drugs resolved on MCTA [60], include chlormezanon (**10**) (also on "Resolvosil" [74]), oxapadol (**11**), ketamin (**12**) and mianserin (**13**). Poly[N-(S)-acryloylphenylalanine ethyl ester] has been successfully used for optical resolution of the chiral sulphur compound **14** and of chlorthalidone (**15**) [both with toluene/dioxan (1:1) as mobile phase] [60].

10   **11**

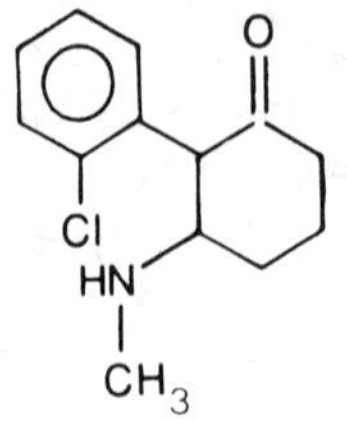

12

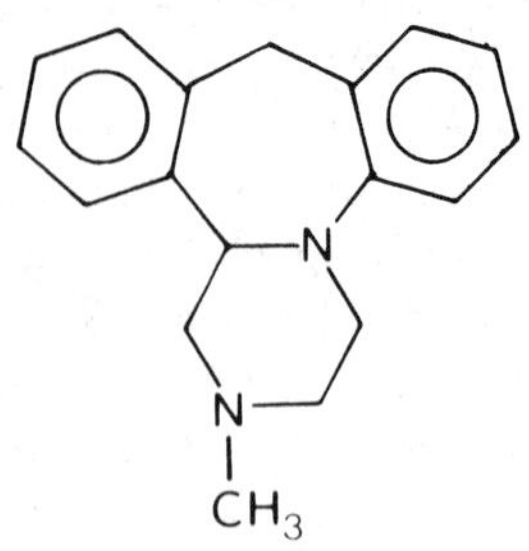

13

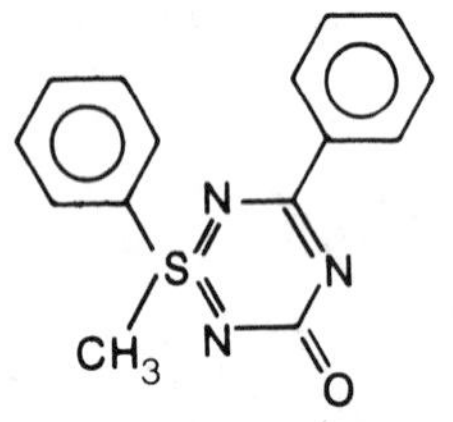

14

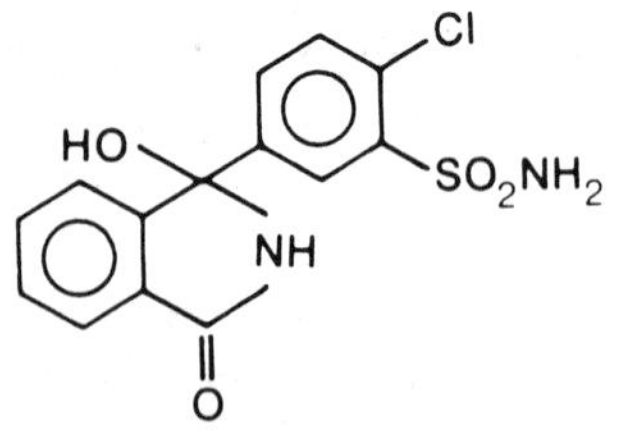

15

8.3.2   Protolytic (charged) pharmaceuticals

A number of LC strategies have been attempted for these compounds, but two modes are particularly attractive as they do not require derivatization. One is based on metal ion complexation and has been successful in only a limited number of cases. The other, more general approach, is an application of ion-pair chromatography under chiral conditions, in two different ways, viz. (i) with an achiral sorbent and an optically active counter-ion (separation of diastereomeric ion-pairs), and (ii) with a chiral sorbent and an achiral counter-ion (separation of enantiomeric ion-pairs, cf. Section 7.1.4.2). Apart from these methods, successful resolutions have also been achieved by other means in aqueous as well as non-protolytic media.

Most GC methods require derivatization, but some very simple and elegant techniques are available.

8.3.2.1   *Amphoteric compounds*

A. Penicillamine. This β-mercapto α-amino acid, **16,** is an excellent example of the different pharmacological effects exerted by enantiomers. Whereas the D-form is a valuable pharmaceutical used against rheumatoid arthritis [75] and has been clinically tested against some other illnesses, notably Wilson's disease [76], the L-enantiomer is highly toxic [77]. It is therefore essential to have access to analytical methods for accurate and direct determination of the optical purity of the drug. This problem has been solved by two elegant chromatographic methods.

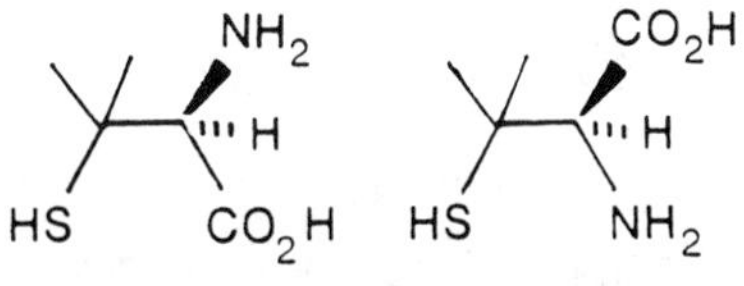

16

(1) Gas chromatography [69]. Esterification of **16** with 2-propanol, followed by reaction with phosgene, yields the thiazolidin-2-one (**17**) in good yield (Scheme 8.6).

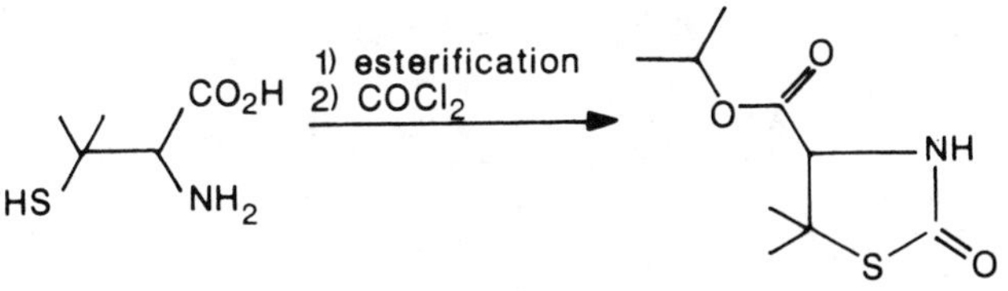

1) esterification
2) COCl$_2$

17

Scheme 8.6 — Derivatization of penicillamine for gas chromatography.

GC on a capillary column coated with the CSP XE-60–L-valine-(R)-α-phenylethyl-amide (Chrompack) at 170°C, with hydrogen as carrier, affords excellent resolution. The non-methylated analogue (cysteine) is also very well resolved (Fig. 8.13). On this CSP the α values recorded were 1.085 for penicillamine and 1.079 for cysteine. In both cases the L-enantiomer was the first eluted.

(2) Liquid chromatography [79]. Reaction of **16** with formaldehyde yields the dimethylthiazolidinecarboxylic acid, **18**, (Scheme 8.7). Separation of the enantiomers was effected by CLEC on a column packed with LiChrosorb RP-8 (Merck) coated with the Cu(II) complex of (2S,4R,2'RS)-4-hydroxy-1-(2'-hydroxydodecyl)-proline **19** [80], and a mobile phase composed of methanol–water (12:88 v/v)

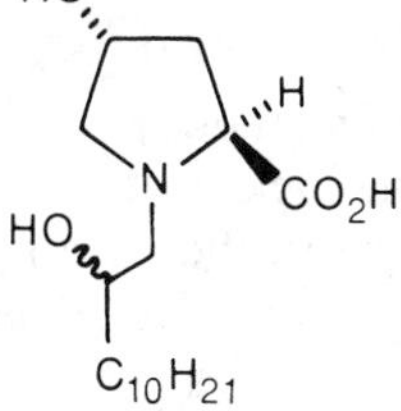

19

containing 0.1mM copper(II) sulphate and adjusted to pH 4.5 (with phosphoric acid). A column temperature of 50°C was used. The chromatographic performance is

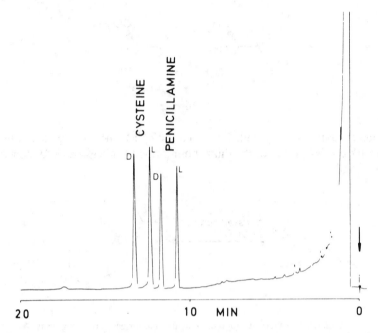

Fig. 8.13 — Optical resolution of D.L-penicillamine and D.L-cysteine after conversion into the thiazolidin-2-one derivatives. (Reprinted, with permission, from W. A. König, E. Steinbach and K. Ernst, *J. Chromatog.*, 1984, **301**, 129. Copyright 1984, Elsevier Science Publishers B.V.).

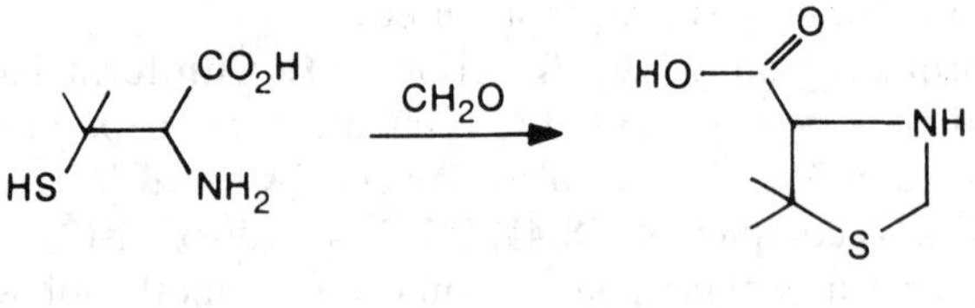

Scheme 8.7 — Derivatization of penicillamine for CLEC.

shown in Fig. 8.14.   The two methods permit determination of less than 0.1% of contaminating enantiomer. The somewhat higher sensitivity and precision of the GC method should be balanced against the risk of partial racemization induced by the higher temperature necessary for derivatization [81].

B. Dopa and derivatives. L-Dopa, L-3-(3,4-dihydroxyphenyl)alanine, is commonly used for the treatment of Parkinson's disease and acts as the precursor of dopamine, which is deficient in patients suffering from this illness. D-Dopa, however, is toxic [82,83] and its presence in the drug has to be monitored because of the rather large amounts of the drug that are usually taken in L-dopa therapy.

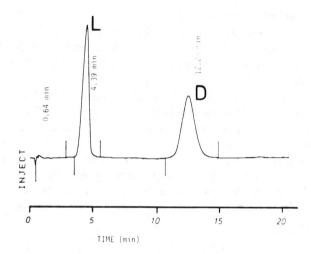

Fig. 8.14 — Optical resolution of D.L-5,5-dimethylthiazolidinecarboxylic acid derived from D.L-penicillamine. 4×125 mm column; flow-rate 3 ml/min; UV 235 nm. (Reprinted, with permission, from E. Busker, K. Günther and J. Martens, *J. Chromatog.*, 1985, **350**, 179. Copyright 1985, Elsevier Science Publishers B.V.).

An LC method, utilizing the principle of CLEC and mobile phase addition of the chiral constituent (cf. Section 7.3 as well as the technique described above), has proved to be very useful for this purpose [84]. The chromatographic system consists of a C_{18} column (4.6×250 mm) which has been equilibrated with a mobile phase composed of L-phenylalanine (6mM) and copper(II) sulphate (3mM) in water. Eluted species are detected by UV at 280 nm.

Two chemically related compounds, methyldopa and carbidopa, the L-forms of which are used as inhibitors of the decarboxylative enzyme, as well as tryptophan, can also be determined by the same method. As seen from Table 8.7, which also gives

Table 8.7 — Data obtained for the optical resolution of dopa and dopa–analogues by CLEC. (Reprinted, with permission, from L. R. Gelber and J. L. Neumeyer, *J. Chromatog.*, 1983, **257**, 317. Copyright 1983, Elsevier Science Publishers B.V.)

Compound	k'_D	k'_L	α	R_s	Mobile phase composition		
					L-Phe(mM)	Cu(II)(mM)	MeOH(%)
Dopa	1.7	2.4	1.4	2.9	6	3	0
Methyldopa	1.9	2.6	1.2	1.9	12	6	0
Carbidopa	7.4	9.5	1.2	3.4	6	3	0
Tryptophan	5.8	7.0	1.2	2.6	8	4	10

the optimized mobile phase compositions, all D-enantiomers are eluted prior to the L-forms . The α values (1.2–1.4) are sufficient for complete resolution. Standard calibration curves give correlation coefficients >0.986 (for dopa 0.996) and the relative standard deviations are <1%.

8.3.2.2 Basic (cationic) compounds

Many important drugs contain aliphatic amino groups and are therefore charged at physiological pH. Among these are many anticholinergic alkaloid derivatives, amino-alcohols with β-receptor blocking or stimulating action, and a variety of chiral local anaesthetics.

The number and diversity of important chiral basic drugs has led to intensive research on their direct chromatographic optical resolution. Selected examples of useful bioanalytical procedures will be given after some general comments on the different merits of the various strategies.

The different physiological effects of the enantiomers of β-adrenergic blocking agents have stimulated intensive research on chromatographic optical resolution of such compounds. Most of them possess the general structure **20**. It is known that the (S)-form is often 50–500 times more effective than the (R)-enantiomer [85] and that the latter may also be toxic.

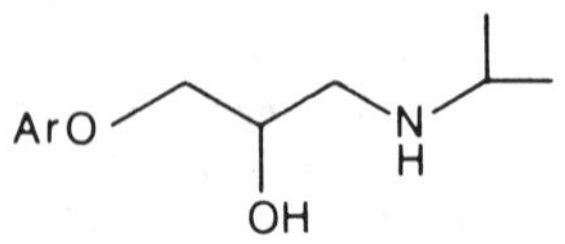

20

Ar=

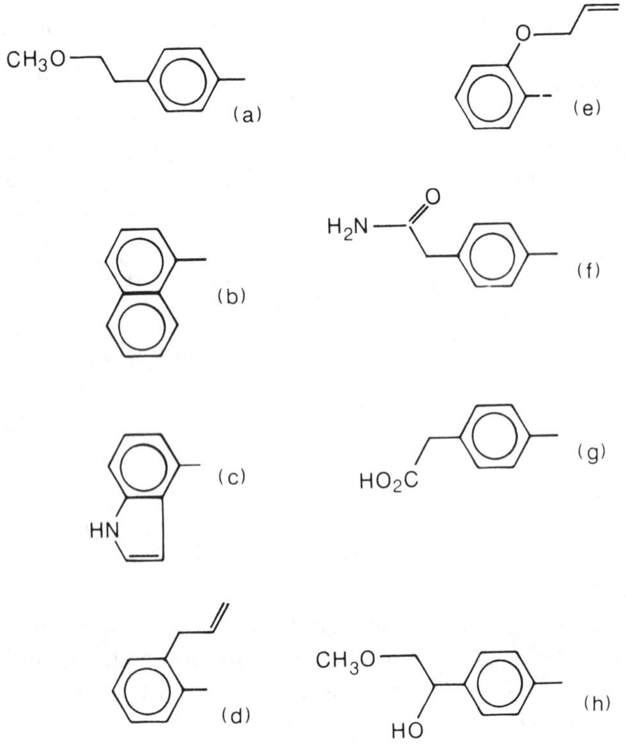

(a)

(b)

(c)

(d)

(e)

(f)

(g)

(h)

These compounds are too polar to be subjected to GC without derivatization, which is also required by some LC methods. The favoured derivatization process is a conversion into the oxazolidones by reaction with phosgene [86]. In some cases non-cyclic carbamates, formed by reaction with an isocyanate, have been used. The reactions are shown in Scheme 8.8. Note that the oxazolidinones are easily cleaved by dilute alkali to regenerate the amino-alcohol without racemization.

Some experimental procedures, useful for a determination of enantiomer composition of β-blockers, are given below. The first of these, based on GC of the oxazolidinone derivative of metoprolol (**20a**) and two metabolites (**20g**) and (**20h**) [87], used a capillary column (0.25 mm × 18 m, Duran glass) coated with the XE-60–L-valine-(R)-α-phenylethylamide polymer described by König and co-workers [88,89]. The reactions and the structures of the actual compounds are given in Scheme 8.8.

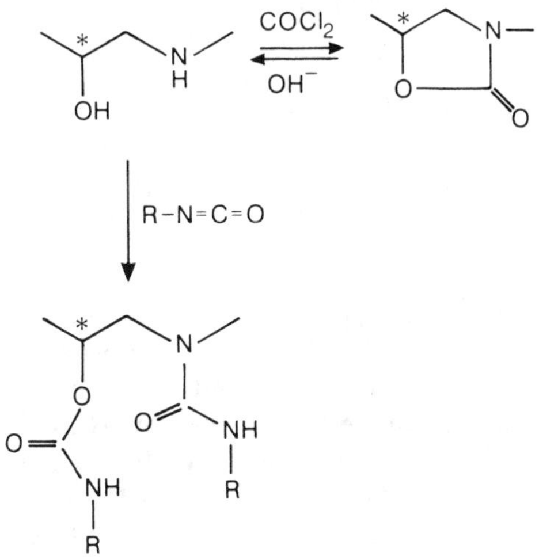

Scheme 8.8 — Structures of common β-blockers and the reactions used to produce oxazolidinone and carbamoyl derivatives.

Derivatization of the sample is typically performed by dissolution of 1 mg in 1 ml of buffer (pH 12; ionic strength 1) and mixing with 1 ml of dichloromethane, then addition of 30 μl of 2M phosgene solution in toluene and shaking of the mixture for 10 min. After separation of the phases, an aliquot of the organic phase is evaporated and the residue dissolved in a suitable amount of dichloromethane.

The oxazolidinone obtained from the α-hydroxylated metabolite (**20h**) is converted into the corresponding trimethylsilyl ether by reaction with N,O-bis(trimethylsilyl)acetamide (BSA) and the other metabolite, metoprolol acid (**20g**), is esterified with diazomethane by standard techniques.

The GC separation is preferably done at 195°C with hydrogen (at a pressure of 100 kPa) as carrier gas. The α values obtained are about 1.03, giving resolution that is almost or quite complete. The elution order has not been reported.

The second procedure [90], used for propranolol (**20b**), is based on LC and makes use of an (R)-N-(3,5-dinitrobenzoyl)phenylglycine column (Pirkle Type 1-A, 5 μm, Regis Co., 4.6×250 mm).

Samples for derivatization are obtained from human blood plasma [spiked with internal standard, i.e. 100 μl of a 20 μg/ml solution of pronethanol (**21**) in methanol, per tube] by extraction with diethyl ether after the addition of carbonate buffer to give pH 10. After collection of the ether phase (10 ml) and cooling to 0°C, phosgene (10 μl of a 12.5% solution in toluene) is added and the mixture vortexed for 30 min before centrifugation. The ether phase is evaporated in a stream of nitrogen and the residue is dissolved in 50 μl of methylene chloride.

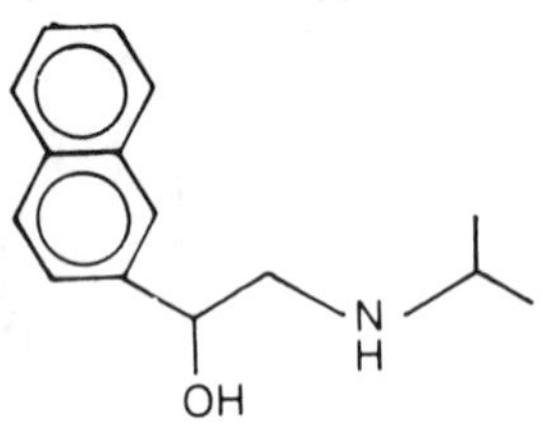

21

For the LC separation a mobile phase comprised of hexane/2-propanol/aceto-nitrile (96:3:1 v/v) and a flow-rate of 2 ml/min are used. Detection is performed fluorimetrically with excitation at 290 nm and emission measurement at 335 nm. Propranolol is detectable in whole blood, down to 10 ng/ml. The α value of the oxazolidinone of propranolol is 1.09. The elution order (for the oxazolidinones) is (R,S)-pronethanol (one peak), (S)-propranolol and (R)-propranolol, respectively.

The third procedure uses LC with a "Resolvosil" BSA-silica column coupled to a thermospray–MS system [91]. By this method bopindolol (**22**) was separated into enantiomers without derivatization. Pindolol (**20c**) was first converted into a bis-isopropylcarbamoyl derivative [R=(CH₃)₂CHNHCO–] (Scheme 8.8) by reaction with isopropyl isocyanate.

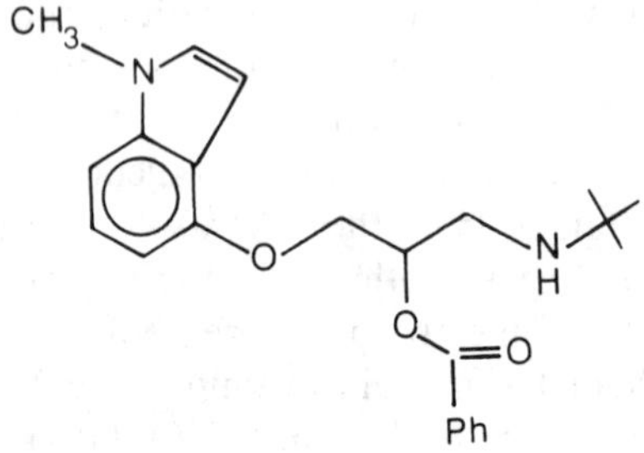

22

Derivatization of pindolol is performed by heating 3 mg of the compound with 20 μl of dichloromethane, 300 μl of triethylamine and 400 μl of isopropyl isocyanate at 100°C for 1.5 hr. After cooling, the solvent and excess of reagents are expelled with a stream of nitrogen. The residue is dissolved in a small amount of 2-propanol.

For UV detection, a mobile phase comprised of phosphate buffer, with 2-propanol as retention modifier may be used, e.g. for bopindolol, 0.1M phosphate buffer, pH 7.0, 0.5% 2-propanol, $\alpha=1.81$; for the pindolol derivative, 0.1M disodium hydrogen phosphate, 5% 2-propanol, $\alpha=1.46$. With the MS interface the phosphate buffer is replaced by 0.05M ammonium formate+2% 2-propanol (suitable for the pindolol derivative).

Other methods used for direct optical resolution of β-blockers include ion-pair chromatographic techniques with chiral counter-ions [92–94] and the use of an α_1-acid glycoprotein (EnantioPac®) column [95]. Very impressive results were also obtained by Okamoto *et al.* [96] with the use of a silica-based cellulose tris(3,5-dimethylphenylcarbamate) column. On this column five β-blockers (alprenolol **20d**, oxyprenolol **20e**, propranolol **20b**, pindolol **20c** and atenolol **20f**) were completely resolved without derivatization, by elution with hexane/2-propanol (9:1). Figure 8.15 shows some of the results obtained. The chromatographic separation can easily be scaled up. Thus, with a preparative column (20×500 mm), 100 mg of **20b**, 150 mg of **20d** and 400 mg of **20e**, respectively, were completely resolved, each in a single run.

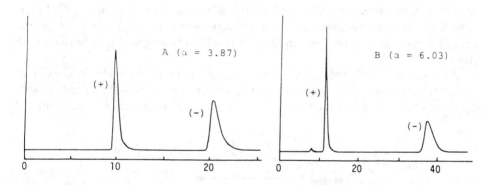

Fig. 8.15 — Optical resolution of (a) alprenolol and (b) oxyprenolol on a cellulose tris(3,5-dimethylphenylcarbamate) (silica supported) column. (Reprinted, with permission, from Y. Okamoto, M. Kawashima, R. Aburatani, K. Hatada, T. Nishiyama and M. Masuda, *Chem. Lett.*, 1986, 1237. Copyright 1986, Chemical Society of Japan).

Compounds of this type are further resolvable with high α-values on the cellobiohydrolase-derived sorbent (CBH-I) (Section 7.1.4.5). An example of the performance of this protein-based column is given in Fig. 8.16.

Optical resolution of **20a** and of **20b** (as the oxazolidinones) has also been achieved on a semipreparative scale by LC on MCTA [98].

The structurally related ephedrines (β-amino-alcohol structure) have also been studied. Direct resolution of the enantiomers of ephedrine and its metabolites (as

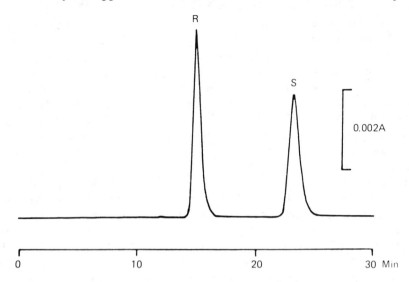

Fig. 8.16 — Separation of the propranolol enantiomers (CBH-I column (4.6×250 mm), mobile phase: 10m*M* sodium acetate buffer pH 4.7 with 0.5% of 2-propanol, flow rate: 0.3 ml/min). (Reprinted, with permission, from P. Erlandsson, I. Marle, L. Hansson, R. Isaksson, C. Pettersson and G. Pettersson, *J. Am. Chem. Soc.*, 1990, **112**, 4573. Copyright 1990, American Chemical Society.)

N,O-PFP derivatives) by GC on "Chirasil-Val" was described in 1978 [99]. Later, ephedrine, pseudoephedrine and norephedrine were optically resolved as the oxazolidinone derivatives on an XE-60–L-valine-(R)-α-phenylethylamide CSP [100].

An interesting LC separation of ephedrine has been described by Wainer *et al.* [101]. By cyclization with 2-naphthaldehyde to form a naphthyl-substituted oxazolidine, an efficient π-donating group was introduced (Scheme 8.9) and optical

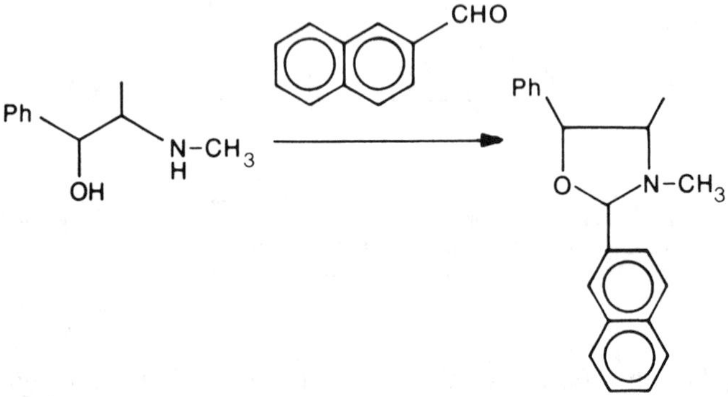

Scheme 8.9 — The cyclization of ephedrine by reaction with 2-naphthaldehyde.

resolution could be achieved (although not completely) by the use of an (R)-N-(3,5-dinitrobenzoyl)phenylglycine column (Pirkle Type 1-A, 5 µm, Regis Co., 4.6×250 mm). The (1R,2S)-enantiomer is eluted before the (1S,2R) enantiomer.

Optical resolutions, without derivatization, of some basic drugs, including ephedrines, atropines and some local anaesthetics, by the use of "EnantioPac" columns and ion-pair techniques, have been described [95].

8.3.2.3 *Acid (anionic) compounds*

Many acidic, carboxyl-substituted, drugs are well resolved by LC without derivatization. Particularly useful are the chiral columns based on proteins as stationary phases, because of the ease with which retention and resolution are regulated. Typical examples are the 2-aryl-substituted propionic acids, **23**, used as analgesics, of which many have been resolved on BSA-, AGP- as well as OVM-based columns.

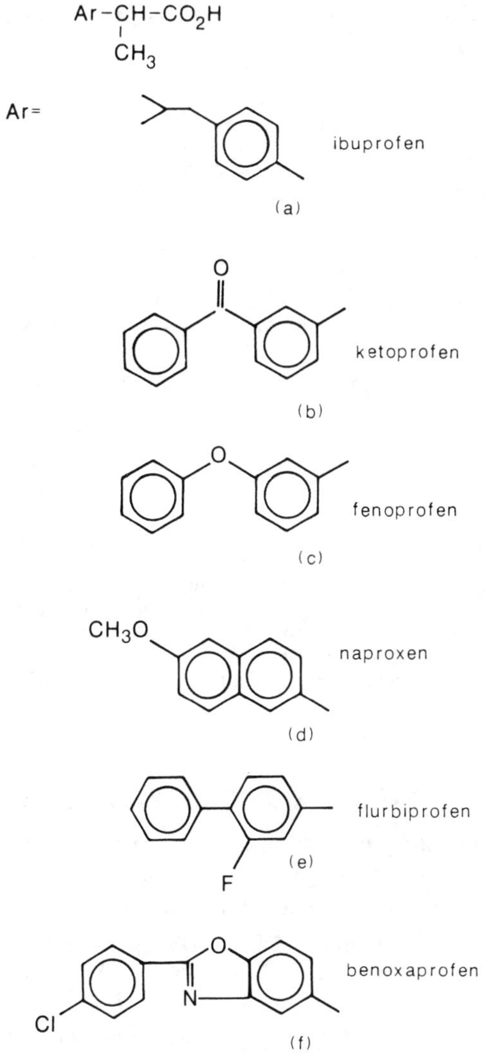

$$Ar-CH-CO_2H$$
$$|$$
$$CH_3$$

Ar =

(a) ibuprofen

(b) ketoprofen

(c) fenoprofen

(d) naproxen

(e) flurbiprofen

(f) benoxaprofen

23

By conversion of the free carboxyl group into amide derivatives, good results have also been obtained with the use of other types of columns such as those based on covalent (R)-*N*-(3,5-dinitrobenzoyl)phenylglycine (Regis 'Hi-Chrom Reversible') as well as cellulose tris(phenylcarbamate)-coated macroporous silica (Daicel 'Chiralcel OC') [102,103]. For analytical purposes, however, the latter column was found to be less well suited, despite its good selectivity, owing to the low column efficiency. As their napthylmethylamide derivatives in particular, ibuprofen (**23a**) and flurbiprofen (**23b**) were well optically resolved on the Pirkle-type column with the use of 7.5% dioxan in hexane, permitting in each case a determination of ca. 0.1% contamination by the other enantiomer.

In Table 8.8 the chiral LC techniques used for optical resolution of some common acidic drugs are summarized.

Table 8.8 — Techniques used for optical resolution of some anti-inflammatory agents of the α-arylpropionic acid type

Compound	Structure	Derivative	Column/Mobile phase	References
Ibuprofen	23a	—	AGP/buffer	[104]
		Amide (with NMA)	PG/7.5% dioxan in hexane	[102,103,105]
Flurbiprofen	23e	Amide (with NMA)	PG/3% 2-propanol in hexane	[103]
		Amide (with NMA)	CTPC/3% methanol in hexane	[103]
Benoxaprofen	23f	Amide (with NMA)	PG/2-propanol in hexane	[102]
Fenoprofen	23c	Amide (with NMA)	PG/2-propanol in hexane	[102]
Naproxen	23d	Amide (with NMA)	PG/2-propanol in hexane	[102]
		—	BSA/buffer	[63]
Ketoprofen	23b	—	BSA/buffer	[63]

Abbreviations used: AGP=α$_1$-acid glycoprotein column (Enantiopac)
BSA=silica-based bovine serum albumin column (Resolvosil)
CTPC=cellulose tris(phenylcarbamate) coated onto silica (Chiralcel OC)
NMA=1-napthylmethylamine
PG=covalent (R)-N-(3,5-dinitrobenzoyl)phenylglycine (Hi-Chrom Reversible)

8.4   STUDIES OF MICROBIAL AND ENZYMATIC REACTIONS

With the progress of biotechnology, a growing interest in the use of enzymes and micro-organisms as catalysts for organic chemical conversions has emerged. Of particular importance in this respect is the possibility of achieving transformations with a high degree of stereoselectivity, to produce optically active compounds. Although there is already a wealth of empirical knowledge concerning the use of enzymes and cells for such purposes, the field and its potential are enormous. In particular, the results from microbial reactions are highly unpredictable and the need for small-scale screening work seems almost unlimited. Such studies were earlier hampered by a lack of really adequate techniques to monitor the progress of a stereoselective reaction. With the new developments in chiral chromatography it is now possible to determine, in a very simple way, the exact enantiomeric composition in a minute sample taken from an enzyme-catalysed reaction. The chromatographic peak areas are measured by an electronic integrator interfaced with the detector. By such techniques the progress of the reaction and its stereochemistry can be readily monitored, with a very small volume of sample.

For such purposes both GC and LC separations on chiral stationary phases have now come into use. These applications, where a minimum of sample treatment is needed, are given below.

8.4.1 Enzymatic and microbial alkene epoxidation

The known carcinogenic action of certain aromatic hydrocarbons has led to intensive research on epoxidation reactions catalysed by liver microsomes. These cell orga-nelles contain a cytochrome P-450-dependent mono-oxygenase, the metabolic function of which is to transform a reactive double bond into an oxirane function. The latter, as a highly reactive intermediate, is supposed to be the actual xenobiotic, having a high carcinogenic and mutagenic potential [106].

It has been shown that such microsomal epoxidations, which are readily per-formed *in vitro*, are highly stereoselective [107]. However, before the technique of chiral GC was available, no precise methods to study these processes existed.

Because epoxides of lower olefins are relatively volatile, they can be subjected to GC at a rather low column temperature, without derivatization. The first study by chiral GC of the stereochemistry of microsomal epoxidation of a series of olefins was made in 1984 [107]. The experimental description given below, taken from [107], and in which the final analysis is performed by complexation chromatography, is illustrative of this technique.

The olefin (3–13 μmole) is incubated with microsomes (ca. 1.5 mg of protein/ml) at pH 7.4 and 37°C in 0.5 ml of 0.15M phosphate buffer containing NADP$^+$ (1.0mM), isocitrate dehydrogenase (0.1 I.U.), racemic isocitrate (8mM) (for recy-cling of the coenzyme), magnesium chloride (5mM) and 2-(trichloromethyl)oxirane (4–9mM) (inhibitor of epoxide hydrolase). After 0.5–2 hr the liberated epoxide is injected into a capillary column coated with 0.125M nickel bis[3-heptafluorobutyryl-(1R)-camphorate] in OV 101 (40 m$\times$0.25 mm, column temperature 70°C, headspace analysis technique).

The elution orders and the absolute configuration of the preferentially formed enantiomer were determined by chromatography of oxiranes of known absolute configuration. In all the cases investigated, the enzymatic reaction favoured the (S)-enantiomer. The stereochemical result from an epoxidation of butadiene to vinylox-irane is given in Fig. 8.17.

Certain bacteria have been found to utilize alkenes to produce epoxides in a stereoselective reaction [108–112]. Recently, various *Mycobacterium* strains were investigated with respect to production of optically active epoxides from a series of 1-alkenes [113]. The optical purity was determined by complexation gas chromato-graphy [113] with a capillary column coated with 0.12M Ni(II)bis[3-heptafluoro-butyryl-(1R)-camphorate] (A) or with 0.1M Co(II)[bis(3-heptafluorobutyryl-(1R)-camphorate] (B) in OV 101 (methyl silicone).

The epoxidation of propene was carried out in a fermenter at 30°C and pH 6.9 with a continuous supply of air containing 2% of propene to the microbial suspen-sion. A rapid small-scale screening procedure was also used. Here, a suspension of

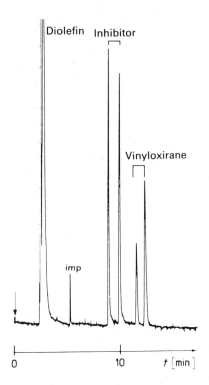

Fig. 8.17 — Microsomal enantioselective epoxidation of butadiene (imp=impurity). The peak
ratio of vinyloxirane gives a 40% e.e. of the (S)-enantiomer. (Reprinted, with permission, from
V. Schurig and D. Wistuba, *Angew. Chem.*, 1984, **96**, 808. Copyright 1984, Verlag Chemie
GmbH).

cells (20–50 mg dry weight) in 50mM phosphate buffer (3 ml) was stirred in a septum-
sealed bottle after the addition of 20 μl of the alkene. The reaction was followed by
periodic removal of 100–200 μl of gas phase, and its direct injection into the gas
chromatograph.

The results show that most of the organisms investigated produced high optical
yields of the same enantiomer.

The stereochemistry of the reaction can be described as follows:

$$X = H \qquad CH_3 \qquad Cl$$
$$(R) \qquad (R), \qquad (S)$$

Absolute configuration

The enantiomeric purities obtained with the use of the most efficient *Mycobacterium* strains were 93–97% (X=H), 83–95% (X=CH$_3$) and 96–98% (X=Cl). The
value obtained for each strain was accurate to within 1%.

By the use of columns containing chiral selectors of opposite configuration [(R) and (S) metal chelates], a completely reversed elution pattern for the enantiomers was found. By this elegant and useful technique it is possible to verify the purity and identity of the eluted species by comparison of integration data.

In Fig. 8.18 an example of the chromatograms is given, showing the very fast analysis of the stereochemistry of a microbiological epoxidation reaction.

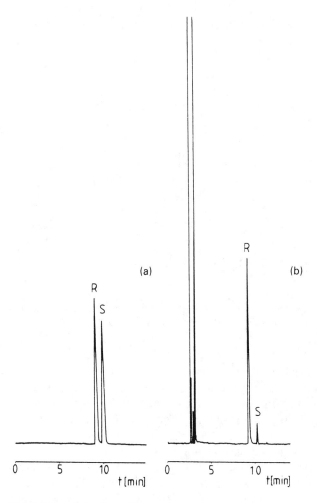

Fig. 8.18 — Chiral GC analysis of epoxide enantiomeric yield from a microbiological reaction. (a) Racemic, (b) microbiological 1,2-epoxypropane. (Reprinted, with permission, from A. Q. H. Habets-Crützen, S. J. N. Carlier, J. A. M. de Bont, D. Wistuba, V. Schurig, S. Hartmans and J. Tramper, *Enzyme Microb. Technol.*, 1985, **7**, 17. Copyright 1985, Butterworth Ltd.)

8.4.2   Enantioselective microbial amide and ester hydrolysis

In connection with studies of the optical resolution of a series of racemic *N*-aroylated amino-acids by LC on a "Resolvosil" column, it was accidentally found that the peak area ratio very significantly deviated from unity if the buffer solutions of the compounds to be analysed had been standing at room temperature for 24 hr [114]. This initiated more systematic studies of the microbial process and it was found that a preferential degradation of the L-enantiomer by amide bond hydrolysis occurred in N-benzoylalanine [114] as well as N-(p-nitrobenzoyl)serine [115]. Similarly, enantio-selective hydrolysis of ester bonds by micro-organisms could be demonstrated, in which case the enantiomeric composition of the ester substrate and of the product could be studied simultaneously [116]. The technique used to monitor the progress of the reaction is shown in Fig. 8.19.

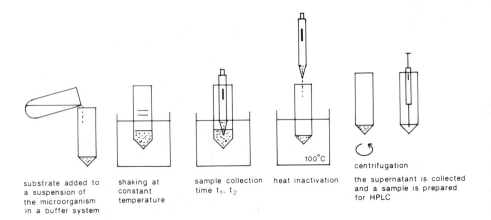

Fig. 8.19 — An outline of the very simple technique by which enantioselective microbial reactions can be studied kinetically by reversed-phase chiral LC.

An aliquot of a few hundred μl is taken from the reaction mixture and is rapidly heated to destroy the enzyme activity, the dead cells are separated by centrifugation, and a portion of the supernatant liquid is injected into the column.

Table 8.9 gives the results obtained for some of the reactions studied.

8.4.3 Asymmetric ketone reduction by yeast organisms

The technique described in Section 8.4.2 has also been used to study an asymmetric microbial synthesis. In this case the substrate is an achiral aromatic ketone and an asymmetric reduction is achieved by an NAD(H)-dependent alcohol dehydrogenase present in the yeast organism. The reduction is best classified as an enantioface-differentiating reaction (i.e. it is of the same type as that in the epoxidation reaction described in Section 8.4.1), Fig. 8.20.

Analysis of the reaction mixture at different times, by direct injection into a "Resolvosil" column, shows that the reaction takes place with high or even complete stereospecificity, depending on the particular micro-organism. In all cases investigated the stereochemistry follows Prelog's rule, i.e. the (S)-form of the alcohol is preferentially produced.

Table 8.9 — Microbial enantioselectivity in some hydrolysis reactions determined by chiral LC. (Reprinted, with permission, from S. Allenmark, B. Bomgren and H. Borén, *Enzyme Microb. Technol.*, 1986, **8**, 404. Copyright 1986, Butterworth Ltd.).

Micro-organism used	Substrate	Reaction intermediate observed	Stereochemical preference
Nocardia restrictus	N-Acetyl-D,L-tryptophan	—	L- (>95%)
	N-Acetyl-D,L-tryptophan ethyl ester	N-Acetyl-D-tryptophan	D-
	N-Benzoyl-D,L-alanine	Benzoic acid	L- (>98%)
Arthrobacter oxydans	N-Formyl-D,L-tryptophan	—	L-
	N-Acetyl-D,L-tryptophan	—	L-
	N-Acetyl-D,L-tryptophan ethyl ester	N-Acetyl-D-tryptophan	D- (75%)
	N-Benzoyl-D,L-alanine	Benzoic acid	L-
Pseudomonas putida	N-Acetyl-D,L-tryptophan	—	L-
	N-Benzoyl-D,L-alanine	—	L-
Nocardia corallina	N-Acetyl-D,L-tryptophan ethyl ester	N-Acetyl-L-tryptophan	L-

8.4.4 Calculation of enantioselectivity from chromatographic data

From the examples given it is clear that chiral chromatographic techniques should be most valuable tools for detailed studies of stereoselective enzyme-catalysed reactions. It is thus of interest to go into some further detail with respect to definitions of reaction types and the calculations that can be made from basic chromatographic data.

Since there has been some confusion and ambiguity in the terminology of these reactions over the years, a short summary of the classification made by Izumi and Tai [117] is given in Scheme 8.10. Using the general term *stereodifferentiating reactions*, six classes of reactions can be found. The first three, belonging to the group of enantio-differentiating reactions, will be discussed here. It is seen that the first two reactions start from a prochiral substrate and that the differentiation is the result of

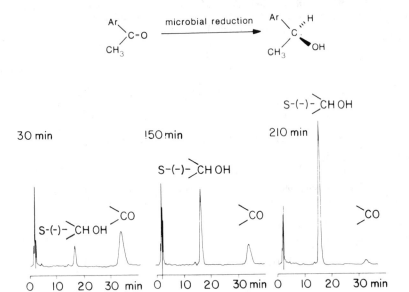

Fig. 8.20a — Example of an asymmetric reduction of an aromatic ketone studied by chiral reversed-phase LC.

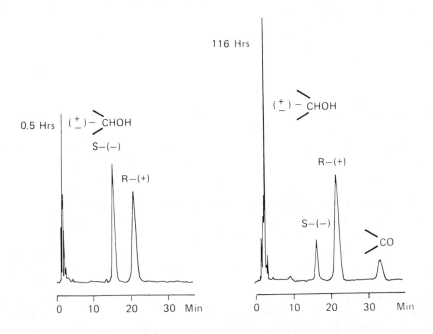

Fig. 8.20b — Reversal of the reaction: Enantioselection in the microbial oxidation of the rac. alcohol [(S)-(−)-form, the favoured substrate]. (Reprinted, with permission, from S. Allenmark and S. Andersson, *Enzyme Microb. Technol.*, 1989, **11**, 177. Copyright 1989, Butterworth Ltd.)

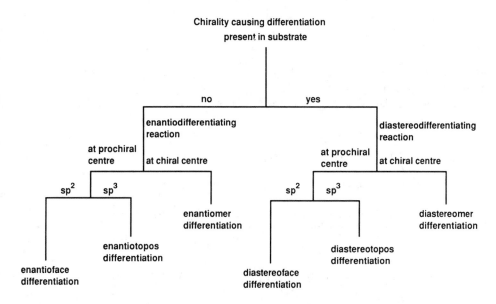

Scheme 8.10 — Different types of stereodifferentiating reactions.

preferential attack by the reagent from one side of the substrate molecule. The situation is different in the third case, since here there is competition between two enantiomeric substrates which leads to a faster conversion of one of them. However, since the enantiomeric ratio of the substrate changes with time, the enantiomer excess of the product will, of course, not be constant but time-dependent.

Thus, while the kinetic treatment of the first two reaction types is quite simple, since it involves only the formation of two enantiomeric products, the case of enantiomer-differentiation is a bit more complex to handle. This reaction type, however, has been exhaustively treated by Sih and his collaborators [118–120], and the following discussion originates from their results.

Let us assume that a racemate, composed of the enantiomers S_1 and S_2, undergoes an enantiomer-differentiating reaction to form the enantiomeric products P_1 and P_2:

$$S_1 \rightleftharpoons P_1$$
$$a_1 - x_1 \qquad x_1$$
$$S_2 \rightleftharpoons P_2$$
$$a_2 - x_2 \qquad x_2$$

If this reaction is kinetically controlled, i.e. the equilibrium is sufficiently far to the right, then at a certain time t, the concentrations are as written above and the enantiomeric excess of the product (ee_p) and substrate (ee_s), respectively, can be formulated as:

$$ee_p = \frac{x_1 - x_2}{x_1 + x_2} \quad \text{and} \quad ee_s = \frac{x_1 - x_2}{2a - (x_1 + x_2)}$$

It is assumed here that $x_1 > x_2$ and that at $t = 0$ $[S_1]_0 = [S_2]_0$, i.e.: $a_1 = a_2 = a$. Thus, the initial concentration of the racemate $= 2a$.

The *enantioselectivity* (E) of the reaction has been defined [118,119] as:

$$E = \frac{\ln([S_1]/[S_1]_0)}{\ln([S_2]/[S_2]_0)} = \frac{(k_{cat}/K_m)_1}{(k_{cat}/K_m)_2} \tag{8.1}$$

It has also been shown [120] that E can be expressed as:

$$E = \frac{\ln[(1-c)(1-ee_s)]}{\ln[(1-c)(1+ee_s)]} = \frac{\ln[1-c(1+ee_p)]}{\ln[1-c(1-ee_p)]} \tag{8.2}$$

$$\text{where } c \text{ (the degree of substrate conversion)} = \frac{ee_s}{ee_s + ee_p} \tag{8.3}$$

However, with the expression for ee_s and ee_p given above, Eq. (8.3) can be rewritten as:

$$c = (x_1 + x_2)/2a$$

Since $x_1 + x_2$ is the sum of the two product enantiomers at time t, and $2a$ the initial substrate (racemate) concentration, it follows that c is readily determined by achiral LC. Further, it is clear from Eq. (8.2) that it is only necessary to determine *either ee_p or ee_s* to be able to calculate E. For an irreversible reaction E is (unlike ee_p or ee_s) time-independent and thereby a true measure of the enantioselectivity.

To date, determination of enantiomeric composition for calculation of E has relied upon the use of NMR-techniques involving chiral lanthanide shift reagents [119]. However, the use of chiral chromatography should greatly facilitate such determinations, especially since it is only necessary to resolve either the substrate or the product. An example, showing the application of a BSA-based chiral LC column for a study of the enantiomeric product composition after the lipase-catalysed hydrolysis of chloroethyl 2-(p-nitrophenyl)propionate [121], is given in Fig. 8.21. Integration over the separated enantiomers of the product yields ee_p with high precision on a very small amount of sample. The enzyme was a commercial *Candida cylindracea* lipase, used at various stages of purification and further treatment. The chromatogram shown on the right is obtained after further processing of the enzyme [120], which results in a significantly increased enantioselectivity. The corresponding E-values are: 51 (stage 2; left) and > 150 (stage 4; right). Another interesting way to modify the reaction process is to use an enantioselective enzyme inhibitor [119]. Ester hydrolyses are very suitable reactions for studies with BSA-based chiral LC

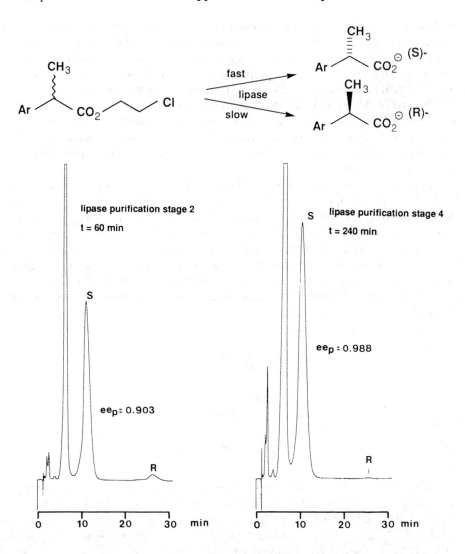

Fig. 8.21 — Determination of enantioselectivity by resolution of the product enantiomers obtained from a lipase-catalysed hydrolysis of 2-chloroethyl 2-(p-nitrophenyl)-propionate on a BSA-based analytical column. The first large peak in the chromatogram is the unresolved ester. The chromatograms (a) and (b) refer to different stages in the lipase purification procedure. (S. Allenmark and A. Ohlsson, unpublished work.)

columns, since the latter permit resolution of a wide range of organic acids into enantiomers [122].

8.5 MISCELLANEOUS APPLICATIONS AND TECHNIQUES

8.5.1 Determination of enantiomerization barriers

As described in Section 5.3, many chiral compounds undergo enantiomerization processes which are fast enough to be conveniently studied by chromatographic or

NMR techniques. In both cases enantiomer interconversion rates are correlated with temperature-dependent peak-coalescence phenomena. Chiral chromatography may play a twofold role, as an analytical tool for direct observation of enantiomerization during chromatography, or as a preparative technique for optical enrichment, to enable polarimetric studies of enantiomer interconversion to be made.

The first of these techniques has been very elegantly applied by Schurig and co-workers [123,124], by use of complexation gas chromatography (cf. Section 6.2). GC is almost ideal for this purpose since the chromatographic temperature-dependence is easily determined, thanks to the precise oven-temperature regulation afforded by modern instruments.

On-column enantiomerization has also been observed in liquid chromatography. During attempts at optical resolution of the members of a series of N,N-dimethyl-thiobenzamides, **24**, on microcrystalline cellulose triacetate (MCTA) at room temperature, it was found [125] that the single peaks obtained for some of the compounds (X=H and F) were not due to insufficient separation, but rather to fast enantiomerization on the column (peak coalescence). This proved to be a general phenomenon for compounds which showed thermal enantiomerization with $\Delta\Delta G^+$ barriers of about 90 kJ/mole. The temperature-dependence of a chromatogram of **24** (X=CH$_3$) is shown in Fig. 8.22.

$$CH_3\underset{N}{\overset{\oplus}{\diagdown}}CH_3$$

24

Optical enrichment of many theoretically interesting compounds for the purpose of polarimetric rate studies have been performed by the use of MCTA. Compounds studied in this field include chiral hydrocarbons such as substituted phenanthrenes [126] and cis,trans-1,3-cyclo-octadiene [127] as well as other twisted 1,3-cycloalka-dienes [128] and polarized alkene systems [129,130].

Recently very interesting techniques, combining chiral liquid chromatographic separation with on-line studies of chiroptical properties of a pure enantiomer after stopped-flow, have come into use [130–132]. The importance of such methods lies in an elimination of the need for *preparative* enantiomer enrichment, which often can be a difficult task, especially if the compound to be investigated is easily racemized. The principle used is to perform the enantiomer separation on a column of relatively high sample capacity at a temperature lower than that of the polarimetric detector cell. When, on elution, the first enantiomer has reached the cell, the flow is stopped and the optical rotation produced is measured as a function of time. By this technique enantiomer half-lives of a few minutes can be readily determined and $\Delta G^{\neq}$-values of the thermal racemization process computed. The data collected in Table 8.10 show the usefulness of the method. It should be remembered that a lowering of the column

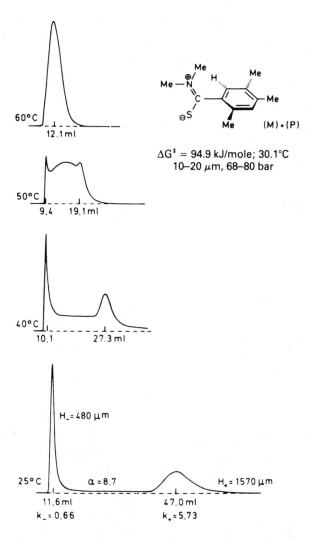

Fig. 8.22 — Illustration of the use of chiral chromatography for a determination of the thermal enantiomerization barrier in a racemic thiobenzamide. (Reprinted, with permission, from A. Eiglsperger, F. Kastner and A. Mannschreck, *J. Mol. Structure*, 1985, **126**, 421. Copyright 1985, Elsevier Science Publishers B.V.).

temperature (for the second substrate given in the table, to −40°C) will usually increase the separation factor, which is actually an additional bonus.

Use has also been made of on-line circular dichroic (CD) detection in combination with the stopped-flow technique. The left part of Fig. 8.23 shows a chromatogram of 0.35 mg of racemic Tröger's base. The chromatographic system consisted of an 8×250 mm triacetylcellulose (10–20 μm particles) column connected to a circular dichrograph equipped with an 8-μl cell of 10-mm pathlength. Elution was carried out with 96% ethanol at a linear flow rate of 0.58 mm/sec. The differential absorbance

Table 8.10 — Data obtained from the use of low-temperature chiral chromatography and stopped-flow, on-line polarimetric determination of the racemization rate of the first eluted enantiomer. T_{col}=column temperature, T_{rac}=detector cell temperature during racemization, $t_{1/2}$=enantiomer half-life. In both cases, a column packed with (+)-poly(tritylmethacrylate)-coated silica was used with methanol as the eluent. (Reproduced from A. Mannschreck, D. Andert, A. Eiglsperger, E. Gmahl and H. Buchner, *Chromatographia*, 1988, **25**, 182, with permission. Copyright 1988, Friedr. Vieweg & Sohn, Verlagsgesellschaft, mbH)

Substrate	k_1'	k_2'	T_{col}(°C)	T_{rac}(°C)	$t_{1/2}$(min)	$\Delta G^{\neq}$ (kJ/mole)
(structure)	0.75	1.52	+15	+13.5	4.3	86±1
(structure)	0.68	1.06	−40	−11	4.4	78.3±0.7

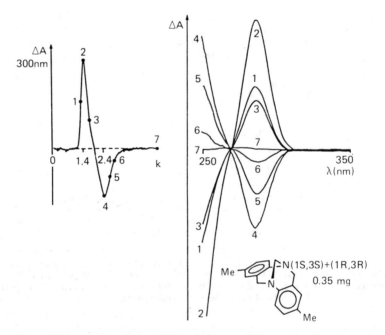

Fig. 8.23 — CD-spectra recordings of Tröger's base at different parts in a chromatogram obtained on elution from a chiral column. (Reproduced from A. Mannschreck, D. Andert, A. Eiglsperger, E. Gmahl and H. Buchner, *Chromatographia*, 1988, **25**, 182, with permission. Copyright 1988, Friedr. Vieweg & Sohn, Verlagsgesellschaft, mbH.)

was measured at 300 nm during the elution. Flow was stopped at the points indicated (nos. 1–7) and the full spectra recorded (right part). It is evident that techniques of this kind will be extremely valuable as analytical tools for stereochemical investigations, since only racemates are needed and accordingly, even enantiomers undergoing fast racemization can be investigated.

8.5.2   Determination of configuration from chromatographic data

It should be evident from the preceding chapters that chiral chromatography offers the possibility of determining the absolute configuration of a compound present in extremely small amount, if the retention data of both antipodes are known and consistent with a general chiral recognition mechanism. We can distinguish between two cases: first establishment of identity from retention data alone, by use of a reference of previously known absolute configuration, and secondly assumption of an enantiomer elution order identical with that for a related compound of known absolute configuration.

The first case represents the use of chiral chromatography only as a sensitive and selective analytical technique for identification purposes. In the second case, on the other hand, the absolute configuration of the compound is determined solely by the mechanism of stereodifferentiation assumed to be exerted by the CSP. Therefore, a detailed understanding of such solute–sorbent interactions is needed. As recently pointed out [134], the number of functional groups present in a given solute will add to the difficulty of postulating a single chiral recognition mechanism. At the present stage, therefore, a determination of absolute configuration of structurally more complex compounds from chromatographic data alone is not quite reliable. In such cases, support from independent techniques such as CD and NMR spectroscopy (cf. Chapter 3) is still required [135]. As evident from the data presented in Table 8.5, however, there appears to be a clear and consistent correlation between elution order and configuration of the variously substituted benzodiazepinones, indicating a common chiral recognition mechanism. As long as large separation factors are obtained, absolute configurations could certainly be deduced with confidence for the members of the series, from the retention data obtained. It is most likely that techniques of this kind will gain importance in the future.

8.5.3   Evaluation of enantiomeric purity from chromatographic partial optical resolution (Mannschreck's method)

Though the procedure for determination of the optical purity of a sample is quite simple in those cases where baseline separation of the enantiomers is obtained, the situation is more difficult when only incomplete separation can be achieved. However, by a combination of UV and polarimetric detection an elegant solution to this problem has been devised by Mannschreck *et al.* [135]. This method is described below.

Let us assume that we are able to measure simultaneously the optical rotation α and the absorbance A of a mixture of enantiomers present in a small volume element, at different times. From the Lambert–Beer law we know that

$$A = A_+ + A_- = \varepsilon b (c_+ + c_-)$$

and from Biot's law that:

$$\alpha = \alpha_+ + \alpha_- = [\alpha] \, l \, (c_+ - c_-)$$

where b and l are the path-lengths in the measurement cells.

For a chromatographic process the time dependence is equivalent to a volume dependence. Therefore, $A(v)$ and $\alpha(v)$ represent chromatograms and a function $C(v)\equiv(\alpha/A)(v)$, where $\alpha/A=[\alpha]\,l\,(c_+-c_-)/\varepsilon b(c_++c_-)$, can be defined.

Because c_+ and c_- are the actual concentrations of the respective enantiomers at a given time in a given volume element dv, it follows that $c_+=dn_+/dv$ and $c_-=dn_-/dv$, and we arrive at the equation:

$$C=\alpha/A=\frac{[\alpha]l}{\varepsilon b}\left(\frac{dn_+-dn_-}{dn_++dn_-}\right)=\frac{[\alpha]l}{\varepsilon b}P'$$ (8.1)

where dn_+ and dn_- are the numbers of moles of the respective enantiomers present in the volume element. Therefore $(dn_+-dn_-)/(dn_++dn_-)$ is identical with the actual enantiomeric purity, P', present in the volume element.

If no separation of the antipodes in the sample occurs (passage through an achiral column), then P' will be constant and equal to the optical purity of the sample, P. The equation then reduces to:

$$C_m=\alpha_m/A_m=\frac{[\alpha]lP}{\varepsilon b}$$ (8.2)

If, however, some separation takes place (passage through a chiral column of partial resolving capacity, then P' will vary with v, the volume of eluate. In the very first or last part of such a chromatogram $P'=1$ (100% enantiomeric purity) and Eq. (8.1) reduces to:

$$C_+=\alpha_+/A_+=\frac{[\alpha]l}{\varepsilon b}$$ (8.3)

Consequently, an expression for P can be derived:

$$P=C_m/C_+$$ (8.4)

Provided α and A are measured simultaneously, the enantiomeric purity, P, of a sample can be determined experimentally from the slopes of plots of α vs. A obtained with an X-Y recorder during chromatography. Thus, C_m is derived from use of an achiral column or by separate α and A determinations on the sample and C_+ (or C_-) is found from the positive or negative slopes obtained on chiral chromatography.

To allow for simultaneous determination of α and A, it is advisable to pass the eluate through the UV-detector twice; a double-cell configuration is then used. A representative example illustrating the technique is given in Fig. 8.24.

Exercises
(1) Neither of the techniques illustrated in Fig. 8.1 gives any information regarding the amount of analyte racemized during work-up procedures. Why? How would you

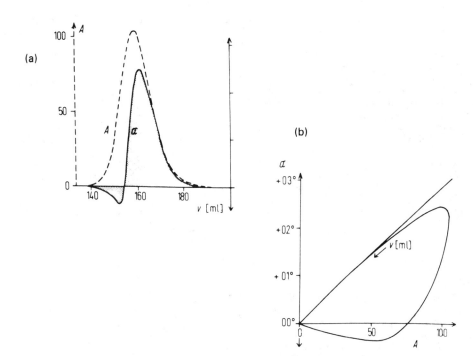

Fig. 8.24 — Experimental results showing an application of Mannschreck's method. (a) Chromatograms showing A and α and (b) x-y recording of $\alpha(A)$ of (+)-1-phenylethanol ($P = 52\%$) obtained after passage through an MCTA column (96% ethanol). (Reprinted, with permission, from A. Mannschreck, M. Mintas, G. Becher and G. Stühler, *Angew. Chem.*, 1980, **92**, 490 (*Int. Ed.*: 1980, **19**, 469. Copyright 1980, Verlag Chemie GmbH).

design a system which could also give this information in one single chromatographic experiment?

(2) Draw a stereoprojection formula which shows how (S)-1-(α-naphthyl)ethanol is formed by hydride transfer to the prochiral ketone shown in Fig. 8.19.

BIBLIOGRAPHY

J. Retey and J. A. Robinson, *Stereospecificity in Organic Chemistry and Enzymology*, Verlag Chemie, Weinheim, 1982.

K. Mori in *Techniques of Pheromone Research*, H. E. Hummel and T. A. Miller (eds.), Springer Verlag, New York 1984, p. 323.

R. Tressl, K.-H. Engel, W. Albrecht and H. Bille-Abdullah, Analysis of Chiral Aroma Components in Trace Amounts, in *Characterization and Measurement of Flavor Compounds*, B. B. Bills and C. J. Mussinan (eds.), ACS, Washington DC, 1985, p. 43.

J. H. Tumlinson, III, Pheromonal Chemists par Excellence, in *Bioregulators for Pest Control*, P. A. Hedin (ed.), ACS, Washington DC, 1985, p. 367.

J. Bojarski, Chromatographic Resolution of Enantiomers and its Pharmaceutical Applications, in *Chromatography '85*, H. Kalász and L. S. Ettre (eds.), Akadémiai Kiadó, Budapest, 1986, p. 343.

R. Däppen, H. Arm and V. R. Meyer, Applications and Limitations of Commercially Available Chiral Stationary Phases for High-Performance Liquid Chromatography, *J. Chromatog.*, 1986, **373**, 1 (*Chromatog. Rev.*, 1986, **31**, 1).

V. Schurig, Current Methods for Determination of Enantiomeric Compositions, Part 3, Gas Chromato-
graphy on Chiral Stationary Phases, *Kontakte (Darmstadt)*, 1986, No. 1, 3.
M. Zief and L. J. Crane, *Chromatographic Chiral Separation*, Dekker, New York, 1988.
I. W. Wainer, *A Practical Guide to the Selection and Use of HPLC Chiral Stationary Phases*, J. T. Baker
Inc., Phillipsburg, NJ, 1988.
I. W. Wainer and D. E. Dryer (eds.), *Drug Stereochemistry, Analytical Methods and Pharmacology*,
Dekker, New York & Basel, 1988.
A. M. Krstulovic (ed.), *Chiral Separations by HPLC: Applications to Pharmaceutical Compounds*, Ellis
Horwood, Chichester, 1989.
C. J. Sih and S.-H. Wu, Resolution of Enantiomers via Biocatalysis, *Top. Stereochem.*, 1989, **19**, 63.
C.-S. Chen and C. J. Sih, General Aspects and Optimization of Enantioselective Biocatalysis in Organic
Solvents: The Use of Lipases, *Angew. Chem., Int. Ed.*, 1989, **28**, 695.
S. Allenmark, Chromatographic Methods for Optical Purity Determination of Drugs, in *Racemates and
Enantiomers in Drug Design and Development*, L. Dalgaard (ed.), *Acta Pharm. Nord.*, 1990, **2**.
J. Martens and R. Bhushan, Importance of Enantiomeric Purity and Its Control by Thin-Layer
Chromatography, *J. Pharm. Biomed. Anal.*, 1990, **8**, 259.
B. Holmstedt, H. Frank and B. Testa (eds.), *Chirality and Biological Activity*, Alan R. Liss, New York,
1990.

REFERENCES

[1] S. Moore and W. H. Stein, *J. Biol. Chem.*, 1951, **192**, 663.
[2] H. Kleinkauf and H. von Doehren (eds.), *Peptide Antibiotics: Biosynthesis and Function*, de
Gruyter, Berlin, 1982.
[3] H. Frank, G. Nicholson and E. Bayer, *J. Chromatog.*, 1978, **167**, 187.
[4] N. E. Blair and W. A. Bonner, *J. Chromatog.*, 1980, **198**, 185.
[5] S. Allenmark, *Tetrahedron Lett.*, 1990, **31**, 1455.
[6] S. Allenmark and S. Andersson, *Chromatographia*, 1991, **31**, 429.
[7] S. Allenmark, in *Chiral Separations*, S. Ahuja (ed.), ACS Symp. Ser., Washington DC, 1991, in the
press.
[8] H. J. Bestmann, W. L. Hirsch, H. Platz, M. Rheinwald and O. Wostrowsk, *Angew. Chem.*, 1980,
92, 492.
[9] R. M. Silverstein, in *Chemical Ecology: Odour Communication in Animals*, F. J. Ritter (ed.),
Elsevier/North Holland, Amsterdam, 1979, p. 133.
[10] M. Cammaerts, A. B. Attygalle, R. P. Evershed and E. D. Morgan, *Physiol. Entomol.*, 1985, **10**,
33.
[11] G. T. Pearce, W. E. Gore, R. M. Silverstein, J. W. Peacock, R. A. Cuthbert, G. N. Lanier and J. B.
Simeone, *J. Chem. Ecol.*, 1975, **1**, 115.
[12] K. Mori, *Tetrahedron*, 1977, **33**, 289.
[13] R. G. Riley, R. M. Silverstein and J. C. Moser, *Science*, 1974, **183**, 760.
[14] Y. Sakito and T. Makaiyama, *Chem. Lett.*, 1979, 1027.
[15] L. Colombo, C. Gennari, G. Poli and C. Scolastico, *Tetrahedron*, 1982, **38**, 2725.
[16] K. Mori and H. Watanabe, *Tetrahedron*, 1985, **41**, 3423.
[17] H. J. Bestmann, A. B. Attygalle, J. Glasbrenner, R. Riemer and O. Vostrowsky, *Angew. Chem.*,
1987, **99**, 784.
[18] W. Franke, V. Heemann, B. Gerken, J. A. A. Renwick and J. P. Vite, *Naturwiss.*, 1977, **64**, 590.
[19] W. Franke, G. Hindorf and W. Reith, *Leibigs Ann. Chem.*, 1979, 1.
[20] W. Franke, W. Reith, G. Bergström and J. Tengö, *Naturwiss.*, 1980, **67**, 149.
[21] W. Franke, W. Reith, G. Bergström and J. Tengö, *Z. Naturforsch.*, 1981, **36**, 928.
[22] B. Koppenhöfer, K. Hintzer, R. Weber and V. Schurig, *Angew. Chem.*, 1980, **92**, 473.
[23] R. Weber and V. Schurig, *Naturwiss.*, 1981, **68**, 330.
[24] V. Schurig, R. Weber, D. Klimetzek, U. Kohnle and K. Mori, *Naturwiss.*, 1982, **69**, 602.
[25] V. Schurig, R. Weber, G. J. Nicholson, A. C. Oehlschlager, H. Pierce, Jr., A. M. Pierce, J. H.
Borden and L. C. Ryker, *Naturwiss.*, 1983, **70**, 92.
[26] R. Isaksson, T. Liljefors and P. Reinholdsson, *Chem. Commun.*, 1984, 137.
[27] H. Allgaier, G. Jung, R. G. Werner, U. Schneider and H. Zähner, *Angew. Chem.*, 1985, **97**, 1052.
[28] E. Küsters, H. Allgaier, G. Jung and E. Bayer, *Chromatographia*, 1984, **18**, 287.
[29] E. Gross and J. L. Morell, *J. Am. Chem. Soc.*, 1971, **93**, 4634.
[30] E. Gross and H.H. Kiltz and E. Nebelin, *Z. Physiol. Chem.*, 1973, **354**, 810.
[31] T. Wakamiya, Y. Ueki, T. Shiba, Y. Kido and Y. Motoki, *Tetrahedron Lett.*, 1985, **26**, 665.

[32] T. Bolte, D. Yu, H.-T. Stuwe and W. A. König, *Angew. Chem.*, 1987, **99**, 362, (*Int. Ed.*, 1987, **26**, 331).
[33] W. A. König, I. Benecke, N. Lucht, E. Schmidt, J. Schulze and S. Sievers, *J. Chromatog.*, 1983, **279**, 555.
[34] T. Koscielski, D. Sybilska, S. Belniak and J. Jurczak, *Chromatographia*, 1984, **19**, 292.
[35] T. Koscielski, D. Sybilska and J. Jurczak, *J. Chromatog.*, 1986, **364**, 299.
[36] A. Mosandl, U. Hener, U. Hagenauer-Hener and A. Kustermann, *J. High Res. Chromatog., Chromatog. Commun.*, 1989, **12**, 532.
[37] E. Guichard, A. Kustermann and A. Mosandl, *J. Chromatog.*, 1990, **498**, 396.
[38] H.-G. Schmarr, A. Mosandl and K. Grob, *Chromatographia*, 1990, **29**, 125.
[39] A. Mosandl and C. Günther, *J. Agric. Fd. Chem.*, 1989, **37**, 413.
[40] A. Mosandl and A. Kustermann, *Z. Lebensm. Unters. Forsch.*, 1989, **189**, 212.
[41] R. Kallenborn, H. Hühnerfuss and W. A. König, *Angew. Chem.*, 1991, **103**, 328.
[42] E. J. Ariëns, W. Soudijn and P. Timmermans, *Stereochemistry and Biological Activity of Drugs*, Blackwell, Oxford, 1983.
[43] E. J. Ariëns, *Trends Pharmacol. Sci.*, 1986, **7**, 200.
[44] G. Wahlström, *Life Sci.*, 1966, **5**, 1781.
[45] G. Wahlström, *Acta Pharmacol. Toxicol.*, 1968, **26**, 81.
[46] H. Buch, W. Buzello, O. Nuerohr and W. Rummel, *Biochem. Pharmacol.*, 1968, **17**, 2391.
[47] G. Blaschke, *Angew. Chem.*, 1984, **92**, 14.
[48] I. Okamoto and T. Shibata, unpublished results (see [41]).
[49] D. W. Armstrong and W. DeMond, *J. Chromatog. Sci.*, 1984, **291**, 411.
[50] T. Shibata, I. Okamoto and K. Ishii, *J. Liquid Chromatog.*, 1986, **9**, 313.
[51] Z.-Y. Yang, S. Barkan, C. Brunner, J. D. Weber, T. D. Doyle and I. W. Wainer, *J. Chromatog.*, 1985, **324**, 444.
[52] W. H. Pirkle, J. M. Finn, J. L. Schreiner and B. Hamper, *J. Am. Chem. Soc.*, 1981, **103**, 3964.
[53] P. J. Wedlund, B. J. Sweetman, C. B. McAllister, R. A. Branch and G. R. Wilkinson, *J.Chromatog.*, 1984, **307**, 121.
[54] A. Kupfer, R. K. Roberts, S. Schenker and R. A. Branch, *J. Pharmacol. Exp. Ther.*, 1981, **218**, 193.
[55] H. Wollweber, H. Horstmann and K. Meng, *Eur. J. Med. Chem. (Chim. Ther.)*, 1976, **11**, 159.
[56] J. A. Tobert, V. J. Cirillo, G. Hitzenberger, I. James, J. Pryor, T. Cook, A. Buntninx, I. B. Holmes and P. M. Lutterbeck, *Clin. Pharmacol. Ther.*, 1981, **29**, 344.
[57] G. Blaschke and J. Maibaum, *J. Pharm. Sci.*, 1985, **74**, 438.
[58] L. H. Sternbach, *J. Med. Chem.*, 1979, **22**, 1.
[59] G. Blaschke and H. Markgraf, *Chem. Ber.*, 1980, **113**, 2031.
[60] G. Blaschke, *J. Liquid Chromatog.*, 1986, **9**, 341.
[61] W. H. Pirkle and A. Tsipouras, *J. Chromatog.*, 1984, **291**, 291.
[62] S. Allenmark, *J. Liquid Chromatog.*, 1986, **9**, 425.
[63] S. Allenmark and S. Andersson, *Chirality*, 1989, **1**, 154.
[64] H. A. Brassfield, R. A. Jacobsson and J. G. Verkade, *J. Am. Chem. Soc.*, 1975, **97**, 4143.
[65] S. Johne, *Pharmacie*, 1981, **36**, 583.
[66] L. D. Colebrook and H. G. Giles, *Can. J. Chem.*, 1975, **53**, 3431.
[67] A. Mannschreck, H. Koller, G. Stühler, M. A. Davies and J. Traber, *Eur. J. Med. Chem. (Chim. Ther.)*, 1984, **19**, 381.
[68] P. Hess, J. B. Lansman and R. W. Tsien, *Nature*, 1984, **311**, 538.
[69] G. Franckowiak, M. Bechem, M. Schramm and G. Thomas, *Eur. J. Pharmacol.*, 1985, **114**, 223.
[70] R. P. Hof, U. T. Ruegg, A. Hof and A. Vogel, *J. Cardiovasc. Pharmacol.*, 1985, **7**, 689.
[71] G. Blaschke and J. Maibaum, *J. Chromatog.*, 1986, **366**, 329.
[72] G. Blaschke, W. Bröker and W. Fraenkel, *Angew. Chem.*, 1986, **98**, 808 (*Int. Ed.*, 1986, **25**, 830).
[73] P. Gjörstrup, H. Harding, R. Isaksson and C. Westerlund, *J. Eur. Pharm.*, 1986, **122**, 357.
[74] S. Allenmark and R. A. Thompson, *Tetrahedron Lett.*, 1987, **28**, 3751.
[75] J. A. Jaffe, *Arthritis Rheum.*, 1970, **13**, 436.
[76] I. Sternlieb and I. H. Scheinberg, *J. Am. Med. Assoc.*, 1964, **189**, 748.
[77] A. Wacker, E. Heyl and P. Chandra, *Arzneim. Forsch.*, 1971, **30**, 395.
[78] W. A. König, E. Steinbach and K. Ernst, *J. Chromatog.*, 1984, **301**, 129.
[79] E. Busker, K. Günther and J. Martens, *J. Chromatog.*, 1985, **350**, 179.
[80] E. Busker and J. Martens, *Z. Anal. Chem.*, 1984, **319**, 907.
[81] H. Frank, W. Woiwode, G. Nicholson and E. Bayer, *Liebigs Ann. Chem.*, 1981, 354.
[82] G. C. Cotzias, P. S. Papavasilou and R. Gellene, *New Engl. J. Med.*, 1969, **280**, 337.
[83] D. C. Poskanzer, *New Engl. J. Med.*, 1969, **280**, 362.

[84] L. R. Gelber and J. L. Neumeyer, *J.Chromatog.*, 1983, **257**, 317.
[85] W. L. Nelson and T. R. Burke, Jr., *J. Org. Chem.*, 1978, **43**, 3641.
[86] O. Gyllenhaal and J. Vessman, *J. Chromatog.*, 1983, **273**, 129.
[87] O. Gyllenhaal, W. A. König and J. Vessman, *J. Chromatog.*, 1985, **350**, 328.
[88] W. A. König, I. Benecke and S. Sievers, *J.Chromatog.*, 1981, **217**, 71.
[89] W. A. König and K. Ernst, *J.Chromatog.*, 1983, **280**, 135.
[90] I. W. Wainer, T. D. Doyle, K. H. Donn and J. R. Powell, *J. Chromatog.*, 1984, **306**, 405.
[91] E. Küsters and D. Giron, *J. High Resol. Chromatog.*, *Chromatogr. Commun.*, 1986, **9**, 531.
[92] C. Pettersson and G. Schill, *J.Chromatog.*, 1981, **204**, 179.
[93] C. Pettersson and G. Schill, *J. Liquid Chromatog.*, 1986, **9**, 269.
[94] C. Pettersson and M. Josefsson, *Chromatographia*, 1986, **21**, 321.
[95] G. Schill, I. W. Wainer and S. A. Barkan, *J. Liquid Chromatog.*, 1986, **9**, 641.
[96] Y. Okamoto, M. Kawashima, R. Aburatani, K. Hatada, T. Nishiyama and M. Masuda, *Chem. Lett.*, 1986, 1237.
[97] P. Erlandsson, I. Marle, L. Hansson, R. Isaksson, C. Pettersson and G. Pettersson, *J. Am. Chem. Soc.*, 1990, **112**, 4573.
[98] R. Isaksson and B. Lamm, *J.Chromatog.*, 1986, **362**, 436.
[99] H. Frank, G. J. Nicholson and E. Bayer, *J. Chromatog.*, 1978, **146**, 197.
[100] W. A. König, K. Ernst and J. Vessman, *J. Chromatog.*, 1984, **294**, 423.
[101] I. W. Wainer, T. D. Doyle, Z. Hamidzadeh and M. Aldridge, *J. Chromatog.*, 1983, **261**, 123.
[102] I. W. Wainer and T. D. Doyle, *J. Chromatog.*, 1984, **284**, 117.
[103] D. M. McDaniel and B. G. Snider, *J. Chromatog.*, 1987, **404**, 123.
[104] J. Hermansson and M. Eriksson, *J. Liquid Chromatog.*, 1986, **9**, 621.
[105] J. B. Crowther, T. R. Covey, E. A. Dewey and J. D. Henion, *Anal. Chem.*, 1984, **56**, 2921.
[106] P. J. van Bladeren, R. N. Armstrong, D. Cobb, D. R. Thakker, D. E. Ryan, P. E. Thomas, N. D. Sharma, D. R. Boyd, W. Levin and D. M. Jerina, *Biochem. Biophys. Res. Commun.*, 1982, **106**, 602.
[107] V. Schurig and D. Wistuba, *Angew. Chem.*, 1984, **96**, 808.
[108] S. W. May and R. D. Schwartz, *J. Am. Chem. Soc.*, 1974, **96**, 4031.
[109] S. W. May, M. S. Steltenkamp, R. D. Schwartz and C. J. McCoy, *J. Am. Chem. Soc.*, 1976, **98**, 7856.
[110] H. Ohta and H. Tetsukawa, *Agric. Biol. Chem.*, 1979, **43**, 2099.
[111] K. Furuhashi, A. Taoka, S. Uchida, I. Karube and S. Suzuki, *Eur. J. Appl. Microbiol. Biotechnol.*, 1981, **12**, 39.
[112] M. J. de Smet, J. Kingma, H. Wynberg and B. Witholt, *Enzyme Microb. Technol.*, 1983, **5**, 352.
[113] A. Q. H. Habets-Crützen, S. J. N. Carlier, J. A. M. de Bont, D. Wistuba, V. Schurig, S. Hartmans and J. Tramper, *Enzyme Microb. Technol.*, 1985, **7**, 17.
[114] S. Allenmark, B. Bomgren and H. Borén, *J. Chromatog.*, 1983, **264**, 63.
[115] S. Allenmark, B. Bomgren and S. Andersson, *Prep. Biochem.*, 1984, **14**, 139.
[116] S. Allenmark, B. Bomgren and H. Borén, *Enzyme Microb. Technol.*, 1986, **8**, 404.
[117] Y. Izumi and A. Tai, *Stereo-differentiating Reactions: the Nature of Asymmetric Reactions*, Kodansha/Academic Press, Tokyo/New York, 1977, Ch. 4.
[118] C. S. Chen, Y. Fujimoto, G. Girdaukas and C. J. Sih, *J. Am. Chem. Soc.*, 1982, **104**, 7294.
[119] Z.-W. Guo and C. J. Sih, *J. Am. Chem. Soc.*, 1989, **111**, 6836.
[120] S.-H. Wu, Z.-W. Guo and C. J. Sih, *J. Am. Chem. Soc.*, 1990, **112**, 1990.
[121] S. Allenmark and A. Ohlsson, to be published.
[122] S. Allenmark and S. Andersson, *Chirality*, 1991, in the press.
[123] V. Schurig and W. Bürkle, *J. Am. Chem. Soc.*, 1982, **104**, 7573.
[124] W. Bürkle, H. Karfunkel and V. Schurig, *J.Chromatog.*, 1984, **288**, 1.
[125] A. Eiglsperger, F. Kastner and A. Mannschreck, *J. Mol. Structure*, 1985, **126**, 421.
[126] H. Scherübl, U. Fritzsche and A. Mannschreck, *Chem. Ber.*, 1984, **117**, 336.
[127] R. Isaksson, J. Roschester, J. Sandström and L.-G. Wistrand, *J. Am. Chem. Soc.*, 1985, **107**, 4074.
[128] L. Andersson, C.-J. Aurell, B. Lamm, R. Isaksson, J. Sandström and K. Stenvall, *Chem. Commun.*, 1984, 411.
[129] U. Berg, R. Isaksson, J. Sandström, U. Sjöstrand, A. Eiglsperger and A. Mannschreck, *Tetrahedron Lett.*, 1982, **23**, 4237.
[130] A. Zul-Quarnain Khan, R. Isaksson and J. Sandström, *J. Chem. Soc. Perkin Trans. II*, 1987, 491.
[131] M. Mintas, Z. Orhanovic, K. Jakopcic, H. Koller, G. Stühler and A. Mannschreck, *Tetrahedron*, 1985, **41**, 229.
[132] C. Roussel and A. Djafri, *Nouveau J. Chim.*, 1986, **10**, 399.

[133] A. Mannschreck, D. Andert, A. Eiglsperger, E. Gmahl and H. Buchner, *Chromatographia*, 1988, **25**, 182.

[134] R. D. Stipanovic, J. P. McCormick, E. O. Schlemper, B. C. Hamper, T. Shinmyozu and W. H. Pirkle, *J. Org. Chem.*, 1986, **51**, 2500.

[135] A. Mannschreck, A. Eiglsperger and G. Stühler, *Chem. Ber.*, 1982, **115**, 1568.

9

Preparative scale enantioseparations — need, progress and problems

9.1 GENERAL CONSIDERATIONS

Classical methods of optical resolution, based on recrystallization of diastereomeric salts or labile complexes, possess the advantage of being easily performed on the large scale. However, unless the total process is very straightforward (which is rarely the case), it is generally not very suitable for industrial scale-up and automation. One of the main reasons for this, of course, is that a large number of recrystallizations will often be needed to yield a diastereomer of acceptable purity. The product thus obtained often contains a little of the opposite enantiomer and has to be further recrystallized to improve the optical purity. The second enantiomer, which must be obtained from the mother liquor fractions from the initial recrystallizations, is seldom obtained in the same yield as the first. Thus, there is also a loss of valuable material in the classical resolution process, which is difficult to counteract unless some complicated recycling techniques, which may be very hard to control, are used.

With this in mind, it is easy to appreciate the advantages of a chromatographic process for large-scale, direct separation of enantiomers. These may be summarized below.

(1) When the chiral stationary phase used is completely immobilized, no loss of valuable optically active materials can occur.
(2) The optical purity of the material used as a stationary phase is not critical, as it affects only the separation factors obtained.
(3) If the chromatographic peaks are fully resolved, both enantiomers can be quantitatively obtained, 100% pure.
(4) Recovery of eluted material can be easily automated.

The first two points are quite important and contribute to make LC a very attractive method for preparative optical resolution. Continuous operation without loss of any valuable material is highly desirable in industrial processes. However, in a scale-up of any type of LC-separation, difficult problems have to be faced. The rather limited capacity of a chromatographic sorbent means that the column load cannot be dramatically increased without affecting the performance. On the other hand, the size of a chromatographic column cannot be increased to very large dimensions without causing other problems, such as those concerned with sample application, creation of undesired void volumes, etc. In chromatography it is always necessary to make compromises. The situation is often illustrated as shown in Fig. 9.1. The

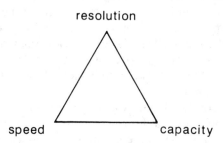

resolution

speed   capacity

Fig. 9.1 — The 'chromatographic triangle' showing that gain in one desired property will be at the expense of the others.

triangle tells us that if we want to increase the capacity, we have to sacrifice speed and/or resolution. In general, for work in the essentially linear part of the sorption isotherm, the amount injected into a column of normal capacity should be less than ca. 1 mg per g of sorbent. Therefore, a preparative column containing 1 kg of sorbent should be able to separate a maximum load of roughly 1 g of injected sample without significant loss of performance. The amount handled can be increased, but only up to the level at which the column efficiency and resolution are still adequate to yield the desired purity. An illustration of the loading capacities for various column sizes is given in Table 9.1

Table 9.1 — Examples of load capacity of columns used for various preparative purposes

Column type	Inner diameter (mm)	Use	Amount of stationary phase (g)	Approximate maximum load (mg)
Analytical	1–5	Isolation of pure samples for MS or IR (0.001–0.1 mg)	0.2–3	0.2–3
Analytical to semi-preparative	6–11	Isolation of pure samples for NMR or elemental analysis (0.1–25 mg)	3–25	3–25
Semipreparative to preparative	10–30	Small scale synthesis (0.1–1 g)	25–100	20–1000
Preparative	20–100	Large scale synthesis (1–100 g)	$100–10^4$	$100–10^4$
Industrial	100–1000	Industrial scale synthesis	$10^3–10^5$	$10^3–10^5$

It is evident that the meaning of preparative LC is very broad, including isolation of one or more particular components by collection of the corresponding fractions of eluate. This may imply isolation of only a few μg of pure product for mass spectrometry, in which case an analytical column can be used, but what is usually meant is use of LC to isolate amounts of several hundred mg or more in a single run. Columns used for sample loads of ca. 0.1–1 g are of normal laboratory size and generally present no particular technical problems. On a scale-up by, say, a factor of 100–1000, which may be required for industrial purposes, it is necessary to find suitable techniques to ensure even distribution of the sample on the top of the column, maintain a constant linear flow-rate throughout the column, and maintain reproducible column performance. This implies controlling the bed compression and eliminating void volumes, as well as mastering various contamination problems. It is not within the scope of this text to treat such matters in detail, but rather to emphasize that chromatographic scale-up is more difficult in practice than it may appear from a merely theoretical point of view.

The following part of this chapter will concentrate on the questions of which are the most desirable properties of a chiral sorbent for direct optical resolution on a true preparative scale, which sorbents are useful on the laboratory scale, and what are their particular merits and disadvantages.

From what has been said above, it is clear that continuous operation of a column will create particular demands on the column packing material. A non-compressible support, such as silica, will be needed unless an organic material of equivalent mechanical properties is available. Next, the chiral sorbent should have a high sample capacity, i.e. the number of enantiodifferentiating sorption sites per g of material should be as high as possible. It is also highly desirable that the sorbent/ mobile phase system gives a high enantiomeric separation factor for the resolution of interest, since a higher α-value will permit a greater column overload. The cost of the chiral sorbent is, of course, also an important factor in making an industrial separation process worthwhile. Further, the sorbent must be chemically inert under the conditions used, i.e. there must be no leakage of stationary phase or occurrence of hydrolytic or other reactions which may change the properties of the sorbent.

Other desirable properties, but less fundamental with respect to a particular separation, are broad applicability of a given sorbent and the possibility of effectively regulating retention times by changing the mobile phase composition.

How are these desirable properties met by the chiral sorbents currently in use for preparative purposes on the laboratory scale? Let us consider the various types of sorbents used.

9.2   CHIRAL SORBENTS WITH LARGE CAPACITY

9.2.1   Sorbents based on polysaccharides and derivatives

As already mentioned in Chapter 7, the most widely used and carefully investigated sorbent in this category is cellulose triacetate, particularly in its microcrystalline form (MCTA). It is a relatively cheap material which can be produced in large quantities by a simple acetylation procedure, but it may differ in properties from batch to batch.

Because it is used in the swollen state, it suffers from some compressibility, although small MCTA particles (5–10 μm) have been successfully used in steel columns at rather high pressures (50–100 bar) [1]. So far, however, preparative separations have been performed only with low or medium pressure systems, giving very low column efficiency. Nevertheless, single-run, g-scale complete optical resolution is possible on columns of moderate size, as illustrated by Fig. 9.2.

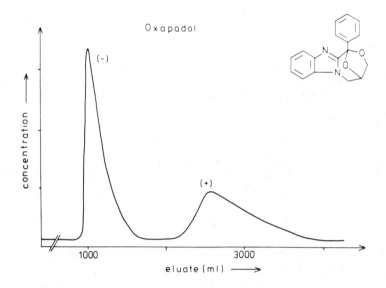

Fig. 9.2 — Separation of 2.1 g of racemic oxapadol into enantiomers by the use of a 38×700 mm column packed with 380 g of MCTA: 95% ethanol was used as eluent at a flow-rate of 90 ml/hr. Note that the time required was almost 48 hr. (Reprinted, with permission, from G. Blaschke, *J. Liquid Chromatog.*, 1986, **9**, 341, by courtesy of Marcel Dekker Inc.).

As a consequence of the successful results obtained by Okamoto *et al.* (cf. Section 7.1.1.2) in depositing CTA and other polysaccharide derivatives onto silica supports, production of improved sorbents of this kind for preparative purposes has recently begun. One important aspect, worth recalling here, is that MCTA can be used for the separation of enantiomers lacking polar groups, such as pure hydrocarbons [2]. This property, at present unique, makes this type of chiral sorbent a very useful tool for obtaining small amounts of optically active hydrocarbons which are otherwise unavailable.

A definite disadvantage with these modified biopolymers is their low compatibility with many common organic solvents. MCTA, for example, is partially dissolved by solvents such as chloroform, dioxan or acetone [3].

The deposition of a microcrystalline cellulose derivative on silica is inevitably associated with the generation of a more amorphous structure even though X-ray diffraction studies performed on these materials [4,5] show that some microcrystalline regions still persist. The effect of such a treatment on chromatographic performances has been carefully investigated by Mannschreck's group [6] who found very great differences in performance between MCTA used alone and when deposited on silica. Under identical conditions (96% ethanol), the k' values were much higher on MCTA. Also, with one exception (hexahelicene), the compounds investigated showed much higher α values on MCTA. However, the k' and α values obtained with the silica-based columns increased significantly with increasing polarity (water content) of the mobile phase. From the point of view of column efficiency, however, the silica-based columns are superior, with plate heights, H, of about 50 μm for the void-volume marker (1,3,5-tri-*tert*-butylbenzene) [5]. Their sample capacity is, on the other hand, lower than that of MCTA columns.

It is clear from Mannschreck's investigations [6] that the load capacity of MCTA is about 0.5 mg/g; with higher loading there is an abrupt decrease in k'. Further, there is a considerable increase in plate height, leading to rapidly decreasing resolution.

Another very important factor in preparative work on MCTA is the flow-rate. It has been shown [3, 7] that to minimize H, according to the van Deemter equation (cf. Section 4.2) a linear flow-rate u of ca. 0.1 mm/sec is required. In the investigation cited here, an axially compressed preparative column (40×243 mm) was used and its void volume (V_0) was found to be 191 ml (with 1,3,5-tri-*tert*-butylbenzene). This means a t_0 value of 40.5 min at the optimal flow-rate (4.72 ml/min). Consequently, it is difficult to accelerate preparative separations on finely ground MCTA, and in most cases they will take several hours or even days.

Because of its availability in large quantities and the ease by which preparative low-pressure columns can be made in the laboratory, MCTA is a very useful material. Isolation of small amounts of enantiomerically pure organic compounds is readily achieved by a simple and inexpensive kind of chiral liquid chromatography. Since such columns are easily interfaced with a polarimetric detector equipped with a standard 1-ml cell of 100 mm pathlength, optically enriched fractions of known sign of rotation are readily obtained from many racemates. Such fractions can subsequently be analysed with an analytical chiral LC system. This will then give the exact enantiomer composition of the fraction investigated, together with the elution order of the enantiomers on the analytical column. The usefulness of such a procedure is shown by Fig. 9.3. Although only partially resolved on the MCTA column (left), the lactam applied to this column gives a fairly pure (−)-enantiomer in the very first fractions taken, as seen from the analytical chromatogram (right) obtained after dilution and reinjection on a BSA-based column [8]. A fraction taken from the later part of the preparative run shows that the content of the (+)-form is *ca.* 70%.

A comprehensive investigation of MCTA with respect to its use as a chiral sorbent has been made [9]. In a mass overload experiment, it was found that the capacity ratio of the more retained enantiomer increased with sample load, showing the complex mechanism of retention on MCTA. It was also shown that column efficiency could be considerably increased at elevated temperatures and at lower flow-rates, indicating a slow mass-transfer rate in the sorbent.

9.2.2 Sorbents based on polyacrylamide and polymethacrylamide derivatives
The sorbents developed by Blaschke and co-workers (cf. Section 7.1.2) have been extensively used for semipreparative and preparative optical resolution of a variety

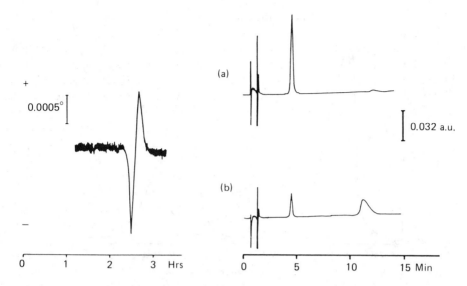

Fig. 9.3 — Enantiomeric purity determination of fractions taken from a preparative, partial separation of *rac*. 3-(*p*-chlorophenyl)butyrolactam (Baclophen-lactam) on MCTA. (Reprinted, with permission, from S. Allenmark and S. Andersson, *Chirality*, 1989, **1**, 154. Copyright 1989, Wiley-Liss, Inc.)

of polar pharmaceuticals [10]. These materials are all relatively cheap and comparatively easy to prepare even on a large scale. There may be undesirable batch to batch variations, however, because of difficulty in exactly reproducing the polymerization conditions.

These sorbents are usually used as swollen polymer beads of particle size ranging between 50 and 100 μm. They are highly compressible and can only be used in columns operating at low pressure. Recently, covalent attachment of the polymer to silica has been achieved, giving a sorbent suitable for analytical HPLC. However, in its present form it apparently suffers from a considerable loss of capacity, making it somewhat unsuitable for scaling-up to preparative work. Although the efficiency of the columns packed with the 50–100 μm polymer particles is rather low, this is often compensated for by large separation factors, making column overload possible. The performance of a column is demonstrated by Fig. 9.4. Here a solute/sorbent ratio of 1:470 is used and the second enantiomer is eluted at elevated temperature (40°C). The load capacity and elution volume used (ca. 4 litres) are comparable with those used for the separation shown in Fig. 9.2.

9.2.3 The Pirkle sorbents

Very impressive preparative optical resolutions have been obtained by Pirkle *et al.* by the use of (R)-*N*-(dinitrobenzoyl)phenylglycine as chiral selector, ionically bound to aminopropyl-silica (cf. Section 7.2.3). In his preparative runs, Pirkle used a relatively cheap, irregularly shaped, totally porous silica of 40 μm mean diameter, which was treated to incorporate 3-aminopropyl ligands [11]. The chiral selector was

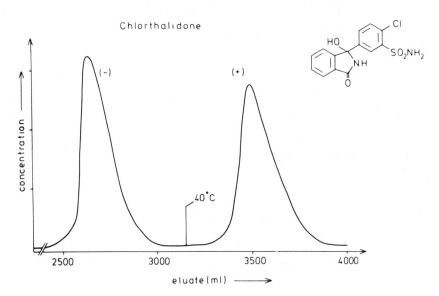

Fig. 9.4 — Separation of 0.53 g of racemic chlorthalidone into enantiomers on a 32×360 mm column packed with 250 g of optically active poly(*N*-acryloylphenylalanine ethyl ester). Toluene/dioxan (1:1) was used as eluent. (Reprinted, with permission, from G. Blaschke, *J. Liquid Chromatog.*, 1986, **9**, 341, by courtesy of Marcel Dekker Inc.).

then adsorbed on the matrix from a non-polar solvent. This sorbent naturally yielded columns with significantly lower efficiency than the analytical columns based on the much more expensive 5 μm silica. However, the overall effect on resolution was claimed to be partially compensated by the somewhat higher α values obtained (as a result of the different type of silica used!).

The sorbent fulfils the requirements of being mechanically stable as well as possessing adequate load capacity. Further, it has broad applicability and is reasonably cheap. The stability has proved to be acceptable as long as the polarity of the mobile phase is not increased too much (which could cause displacement of the selector).

Another attractive feature of this system is that various *N*-substituted aminoacids can be used as chiral ligands on the same basic matrix, giving further possibilities for optimizing a given separation system. For example, the DNB–leucine selector has been found to give superior results in many cases (cf. Section 7.2.3).

The mobile phase used in conjunction with these sorbents has usually been a hexane/2-propanol mixture containing a maximum of 20% of the polar component. Consequently, isolation of the separated enantiomers is very easily performed by evaporation of the solvent from the fractions collected.

In his first series of preparative resolutions [12], Pirkle used a 2×30 inch column (51×760 mm). The loadings were quite high, ranging between ca. 1 and 8 g. For solutes exhibiting α values larger than ca. 1.4 essentially quantitative resolution of several g of sample could be achieved. Figure 9.5 gives a typical example of such a

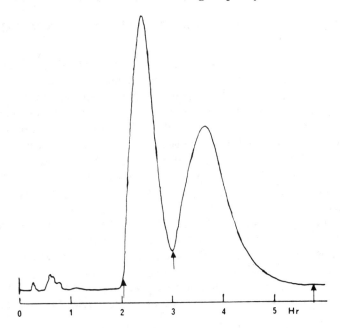

Fig. 9.5 — Chromatogram showing the resolution of 1.6 g of racemic 5-(1-naphthyl)-5-(4-pentenyl)hydantoin on a 2×30 inch column. (Reproduced from W. H. Pirkle and J. M. Finn, *J. Org. Chem.*, 1982, **47**, 4037. Copyright 1982, American Chemical Society).

resolution. Apparatus for continuous operation of this LC-system in a repetitive mode has also been constructed, and yields a resolution capacity of about 0.9 g/hr in certain cases. Recently, a further scaling-up of the system has been made [13]. This achieved resolution of 50 g of racemic methyl N-(2-naphthyl)alaninate in a single run on a large preparative column containing 13 kg of chiral packing.

The (R)-N-3,5-dinitrobenzoylphenylglycine CSP and its (S)-leucine analogue (covalently bound to aminopropyl-silica) were also used by Pirkle for preparative work by a simple flash chromatographic technique [14]. The sorbents were prepared in kg amounts from silica gel of 58 μm particle size. A linear flow velocity of ca. 50 mm/min was obtained by means of nitrogen pressure. By application of 1 g of certain racemic benzodiazepinones and phthalides, it was possible to obtain ca. 300 mg fractions of each pure enantiomer in less than 20 min.

A problem common to all preparative-scale chromatography is that it is not possible to make the column dimensions very large without decreasing the efficiency. Therefore, a certain column overload has to be accepted and the conditions optimized empirically.

Even though a rather dramatic decrease in resolution will result from a demand for increased capacity and throughput, the situation is often not as bad as it may seem. By collection and work-up of appropriate portions of the eluate, high optical purity can be obtained together with satisfactory yield. The remainder of the eluate can be concentrated and re-injected into the column in a later run.

9.3   SOME SPECIAL PROBLEMS ASSOCIATED WITH INCREASED SAMPLE LOAD

It is well known from work on preparative chromatography [15] that deviation from linear sorption isotherms will cause decreased resolution factors in elution chromatography, owing to peak distortion. In chiral chromatography, however, the complex nature of many stationary phases will create other problems not encountered in conventional, achiral chromatography. Normally, the capacity ratio (k'), which is the ratio of the analyte distributed between the stationary and mobile phase [cf. Section 4.1, Eq. (4.2)] will decrease with increasing column load above a certain limit. This is, of course, quite understandable in terms of saturation of the stationary phase, which will cause more of the analyte to reside in the mobile phase.

Now, in the case of chromatography on certain chiral sorbents, we know that the variety of binding sites available represents a scale of different affinities for the analyte, and also that the analyte, when bound to a particular site, might influence binding to another site (co-operative effects, allosteric interactions). Consequently, the effects of an increasing sample load may be more difficult to predict or rationalize. An interesting observation was recently made, however, which may serve as a good example of the complex situation often present with chiral stationary-phase systems.

During studies of preparative separation of an atropisomeric thiazolinone (rac. 3-(2-propylphenyl)-4-methyl-4-thiazolin-2-one) into its enantiomers by chromatography on microcrystalline triacetylcellulose (MCTA), Roussel and collaborators made the observation that an inversion of the capacity factors was obtained with increasing amount of sample injected [16]. Further, separate studies of the concentration dependence on k' of the isolated enantiomers gave the same result. The situation is illustrated in Fig. 9.6. The conclusions drawn from these experiments are that the two enantiomers behave quite independently on the stationary phase, meaning that they do not compete for the same binding sites, but instead bind to different sites in the supramolecular structure of MCTA. As seen in Fig. 9.6, the (+)-enantiomer becomes more retained with increasing load. This can be taken as an indication of an increased availability of sites of higher affinity for this enantiomer owing to a blockade of more easily accessible sites. In this respect the enantiomer itself can be regarded as a mobile phase additive causing increased retention (cf. Sections 7.3.2 and 9.2.1).

It is evident that a sample load dependence of this kind will further complicate the transfer of a chiral separation from an analytical to a preparative scale. It is not yet known whether a similar situation may arise on chiral sorbents prepared from non-microcrystalline cellulose derivatives deposited on silica (Chiralcel), or on sorbents made from synthetic polymers, all of which are of interest for preparative separations. It is not unlikely, however, that this could be the case.

9.4   SEPARATIONS BY CHIRAL MEMBRANES

Enantioselective liquid membrane techniques are in a strict sense outside the scope of this chapter. However, since they are based on the same type of chiral selectors as those used in chromatography, and since they represent an interesting, novel mode

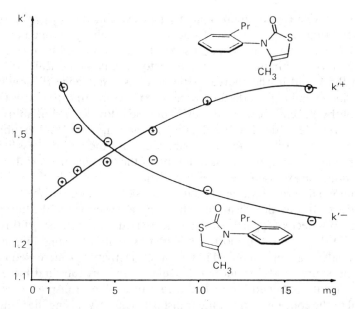

Fig. 9.6 — Variation of the capacity factors ($k'+$ and $k'-$) of the enantiomers of racemic 3-(2-propylphenyl)-4-methyl-4-thiazolin-2-one with the amount injected onto an MCTA column. (Reprinted, with permission, from C. Roussel, J.-L. Stein, F. Beauvais and A. Chemlal, *J. Chromatog.*, 1989, **462**, 95. Copyright 1989, Elsevier Science Publishers BV.)

of preparative chiral separation or, better, preparative enantiomer enrichment, they will be discussed briefly. The general principle of such techniques is given in Fig. 9.7,

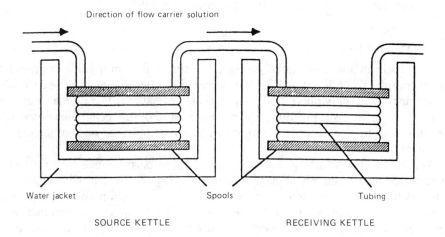

Fig. 9.7 — Illustration of the principle for enantiomer enrichment via transport through an enantioselective liquid membrane. (Reprinted, with permission, from W. H. Pirkle and E. M. Doherty, *J. Am. Chem. Soc.*, 1989, **111**, 4113. Copyright 1989, American Chemical Society.)

which shows a device for continuous enantiomer enrichment as devised by Pirkle [17]. The membrane device consists of silicone rubber tubing wrapped around two spools, each spool being immersed in a separate thermostatted bath containing a methanol/water (4:1) mixture. The racemate to be enantiomerically enriched is dissolved in the liquid in the source kettle and has a tendency to slowly diffuse through the silicone tubing. Dodecane containing the chiral transport agent [here an (S)-N-(1-naphthyl)leucine octadecyl ester] is pumped slowly (*ca.* 1 ml/min) through the tubing in a recycle mode. The chiral transport agent impregnates the tubing walls and increases the transport rate of the analyte from the source vessel into the dodecane stream. This rate increase is different for the two enantiomers, leading to a certain enantiomeric excess in the receiving kettle.

The technique described was applied to a series of N-(3,5-dinitrobenzoyl)amino acid derivatives. It was demonstrated that: (a) a lowering of the temperature increases enantioselectivity, partly owing to decreased achiral transport through the membrane, (*b*) the achiral transport rate increases with increasing lipophilic character of the analyte, giving a reduced net enantioselectivity, (*c*) increased methanol content of the source vessel reduces the achiral transport rate and increases the enantiomeric purity obtained, and (*d*) the rate of analyte transport is essentially proportional to the concentration of the transport agent. By raising the temperature of the receiver liquid (to 50°C) a larger portion of the enriched material can be collected there. All (S)-forms of the compounds investigated were enriched by this technique; the highest rate ratio, 7.6, was found for DNB-leucine butyl amide. This gives an e.e. of 0.77. Typical transport rates at the scale used in the experiments (50 mg of racemate, 100 mg of chiral transport agent) were of the order 0.5–4 mg/hour.

Because of the ease of continuous operation, techniques based on this principle, when optimized, could be most valuable for preparative purposes in the future. Other chiral transport agents, also well known from chromatographic work, have been utilized previously in enantioselective membrane systems, and include, among others, tartaric acid derivatives [18] and cyclodextrins [19].

9.5 CONCLUSION

There are still many problems to be solved in the field of preparative chromatographic optical resolutions, although considerable progress has now been made. While many semipreparative scale separations (of say <0.1 g of material) are relatively straightforward with the aid of one or more of the techniques described, operations on the g or kg scale, requiring an automated process on a large column, are still at the experimental stage.

At present, the Pirkle-type sorbents give the best performance with respect to capacity, speed and column efficiency. A further advantage is that they can be used with volatile organic solvents, which facilitates the isolation of the separated antipodes. The limitation lies in the cost associated with large scale production of these sorbents, as the chiral selectors and coupling reagents needed are quite expensive.

By far the most readily available and cheapest sorbent is MCTA. There is no doubt that heterogeneous acetylation of microcrystalline cellulose can be quite easily performed on a really large scale, and the product is relatively cheap to make and use. The limitation here is probably not the cost of large columns, but rather their low efficiency, slow operation and often poorly reproducible performance.

BIBLIOGRAPHY

L. R. Snyder and J. J. Kirkland, *Introduction to Modern Liquid Chromatography*, 2nd Ed., Wiley, New York, 1979.

C. E. Reese, in *Techniques in Liquid Chromatography*, C. F. Simpson (ed.), Wiley, New York, 1982, p. 97.

C. F. Poole and S. A. Schuette, *Contemporary Practice of Chromatography*, Elsevier, Amsterdam, 1984.

A. Mannschreck, H. Koller and R. Wernicke, Microcrystalline Cellulose Triacetate, a Versatile Stationary Phase for the Separation of Enantiomers, in *Kontakte (Darmstadt)*, 1985, No. 1, 40.

W. H. Pirkle, Chromatographic Separation of Enantiomers on Rationally Designed Chiral Stationary Phases, in *Chromatography and Separation Chemistry*, S. Ahuja (ed.), ACS, Washington DC, 1986, p.101.

B. A. Bidlingmeyer (ed.), *Preparative Liquid Chromatography*, Elsevier, Amsterdam, 1987.

V. R. Meyer, Some Aspects of the Preparative Separation of Enantiomers on Chiral Stationary Phases, *Chromatographia*, 1987, **24**, 639.

E. Grushka (ed.), *Preparative-Scale Chromatography*, Chromatographic Science Series, vol. 46.

M. Zief, Preparative Enantiomeric Separation, *Chromatog. Sci.*, 1988, **40**, 337.

REFERENCES

[1] K. R. Lindner and A. Mannschreck, *J. Chromatog.*, 1980, **193**, 308.

[2] H. Scherübl, U. Fritzsche and A. Mannschreck, *Chem. Ber.*, 1984, **117**, 336.

[3] A. Mannschreck, H. Koller and R. Wernicke, *Kontakte (Darmstadt)*, 1985, No. 1, 40.

[4] T. Shibata, I. Okamoto and K. Ishii, *J. Liquid Chromatog.*, 1986, **9**, 313.

[5] K.-H. Rimböck, M. A. Cuyegkeng and A. Mannschreck, *Chromatographia*, 1986, **21**, 223.

[6] K.-H. Rimböck, F. Kastner and A. Mannschreck, *J. Chromatog.*, 1985, **329**, 307.

[7] H. Koller, K.-H. Rimböck and A. Mannschreck, *J. Chromatog.*, 1983, **282**, 69.

[8] S. Allenmark and S. Andersson, *Chirality*, 1989, **1**, 154.

[9] R. Isaksson, P. Erlandsson, L. Hansson, A. Holmberg and S. Berner, *J. Chromatog.*, 1990, **498**, 257.

[10] G. Blaschke, *J. Liquid Chromatog.*, 1986, **9**, 341.

[11] W. H. Pirkle and J. M. Finn, *J. Org. Chem.*, 1982, **47**, 4037.

[12] W. H. Pirkle, J. M. Finn, B. C. Hamper, J. Schreiner and J. R. Pribish, in *Asymmetric Reactions and Processes in Chemistry*, E. L. Eliel and S. Otsuka (eds.), ACS, Washington DC, 1982, Chap. 18.

[13] W. H. Pirkle, presented at 11th Int. Symp. on Column Liquid Chromatography, Amsterdam 1987.

[14] W. H. Pirkle, A. Tsipouras and T. J. Sowin, *J. Chromatog.*, 1985, **319**, 392.

[15] A. M. Katti and G. Guiochon, *J. Chromatog.*, 1990, **499**, 21.

[16] C. Roussel, J.-L. Stein, F. Beauvais and A. Chemlal, *J. Chromatog.*, 1989, **462**, 95.

[17] W. H. Pirkle and E. M. Doherty, *J. Am. Chem. Soc.*, 1989, **111**, 4113.

[18] V. Prelog and M. Dumic, *Helv. Chim. Acta*, 1986, **69**, 5.

[19] D. W. Armstrong and H. L. Jin, *Anal. Chem.*, 1987, **59**, 2237.

10

Future trends

It is evident that the rapid development of chromatographic techniques in general will also lead to improvements in the applicability of optical resolution by chromatography. In the following section, some areas of particular interest for the future are discussed.

10.1 NEW DETECTOR SYSTEMS

Detectors often combine high sensitivity with a particular selectivity. Typical examples are the thermionic ionization detectors (TID), specific for nitrogen and phosphorus, as well as the electron capture detectors (ECD) used in GC, and the fluorimetric and electrochemical detectors now common in LC. Combination of such detectors with chiral separation columns will further increase the applicability of analytical-scale optical resolution. One obvious reason is that the risk of accidental peak overlap in chromatograms of more complex samples, leading to erroneous results with respect to enantiomer composition, will be minimized.

The minimum detectable quantity of nitrogen with a TID is of the order of 10^{-13} g/sec, which in practice means that it can be used for determination of nitrogen compounds at sub-picomole levels [1]. Such detection techniques in chiral GC enable determination of enantiomer composition with ultra-small sample amounts, which is of great potential importance in many fields of bioanalysis. So far, however, this type of *detector selectivity* has not been very much explored in chromatographic optical resolution.

The ease and versatility of use of GC–MS, however, has encouraged its application in combination with chiral capillary columns. Frank *et al.* [2] showed as early as 1978 that the thermal stability of a 'Chirasil-Val' column permitted its use in a system coupled to a mass spectrometer and demonstrated its importance in studies of metabolism and related fields. It may be instructive to show part of their results in some detail here.

As mentioned previously (cf. Section 8.3.2.1) L-dopa, as precursor of the catecholamines, is important not only as an endogenous compound but also as a drug for the treatment of Parkinson's disease, whereas the D-form is not metabolized and is only slowly absorbed. Monitoring of L-dopa is therefore often required and can be performed by the use of 'Chirasil-Val' after derivatization. Figure 10.1a shows the

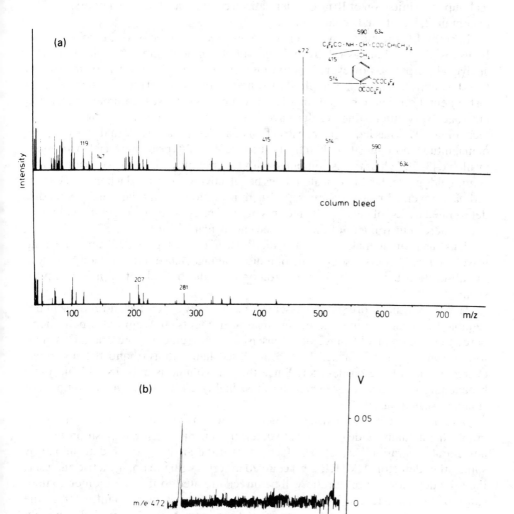

Fig. 10.1 — Chiral GC–MS of L-dopa O,N-PFP isopropyl ester. (a) Mass spectrum of the GC peak. (b) Single ion monitoring (at m/z 472). (Reprinted, with permission, from H. Frank, G. J. Nicholson and E. Bayer, *J. Chromatog.*, 1978, Elsevier Science Publishers, B.V.)

mass spectrum of the O,N-PFP isopropyl ester of L-dopa after passage through the column, and also that of the normal column bleed. The peak at m/z 472 can be used for single ion monitoring (SIM) of the eluate. An illustration of the high sensitivity thus achieved is given by Fig. 10.1b, which shows the chromatogram of only 30 pg of L-dopa, with detection by this technique. As the signal-to-noise ratio is ca. 10, it is evident that determination of trace amounts is possible by this powerful method. A scale-up to an injection of 10 ng of a derivatized commercial L-dopa sample showed it to contain 2.9% of the D-enantiomer.

In chiral LC the most obvious enantiomer-selective detector is the polarimeter. Its use is ideal as long as relatively large sample amounts are available. For most analytical purposes, however, it is too insensitive even with the use of microcells (40 μl volume). The same applies to other chiroptical detectors, such as CD-instrumentation. Though any dramatic improvements in these techniques are hard to envisage, very interesting results have been found recently in the use of laser technology. This has led to the construction of a micropolarimeter that is two orders of magnitude more sensitive than the best conventional-type commercial instrument available [3–5]. When this is combined with LC, not only can many optically active compounds present in biological samples be selectively monitored when eluted from ordinary reversed-phase columns [6–10], but elution orders can also be readily determined by LC of very small amounts of racemates on analytical chiral columns. A commercial instrument of this type became available in 1988.

This instrument makes use of a diode laser operating at 820 nm. The long wavelength is of course, something of a disadvantage, since in the region, far from the absorption band of the chiral chromophore, the optical rotation is relatively small.

However, other more suitable laser sources are very expensive and at present not suitable for incorporating into commercial instruments. It should also be remembered that the main advantage of the laser-based instrument is the dramatic signal-to-noise improvement caused by the increased light intensity and the low noise characteristics of the diode-laser. Since the cell volume is only 18.5 μl, no band broadening will occur even when the most highly efficient analytical columns of standard size are used.

This development of polarimetric detectors for analytical-scale liquid chromatography has actually made it possible to determine enantiomeric composition without any chiral separation [11,12]. The fact is that if the analyte can be determined by some other detector (UV, RI, etc.) coupled in series with the polarimetric analyser, then the detector response ratio will be directly related to the enantiomer composition of the analyte. To use this technique, however, a pure enantiomer of the analyte is needed for the construction of a calibration curve. It seems likely, though, that the technique should be highly useful in many cases where a method not requiring any chiral separation is considered preferable. It has recently been used for studies of enantioselectivity in enzyme-catalysed reductions of some bicyclic ketones [13].

Another application area of these diode-laser-based polarimetric detectors is the determination of natural optically active components in complex mixtures. An illustrative example is given in Fig. 10.2, which clearly shows the selectivity obtained in optical rotation monitoring.

Since the detection level obtained by the use of laser-based polarimetric detectors is in the low ng range [14] and the information content obtained from optical rotation

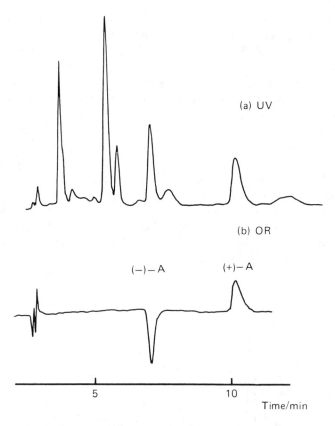

(a) UV

(b) OR

(−)−A   (+)−A

5   10

Time/min

Fig. 10.2 — UV (*a*) and diode-laser-based polarimetric detection (*b*) of the enantiomers (15 mg of each) of 2,2-dimethyl-1-phenyl-1-propanol in a complex reaction mixture. Daicel OD column; hexane/2-propanol (95/5). (Reprinted, with permission, from D. M. Goodall, D. K. Lloyd and Z. Wu, in *Recent Advances in Chiral Separations*, D. Stevenson and I. D. Wilson, eds., Proc. Chromatog. Soc., Int. Symp. Chiral Separ., Univ. of Surrey, 1989. Copyright 1990, Plenum Press.)

measurement is high, a combined, sequential UV and polarimetric detection technique has been tried for identification purposes [15]. It was applied to a series of PTH-amino acids where the amino acid side chain was not expected to influence the UV molar absorptivity. Therefore, the measured optical activity-to-UV ratio can be assumed to be proportional to the specific rotation of each eluted PTH-amino acid in the given mobile phase. A computerized data acquisition system was used, permitting readings from the detectors every 0.1 sec. At pH 5 the optical activity-to-UV ratios determined in this way ranged from -15.5 ± 0.8 (thr) to $+4.8\pm0.1$ (met). It was concluded that the precision obtained in these ratios was sufficient to permit identification of all the twenty PTH-amino acids used, when separated in an isocratic 10-min run with an acetate buffer of pH 5.0 as the mobile phase. This means that when k'-values were too close, identification could be made by the optical activity-to-UV ratio. The technique is expected to facilitate the Edman degradation used in protein and peptide sequencing, which so far has relied upon identification of the PTH-amino acids only from chromatographic retention data.

A number of possibilities exist for future improvement of microanalytical scale optical resolution by LC. First of all, achiral pre-column as well as post-column derivatization can be used to enhance the sensitivity of detection and/or the retention properties. Secondly, there is a rapid development in column miniaturization and in new techniques of detection.

Electrochemical detectors, introduced in the early 1970s by Adams and Kissinger [16], have been developed into powerful tools for bioanalytical research [17]. Their successful combination with reversed-phase LC separation methods makes them highly attractive for certain applications of enantioseparation where aqueous phase conditions are used. The first monitoring of an optical resolution by electrochemical detection was performed by Allenmark *et al.* [18], who studied the resolution of two aromatic amino-acids (5-hydroxytryptophan and 3-hydroxykynurenine) on a BSA-agarose column. Figure 10.3 shows an example of the chromatographic performance when a 7 μm 'Resolvosil' column is used.

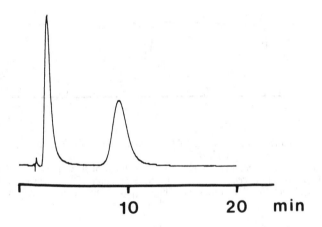

Fig. 10.3 — Electrochemical monitoring of the enantiomers of D,L-3-hydroxykynurenine eluted from a 'Resolvosil' column.

Similarly, application of luminescence detectors can be expected to gain importance in the future. The techniques already available for transformation of amines and amino-acids into fluorescent derivatives are highly useful in chiral chromatography. Thus, dansylamino-acids undergo optical resolution on several types of chiral LC columns [19–21] and can then be detected in trace amounts by various luminescence techniques. Perhaps one of the potentially most interesting is chemiluminescence [22], by which it has been shown that less than one fmole of dansylamino-acid can be determined [23,24]. Compared to conventional photoexcitation (as used in fluorescence detection), chemically induced excitation has the advantage of giving a much lower background, which significantly improves the signal-to-noise ratio, but also has the disadvantage of requiring post-column addition of reagents (usually hydrogen peroxide and an oxalate) together with a highly efficient mixing and rapid transfer into the detector cell. However, the extreme selectivity and sensitivity of the technique will probably make it very important for detection of ultra-small quantities of LC-separated enantiomers in future applications.

10.2   COLUMN IMPROVEMENTS

The introduction of the highly flexible, internally polymer-coated fused-silica capillary columns in GC has improved convenience without diminishing column efficiency. The surface of these columns is quite inert, owing to its low content of alkali-metal oxides. While this inertness means that special deactivation procedures to reduce tailing phenomena are generally not required prior to coating with the stationary phase, it may also cause problems in retaining certain phases. Numerous studies have been made in the field of capillary column treatment and coating, and considerable improvements are still to be expected. Today many of the most useful non-chiral stationary phases for glass capillary columns are also available on fused-silica columns. Of great importance in this respect are the techniques already developed and those still under development for the preparation of immobilized (also called 'bonded' or 'non-extractable') phases. Compared to a stationary phase film, such phases are completely resistant to extraction by organic solvents. Such 'immobilizations', which are generally effected in such a way as to cause a combination of cross-linking and surface bonding, usually have a number of advantages. These are a much lower 'bleed' at higher temperatures, the possibility of preparing films thicker than those normally obtainable by conventional coating techniques, and easy solvent rinsing of the column; consequently, columns with such immobilized phases are important in supercritical fluid chromatography.

Of the chiral GC columns commercially available to date, 'Chirasil-Val' and the Sumipax CC OA series (see Appendix) are produced in fused-silica versions. Thanks to the polysiloxane anchoring of the chiral selector in 'Chirasil-Val', these columns can also be used at high temperatures (up to ca. 200°C).

With the development of techniques to achieve more thermally stable chiral capillary columns, an increased use of chiral GC/MS combinations for various analytical and bioanalytical purposes can be expected in the future.

So far, most work in the field of chiral LC has been concentrated on the synthesis and evaluation of new and improved CSPs and chiral mobile phase additives. As pointed out previously, in LC it is generally much easier to improve enantiomeric separation by changing the mobile phase, whereas in GC high enantiomeric resolution is obtained by virtue of high column efficiency, despite the generally rather low α-values. Many efforts have been made to increase column efficiency in reversed-phase LC, mainly by trying to transfer the advantages of microbore open tubular columns to LC-technology [25]. Successful use of such columns, however, requires a system equipped with injectors and detectors of extremely small volumes (in the low nl range for the injector). Very high column efficiencies have been obtained in well-designed chromatographic systems [26], but the low flow-rates required make the processes highly time-consuming.

Microbore LC may be expected to be useful for obtaining higher resolution and improved peak capacity for chiral separations. With the current conventional techniques, α-values of at least 1.2 are normally required for complete separation. If this value could be reduced to, say 1.1 or even better 1.05, it would mean that a highly increased number of complete (baseline) optical resolutions could be achieved by a given CSP. In combination with selective and sensitive detector systems (e.g. based on laser-induced fluorescence, chemiluminescence, electrochemistry or mass spectrometry) this would significantly widen the analytical capability.

Some interesting attempts in this direction have recently been made. In one of these, microcapillaries (fused-silica, 0.26×106 mm, or glass-lined stainless-steel, 0.30×144 mm) packed with $3\,\mu$m C_{18}-silica were used for separation of twelve pairs of dansylamino-acids in a single run with 12.5mM β-CD as a chiral mobile phase additive [27]. All the D-enantiomers were eluted before the L-forms and enantiomeric separation factors were in the range 1.5–1.12.

A similar study of dansylamino-acids, but with the use of bonded β-CD ($5\,\mu$m spherical particles) in fused-silica capillary tubing (0.25×1900 mm), was recently made [28]. Because of the large theoretical plate numbers obtained with this system, resolution factors R_s greater than 2.00 were found for many of the compounds. Thus, dansyl-D,L-valine gave the highest resolution, R_s=2.83 at an α-value of 1.39.

In the preparative field, chiral sorbents with high α values are necessary, as we have seen, in order to make it possible to overload the column to a certain extent. Values of α exceeding 10 have already been obtained with certain sorbent/solute/solvent combinations. Some examples are given in Table 10.1.

Table 10.1 — Some unusually high enantiomeric separation factors obtained in chiral liquid chromatography

Sorbent	Solute	Solvent	α	Reference
MCTA	3-(N'-methylpiperidine-2'-thion-3'-yl)-1-methylindole	ethanol (96%)	16.7	[29]
MCTA	1-(1-N,N-dimethylcarbamoylethyl)-2-methylindole	ethanol (96%)	12.7	[29]
BSA-silica	D,L-kynurenine (3-(2-aminobenzoyl)alanine	phosphate buffer	39	[30]
BSA-silica	N-(p-nitrobenzoyl)-D,L-alanine	phosphate buffer	17	[31]
(S)-N-(2-naphthyl)alanine-silica	N-(3,5-dinitrobenzoyl)leucine butylamide	10% 2-propanol in hexane	15.4	[32]

Theoretically, a doubling of $\Delta\Delta G$ means that the related α value is squared (cf. Sections 4.1 and 5.4). An increase in $\Delta\Delta G$ by a factor of two is obtained if a solute can interact with two (identical) binding sites on the sorbent instead of only one. If a racemate is derivatized by means of a bifunctional reagent in such a way that the bis-derivative (of the reagent) possesses two end-groups formed by the racemate, the three possible derivatives will be (R,R), (R,S) (the meso-form) and $(S–S)$ in a 1:2:1 ratio. Suppose now that the (S)-enantiomer is the most strongly retained on a particular column and that the α-value is 10. If the two binding end-groups of the (S,S)-derivative can interact simultaneously and independently (a flexible derivative is assumed) with two selectors or identical binding sites of the chiral sorbent, the α value (relative to the enantiomeric (R,R)-derivative) will be 100, owing to the doubling of $\Delta\Delta G$.

Actually, such high α values of bis-derivatives have been experimentally observed [33]. Pirkle *et al.* were able to obtain an α value of 121 for the enantiomeric bis-derivative, **1**, formed by coupling of N-(3,5-dinitrobenzoyl)-D,L-leucine to 1,10-diaminodecane. The sorbent used was the N-(2-naphthyl)valine phase **2**.

1

2

It is readily appreciated that with such high separation factors, optical resolutions might be achieved in a very simple batch adsorption process. A solution of the racemate could be filtered through the chiral sorbent contained in a Büchner funnel. After washing, desorption from the sorbent could be easily achieved by change of solvent. The first and second filtrate should then contain the (R,R) and (S,S) enantiomers in high optical enrichment. Provided the bonds are easily hydrolytically cleaved, the original compounds in high optical purity could be liberated from the derivatives.

The optimum length of the bridging chain in these *bis*-derivatives has been determined from an extensive study of a homologeous series [34]. It was found that the separation factor for the (R,R)-/(S,S)-enantiomeric pair reached a maximum value when the number of interconnecting methylene groups was eight (cf. compound **1**).

Finally, the importance of reducing the content of free silanol groups on a chiral silica sorbent has been clearly demonstrated by Pirkle [35]. A significantly improved column could be obtained by a simple procedure whereby a dichloromethane solution of a silanol end-capping reagent (hexamethyldisilazane or a mixture of trichloromethylsilane and triethylamine) was pumped through the column at a rate of 0.5 ml/min. The effect of this treatment was mainly shown as an increased enantioselectivity (higher α-values). As expected from theory, the greatest improvement was found with the use of relatively nonpolar mobile phase systems and with columns having the most incomplete surface coverages of their sorbents.

The role of the end-capping from the retention mechanistic point of view is not yet certain. However, the interesting observation that in most cases a pronounced effect in the form of increased k_2'-values was obtained, supports the assumption that the terminal chiral selector has been made more available for direct interaction with the analyte. Since the sorbent investigated was a C-11-bonded (S)-undecanyl-N-(2-naphthyl)alaninate (i.e. a sorbent containing long strands), it is very likely some of these are backfolded by interaction with the free silanol groups. A major effect of end-capping would then be a "straightening" of the strands making the terminal chiral selectors more available for interaction as illustrated by Fig. 10.4.

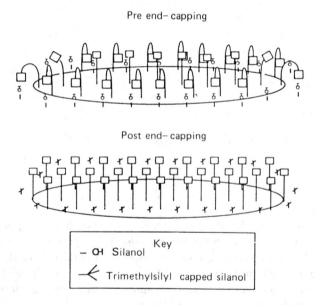

Fig. 10.4 — Illustration of the effect of end-capping on the orientation of strands of bonded chiral selector on a silica surface. (Reprinted, with permission, from W. H. Pirkle and R. S. Readnour, *Chromatographia*, 1991, **31**, 129. Copyright 1991, Friedr. Vieweg & Sohn Verlagsgesellschaft mbH.)

10.3 SUPERCRITICAL FLUID CHROMATOGRAPHY

For some time there has been a general interest in supercritical fluid chromatography (SFC) because of the advantages inherently present in this technique [36]. Since supercritical fluids have physical properties intermediate between those of liquids and gases, they are of interest for use as mobile phases, and SFC is often regarded as a hybrid of GC and LC. The viscosity of such fluids is lower than that of ordinary liquids, which makes possible the rapid diffusion of solutes in the mobile phase. This will, of course, influence the chromatographic flow-rate that can be used in order to achieve a given resolution. Another obvious advantage is the fast evaporation of the mobile phase at atmospheric pressure. Consequently, the most promising features of SFC are the possibilities of fast chromatography without loss of resolution, easy interfacing with mass spectrometry and GC detectors, and fast and simple solute recovery from fractions collected in preparative work.

The most widely studied solvent in SFC is carbon dioxide [37]. Just as in LC, the mobile phase solvent power can be adjusted by addition of a modifier.

SFC is still under development for a variety of applications and has recently been used for some chromatographic optical resolutions, and compared to that of LC with the same CSP. It was found by Hara *et al.* [38] that the enantiomers of some racemic *N*-acetylamino-acid *tert*-butyl esters could be completely separated in less than 4 min on an (*N*-formyl-L-valyl)aminopropyl-silica CSP with supercritical carbon dioxide and a small amount of methanol as modifier.

In other studies, Pirkle-type CSPs were used in SFC (and in subcritical fluid chromatography) for optical resolution of some racemic amides [39] and phosphine oxides [40]. Very similar retentions, stereoselectivities and efficiencies were obtained as for LC, which implies that the techniques can be used interchangeably.

Several studies of other chiral phases in super- and subcritical fluid chromatography have been carried out recently [41–48] and the topic has been reviewed [49].

10.4 ELECTROKINETIC CAPILLARY CHROMATOGRAPHY

Although electrophoretic techniques are usually treated as a subject separate from chromatography, great similarities exist especially when capillary methods are used. Here separation can sometimes be regarded as chromatography driven by an applied electric field rather than by an applied pressure. Accordingly, capillary electrophoresis might also have been given the term electrokinetic capillary chromatography.

The impressive resolution often achieved in capillary electrophoresis, together with the possibility of detecting extremely small amounts by electrochemical [50] or laser-induced fluorimetric [51] techniques, has led to an increased interest in this area. It is therefore not surprising that interest also has been focused on applications to the problem of enantiomer separation.

The first report about a chiral separation based upon a capillary electrophoretic technique appeared in 1988 [52]. The principle used was based on incorporation of a cyclodextrin into the polyacrylamide matrix occupying the capillary. The technique was applied to a series of dansylamino acids, which were detected on-column by means of a microbore UV system. With the use of α-, β- and γ-cyclodextrin, respectively, in the gel matrix under otherwise identical conditions, only a small retardation and no resolution were obtained in the first case. This is as expected, since the α-CD cavity is too small to incorporate the analyte. The largest retardation and highest resolution were obtained with β-CD, although the γ-isomer also yielded chiral separation. The cyclodextrin did not alter the order of retention of the dansylamino acids investigated. Since the L-form within each enantiomeric pair forms the weakest complex with the cyclodextrin, all D-forms were found to be the most retained. The system operated with amazing efficiency: at a field strength of ca. 700 V/cm, up to 100 000 theoretical plates could be obtained in the 15-cm capillary used. Thus, racemic dansyl-serine was resolved with an R_s of 6.4, although the α-value obtained was only 1.12.

Quite another principle was used by Hara and collaborators [53] who utilized incorporation of the analyte into chiral micelles. The equilibria forming the basis of the separation are outlined in Scheme 10.1. The chiral micelle-forming compound is the sodium salt of N-dodecanoyl-L-valine (SDVal), which will undergo self-aggregation in a neutral buffer since it exists there as an anionic amphiphile. To reduce the critical micelle concentration (cmc), however, sodium dodecyl sulphate (SDS) was used as a second component (1/1 ratio) for the formation of co-micelles. Now these chiral, charged micelles have a tendency to incorporate a substrate in a rapid dynamic equilibrium process. Since the mixed micelle is chiral, it can be expected to incorporate the enantiomers of a racemic analyte to different extents. Further, under an electric field there will be a difference in the migration rate through the capillary between the micellar phase and the aqueous phase. Consequently, the separation process is based on the differential distribution between these two phases, which are moving relative to one another. Hence, the process is a partition mode of chromatography, although driven by an electric field.

When this principle was applied to a mixture of four racemic amino acids, as the N-(3,5-dinitrobenzoyl)-O-isopropyl ester derivatives, it was found that, under the conditions used, the analytes separated in order of increasing hydrophobicity, which means that, as expected, the more hydrophobic compounds spent more time in the micelle during the process. As shown by Fig. 10.5, a fairly good resolution of each pair of enantiomers is also achieved. All D-forms consistently migrate through the column faster than the L-forms, showing that the latter have greater affinity for the chiral micelles.

These examples show that capillary electrophoretic techniques may become powerful tools for enantiomer analysis, particularly at low analyte levels, in the near future.

10.5 COMPUTER-AIDED OPTIMIZATION OF MOBILE-PHASE SYSTEMS FOR CHROMATOGRAPHIC CHIRAL RESOLUTIONS

The difficulty in predicting the results of an attempt to separate a pair of enantiomers by chromatography makes optimization of the conditions for such a separation a very arduous task. Moreover, this problem increases with the number of variables that can be changed in the phase system. Therefore, separations based on the use of more complex CSPs like biopolymers are the most difficult to optimize. In fact, for a majority of the chromatographic optical resolutions described in the literature no systematic attempts to improve the result were undertaken.

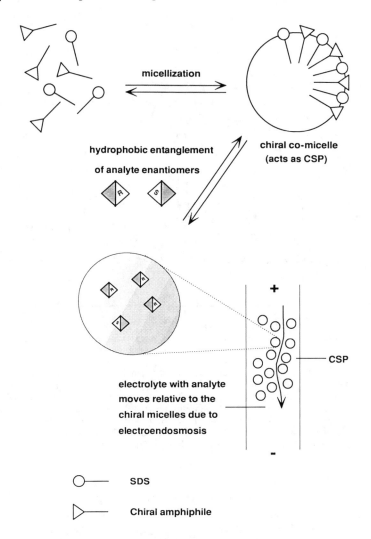

Scheme 10.1 — Basic principle of the separation of enantiomers by use of a chiral micellar phase and electrokinetic chromatography.

The "optimal" conditions to be found are somewhat arbitrary. A chromatographic response function (CRF), describing the chromatogram in terms of resolution and analysis time according to Eq. (10.1) has been proposed [54,55]:

$$CRF = (R_S/1.5)^2 - 0.25(T_L - 16) - 0.05(7 - T_F) \tag{10.1}$$

Here T_L and T_F denote the retention times of the last and first enantiomer, respectively. The factors 0.25 and 0.05 are found empirically.

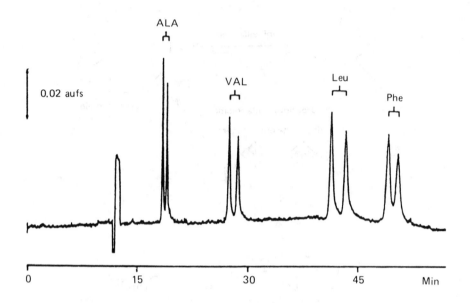

Fig. 10.5 — Electrokinetic capillary chromatography of a mixture of the N-(3,5-dinitrobenzoyl) O-isopropyl ester derivatives of the amino acids alanine, valine, leucine and phenylalanine. Column: fused silica tubing (50-μm i.d., 40-cm length); micellar solution: 12.5mM SDVal and SDS each in 25mM borate/50mM phosphate buffer (pH 7.0) containing 10% of methanol; applied voltage: *ca.* 10.5 kV, current: 26 μA, detection: UV at 254 nm; temperature *ca.* 20°C. (Reprinted, with permission, from A. Dobashi, T. Ono, S. Hara and J. Yamaguchi, *Anal. Chem.*, 1989, **61**, 1984. Copyright 1989, American Chemical Society.)

In principle, determining the CRF as a function of the mobile phase composition allows a CRF contour map to be obtained, in which the maximum CRF peak corresponds to the optimum mobile phase. In the modified sequential simplex (MSS) approach, the CRF values for a set of three different conditions are calculated. The next condition to be used is then determined by the result thus found. In this way a diagram can be constructed leading to a very narrow area where an optimum CRF is found. This kind of optimization has been carried out as a two-parameter problem where pH and organic modifier content of the mobile phase have been varied. An example, illustrating the procedure, is shown in Fig. 10.6. It is likely that such computer-assisted multivariate analysis techniques for optimization of chiral resolutions in complex phase systems will gain in importance in the future.

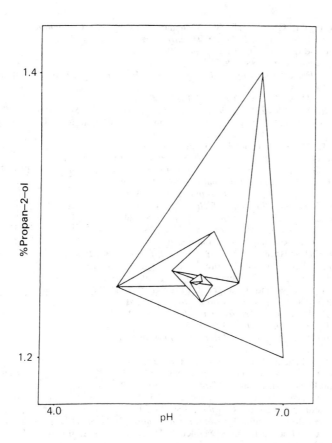

Fig. 10.6 — Optimization of the resolution of rac. oxamniquine on an AGP-based column by use of an MSS algorithm. Note the very distinct optimum that can be found in this way. (Reprinted, with permission, from T. A. G. Noctor, A. F. Fell and B. Kaye, in *Chiral Liquid Chromatography* (W. J. Lough, ed.), Blackie, Glasgow & London 1989, p. 239.

BIBLIOGRAPHY

C. F. Poole and S. A. Schuette, *Contemporary Practice of Chromatography*, Elsevier, Amsterdam, 1984.

P. Kucera (ed.), *Microcolumn High Performance Liquid Chromatography*, Elsevier, Amsterdam, 1984.

R. P. W. Scott (ed.), *Small Bore Liquid Chromatography Columns*, Wiley, New York, 1984.

M. Novotny and D. Ishii (eds.), *Microcolumn Separations*, Elsevier, Amsterdam, 1985.

I. S. Krull, Recent Advances in New and Potentially Novel Detectors in High-Performance Liquid Chromatography and Flow Injection Analysis, in *Chromatography and Separation Chemistry*, S. Ahuja (ed.), ACS, Washington DC, 1986, p. 137.

R. D. Smith, B. W. Wright and H. R. Udseth, Capillary Supercritical Fluid Chromatography and Supercritical Fluid Chromatography–Mass Spectrometry, in *Chromatography and Separation Chemistry*, S. Ahuja (ed.), ACS, Washington DC, 1986, p. 260.

H. Parvez, M. Bastart–Malsot, S. Parvez, T. Nagatsu and G. Carpentier (eds.), *Electrochemical Detection in Medicine and Chemistry, Progress in HPLC*, Vol. 2, VNU Science Press, Utrecht, 1987.

REFERENCES

[1] J. A. Lubkowitz, J. L. Glajch, B. P. Semonian and L. B. Rogers, *J. Chromatog.*, 1977, **133**, 37.
[2] H. Frank, G. J. Nicholson and E. Bayer, *J. Chromatog.*, 1978, **146**, 197.
[3] R. E. Synovec and E. S. Yeung, *Anal. Chem.*, 1985, **57**, 2606.
[4] E. S. Yeung, *Chem. Anal.*, 1986, **89**, 204.
[5] E. S. Yeung, in *Detectors for Liquid Chromatography*, E. S. Yeung, ed., Wiley–Interscience, New York, 1986.
[6] D. K. Lloyd and D. M. Goodall, *Chirality*, 1989, **1**, 251.
[7] D. M. Goodall, D. K. Lloyd and Z. Wu, in *Recent Advances in Chiral Separations*, D. Stevenson and I. D. Wilson, eds., Proc. Chromatog. Soc., Int. Symp. on Chiral Separations, Univ. Surrey, England, 1989.
[8] D. K. Lloyd, D. M. Goodall and H. Scrivener, *Anal. Chem.*, 1989, **61**, 1238.
[9] Z. Wu, D. M. Goodall and D. Lloyd, *J. Pharm. Biomed. Anal.*, 1990, **8**, 357.
[10] Z. Wu, D. M. Goodall, D. Lloyd, P. Massey and K. C. Sandy, *J. Chromatog.*, 1990, **513**, 209.
[11] B. D. Scott and D. L. Dunn, *J. Chromatog.*, 1985, **319**, 419.
[12] B. H. Reitsma and E. S. Yeung, *J. Chromatog.*, 1986, **362**, 353.
[13] J. Leaver and G. Foster, *Biotechnol. Tech.*, 1989, **3**, 179.
[14] D. R. Bobbitt and E. S. Yeung, *Appl. Spectrosc.*, 1986, **40**, 407.
[15] K. C. Chan and E. S. Yeung, *J. Chromatog.*, 1988, **457**, 421.
[16] P. T. Kissinger, C. J. Refshauge, R. Dreiling and R. N. Adams, *Anal. Lett.*, 1973, **6**, 465.
[17] P. T. Kissinger, C. S. Bruntlett and R. S. Shoup, *Life Sci.*, 1981, **28**, 455.
[18] S. Allenmark, B. Bomgren and H. Borén, *J. Chromatog.*, 1982, **237**, 473.
[19] Y. Taphui, N. Miller and B. Karger, *J. Chromatog.*, 1981, **205**, 325.
[20] B. Feibush, M. J. Cohen and B. Karger, *J. Chromatog.*, 1983, **282**, 3.
[21] S. Allenmark and S. Andersson, *J. Chromatog.*, 1986, **351**, 231.
[22] K. Imai and R. Weinberger, *Trends Anal. Chem.*, 1985, **4**, 170.
[23] G. Mellbin, *J. Liquid Chromatog.*, 1983, **6**, 1603.
[24] G. Mellbin and B. E. F. Smith, *J. Chromatog.*, 1984, **312**, 203.
[25] M. Novotny, *Anal. Chem.*, 1981, **53**, 1294A.
[26] D. Ishii and T. Takeuchi, *Adv. Chromatog.*, 1983, **21**, 131.
[27] T. Takeuchi, H. Asai and D. Ishii, *J. Chromatog.*, 1986, **357**, 409.
[28] S. M. Han and D. W. Armstrong, *J. Chromatog.*, 1987, **389**, 256.
[29] I. Nilsson and R. Isaksson, *Acta Chem. Scand.*, 1985, **B39**, 531.
[30] S. Allenmark and S. Andersson, *Chirality*, 1989, **1**, 154.
[31] S. Allenmark, B. Bomgren and H. Borén, *J. Chromatog.*, 1984, **316**, 617.
[32] W. H. Pirkle, T. C. Pochapsky, G. S. Mahler, E. D. Corey, D. S. Reno and D. M. Alessi, *J. Org. Chem.*, 1986, **51**, 4991.
[33] W. H. Pirkle and T. C. Pochapsky, *J. Chromatog.*, 1986, **369**, 175.
[34] W. H. Pirkle and T. C. Pochapsky, *Chromatographia*, 1988, **25**, 652.
[35] W. H. Pirkle and R. S. Readnour, *Chromatographia*, 1991, **31**, 129.
[36] W. P. Jackson, B. E. Richter, J. C. Fjeldsted, R. C. Kong and M. L. Lee, in *Ultrahigh Resolution Chromatography*, S. Ahuja (ed.), ACS, Washington DC, 1984, p. 121.
[37] L. G. Randall, in *Ultrahigh Resolution Chromatography*, S. Ahuja (ed.), ACS, Washington DC, 1984, p. 135.
[38] S. Hara, A. Dobashi, K. Kinoshita, T. Hondo, M. Saito and M. Senda, *J. Chromatog.*, 1986, **371**, 153.
[39] P. Macaudière, A. Tambuté, M. Caude, R. Rosset, M. A. Alembik and I. W. Wainer, *J. Chromatog.*, 1986, **371**, 177.
[40] P. Mourier, E. Eliot, A. Tambuté, M. Caude and R. Rosset, *Anal. Chem.*, 1985, **57**, 2819.
[41] P. A. Mourier, P. Sassiat, M. H. Caude and R. Rosset, *J. Chromatog.*, 1986, **353**, 61.
[42] P. Macaudière, M. Caude, R. Rosset and A. Tambuté, *J. Chromatog.*, 1987, **405**, 135.
[43] J. S. Bradshaw, S. K. Aggerwal, C. A. Rouse, B. J. Tarbet, K. E. Markides and M. L. Lee, *J. Chromatog.*, 1987, **405**, 169.
[44] P. Macaudière, M. Caude, R. Rosset and A. Tambuté, *J. Chromatog.*, 1988, **450**, 255.
[45] P. Macaudière, M. Lienne, M. Caude, R. Rosset and A. Tambuté, *J. Chromatog.*, 1989, **467**, 357.
[46] A. Dobashi, Y. Dobashi, T. Ono, S. Hara, M. Saito, S. Higashidate and Y. Yamauchi, *J. Chromatog.*, 1989, **461**, 121.
[47] W. Roeder, F. J. Ruffing, G. Schomburg and W. H. Pirkle, *J. High Res. Chromatog.*, *Chromatog. Commun.*, 1987, **10**, 665.

[48] W. Steuer, M. Schindler, G. Schill and F. Erni, *J. Chromatog.*, 1988, **447**, 287.
[49] P. Macaudière, M. Caude, R. Rosset and A. Tambuté, in *Chiral Separations*, D. Stevenson and I. D. Wilson, eds., Plenum, New York, 1988, p. 115.
[50] M. D. Oates and J. W. Jorgenson, *Anal. Chem.*, 1989, **61**, 1977.
[51] E. S. Yueng, *Chromatog. Sci.*, 1989, **45**, 117.
[52] A. Guttman, A. Paulus, A. S. Cohen, N. Grinberg and B. L. Karger, *J. Chromatog.*, 1988, **448**, 41.
[53] A. Dobashi, T. Ohno, S. Hara and J. Yamaguchi, *Anal. Chem.*, 1989, **61**, 1984.
[54] T. A. G. Noctor, B. J. Clarke and A. F. Fell, *Anal. Proc.*, 1986, **23**, 441.
[55] A. F. Fell, T. A. G. Noctor, J. E. Mama, and B. J. Clark, *J. Chromatog.*, 1988, **434**, 377.

11

Experimental procedures for the synthesis of chiral sorbents

This last chapter is devoted to the techniques used to obtain some of the commonly applied chiral stationary phases or sorbents and also to certain problems encountered in these procedures. Some of the more basic preparative techniques have been performed in a variety of modifications and the examples given, taken from the literature, are regarded as representative, well-documented procedures.

11.1   TECHNIQUES FOR THE PREPARATION OF CHIRAL SORBENTS BY DERIVATIZATION OF POLYSACCHARIDES

As described in Section 7.1.1.2, many naturally occurring polysaccharides are of interest as starting materials for production of chiral sorbents. Cellulose, however, is the most thoroughly investigated, and preservation of its microcrystallinity is of importance for the chromatographic properties of the derivatives. Therefore, microcrystalline derivatives have to be prepared in a heterogeneous reaction, i.e. under conditions which do not cause dissolution of the material.

11.1.1   Preparation of microcrystalline cellulose triacetate (MCTA)

(a) *Synthesis* [1,2]. Microcrystalline cellulose (e.g. Avicel) (200 g) is suspended in benzene (4 litres), and acetic acid (800 ml), 60% perchloric acid (6 ml) and acetic anhydride (800 ml) are added in that order with stirring. The suspension is then kept at 35°C for 3 days during which time it becomes more transparent but also darker in colour. After centrifugation the mass is suspended in methanol and again isolated by centrifugation. After sufficient washing with methanol to remove excess of reagent, the light brown product is dried at 30°C under vacuum.

(b) *Grinding and sieving* [3]. The dry, brittle material is ground to fine particles with a ball-mill or other suitable device. Size fractionation of the particles is best done by a fan-sieving technique (Zick-Zack Sichter A 100 MZR, Alpine AG, Augsburg, FRG). A fraction corresponding to a 15–25 μm particle-size range is suitable for use in glass columns and low–medium pressure operations.

(c) *Swelling and slurry preparation prior to column packing* [2,3]. The MCTA particles can be swollen and must be treated with an excess of boiling ethanol (95%) for 20–30 min. After this treatment the suspension is cooled to room temperature and centrifuged, or filtered off on a fritted glass disc. The moist material is then suspended in an equal amount of ethanol and ultrasonicated for 2–3 min, followed by degassing under reduced pressure. The slurry is then ready for packing. For calculation purposes, 10 g of dry MCTA will give a final volume of approximately 25 ml in the packed column.

11.1.2   Preparation of silica coated with cellulose triacetate [4,5]

MCTA (ca. 1.1 g), prepared as described under (a) above, is dissolved in chloroform (16.5 ml) and the solution added to 10 g of 3-aminopropyl-silica (10 μm, 500 Å) (e.g. LiChrospher 500-NH$_2$, Merck). After complete mixing by stirring, the solvent is slowly evaporated. The coating procedure is repeated with a further amount (1.1 g) of MCTA. A slurry of the dried material in ethylene glycol/methanol (1:2) is suitable for packing under high pressure into a steel column.

   The preparation of other derivatives of microcrystalline cellulose appears to be difficult to perform in a strictly heterogeneous mode, and there is a resulting loss of crystalline regions in the matrix. Of the ester derivatives, the benzoate and cinnamate phases are commercially available (see Appendix), coated on silica.

   The carbamate derivatives are obtained by reaction with an isocyanate, normally with pyridine as solvent and at elevated temperature (80–100°C).

11.1.3   Preparation of silica coated with cellulose tris(phenylcarbamate) [6]

Microcrystalline cellulose (Avicel, Merck) is reacted with an excess of phenyl isocyanate in pyridine at about 100°C. After cooling, precipitation with methanol gives over 60% of a fraction consisting of the tris(phenylcarbamate) (as shown by IR and NMR). Coating is effected by dissolution of the material (0.75 g) in tetrahydrofuran (10 ml) and adding half this solution (ca. 5 ml) to 3 g of 3-aminopropyl-silica (10 μm, 4000 Å) (LiChrospher SI 4000, Merck) followed by slow evaporation of the solvent. The second half of the tris(phenylcarbamate) solution is used to repeat the coating process.

11.1.4   Preparation of pentyl derivatives of cyclodextrins (Lipodex) [7,8]

The cyclodextrin (α- or β-form) is reacted with a threefold excess of 1-bromopentane and sodium hydroxide in dry dimethylsulphoxide at 25°C for 6 min. This yields 2,6-di-*O*-pentylated cyclodextrin in almost quantitative yield. To obtain the peralkylated compound, the above derivative is refluxed in anhydrous THF with sodium hydride and 1-bromopentane for *ca.* 4 days. To obtain the 3-*O*-acetylated compound, the derivative is instead reacted with acetic anhydride/triethylamine (3 molar equivalents) and 5 mole % of 4-dimethylaminopyridine under reflux for 24 hr. The hexyl derivatives can be prepared in an analogous manner.

   For use in capillary GC columns, the inner surface of the Pyrex glass capillary is first treated with a 0.3–0.5% suspension of silanox in carbon tetrachloride [9] and the column, after removal of excess solvent by a stream of nitrogen, is heated to 300°C for 1 hr with carrier gas flowing. Then, coating of the stationary phase is carried out with a 0.2% solution of the cyclodextrin derivative in dichloromethane by the static procedure [10].

11.2 POLYMERIZATION PROCEDURES USED TO OBTAIN CHIRAL SYNTHETIC POLYMER MATERIALS

The technique developed by Blaschke can be used to obtain a variety of chiral polyacrylamides and polymethacrylamides. The synthesis of one of the most useful representatives of this class is described below [11,12].

11.2.1   Preparation of poly[(S)-N-acryloylphenylalanine ethyl ester]

(a) *Synthesis of the monomer*. A stirred suspension of 0.1 mole of the amino-acid ethyl ester hydrochloride and 0.15 mole of acrylic anhydride and 0.1 g of 4-*tert*-butyl-1,2-dihydroxybenzene (a polymerization inhibitor) in 150 ml of chloroform is cooled to 0–5°C. To this suspension is added a saturated aqueous solution of sodium carbonate at a rate sufficient to keep the pH of the aqueous phase at 8.0. When the pH is constant, the mixture is further stirred at 10–15°C for 1 hr, then the chloroform phase is separated and washed with cold 2M sulphuric acid, followed by water, and subsequently evaporated to ca. 20 ml. This concentrate is then applied to a short column (10×2 cm) of basic alumina and the N-acryloylamino-acid ethyl ester is eluted with chloroform. This procedure removes the polymerization inhibitor. The chloroform is then evaporated at a temperature below 40°C. The product obtained is also recrystallized at low temperature, as otherwise spontaneous polymerization may occur. The ethyl (S)-N-acryloylphenylalaninate thus obtained has m.p.=66°C and $[\alpha]_D^{20}$=+121.7 (c. 3.0, benzene).

The method described is applicable to the synthesis of N-acryloyl- and N-methacryloyl derivatives of a variety of optically active amino-acid esters.

(b) *Polymerization*. To a mixture of (S)-N-acryloylphenylalanine ethyl ester (10.0 g, 40.5 mmole), 1,2-ethanediol diacrylate (0.69 g, 4.05 mmole) and azoisobutyronitrile (66 mg, 0.405 mmole) dissolved in toluene (17 ml) is added a 5% aqueous poly(vinyl alcohol) solution (197 ml) with cooling, and stirring under nitrogen (metal stirrer, ca. 300 rpm). During stirring the temperature is gradually raised to 80°C and is kept there for 5 hr. After cooling, the organic phase is removed by steam distillation and the aqueous phase decanted off. The white polymer beads are thoroughly washed with hot water, methanol, toluene and petroleum ether and finally dried under reduced pressure. The yield is almost 100%.

(c) *Fractionation and swelling*. By sieving, a fraction with particle sizes between 50 and 100 μm can be obtained. The particles are swollen by boiling under reflux for ca. 10 min in the solvent system to be used. After cooling, the solvent is decanted off together with the very fine particles and the beads are resuspended in the same solvent to make a slurry for packing into a glass column.

The other type of polymerization procedure, developed by Okamoto and Yuki [13–16] and utilizing a chiral catalyst, also has broad applicability, for synthesis of chiral polymers of single helicity.

11.2.2   Preparation of poly(triphenylmethyl methacrylate)

(a) *Synthesis of the monomer* [13]. A diethyl ether solution of trityl chloride is added to a suspension of silver methacrylate in diethyl ether and the mixture is stirred for several hours at room temperature. The silver chloride liberated is removed by filtration and the ether evaporated. This gives 75% yield of trityl methacrylate as a crystalline product, of m.p. 101–103°C. The monomer is well soluble in benzene or methanol.

(b) *Polymerization* [14–16]. The monomer (20.0 g, 60.7 mmole) is dissolved in dry toluene (400 ml) under nitrogen and the solution cooled to −78°C. Then a toluene solution of (−)-sparteine (0.34 g, 1.46 mmole) and butyllithium (1.21 mmole) is added by means of a syringe. After reaction for 24 hr, the mixture is added to methanol (4 litres), and the polymer formed is precipitated, and isolated by centrifugation. Grinding of the polymer and washing with tetrahydrofuran (700 ml) yields, after drying in vacuum, 19.4 g (97%) of a product which is then finely ground and fractionated by sieving. Before column packing the material is allowed to swell in the appropriate organic solvent.

It is very important to note that the solubility of this polymer limits the number of organic solvents that can be used as eluents. Aromatic hydrocarbons, chloroform and tetrahydrofuran are not compatible with it. Methanol and mixtures of hexane and 2-propanol are preferable.

11.3   TECHNIQUES USED FOR THE BINDING OF CHIRAL SELECTORS TO SILICA

A large number of methods of functionalizing silica have been developed. In most cases, reactions involving the use of a terminally substituted alkyl trialkoxysilane are employed. These reactions lead to formation of new siloxane bonds with concomitant elimination of alkanol (MeOH or EtOH), according to Scheme 11.1.

The silica derivative will thus contain a stable Si–C bond. A number of groups (R_1) can be introduced in this way, the most useful being shown in Scheme 11.2.

The derivatives shown are easily prepared by the use of commercially available silyl reagents. The synthetic procedures have been extensively investigated and are generally used with only minor modifications. However, the relative importance of certain steps, such as the acid pretreatment of the silica and the silylation under strictly anhydrous conditions, is not entirely clear. Thus, it has been claimed that treatment of the silica with boiling hydrochloric acid will open surface siloxane bonds and create more free silanol groups for reaction. Anhydrous conditions are suggested to be necessary for the formation of a monolayer instead of an adsorbed polymeric film of varying thickness. Dry toluene is the preferred reaction solvent, probably because it enables azeotropic removal of the alcohol formed during the reaction.

Many different strategies for chiral sorbent synthesis by covalent binding to modified silica supports have been developed. Since many of these are of more general applicability, they are worth some consideration. One of the fundamental questions related to chiral bonded phases and their synthesis concerns the means of minimizing non-enantioselective interactions. This leads to questions regarding the synthetic strategy, e.g. is it better to use a solid-phase synthesis approach, starting

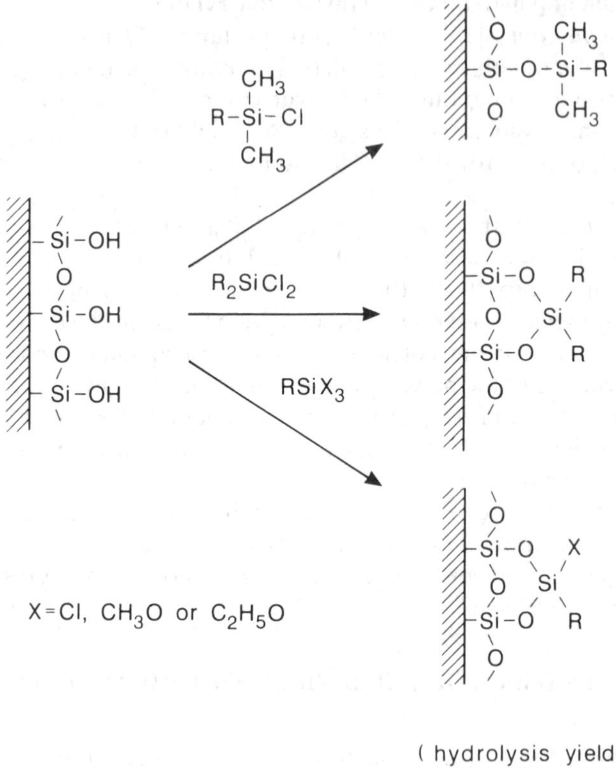

$$CH_3$$
$$R-Si-Cl$$
$$CH_3$$

$$\begin{array}{l} O \quad CH_3 \\ -Si-O-Si-R \\ O \quad CH_3 \end{array}$$

$$-Si-OH$$
$$O$$
$$-Si-OH$$
$$O$$
$$-Si-OH$$

$$R_2SiCl_2$$

$$RSiX_3$$

$$\begin{array}{l} O \\ -Si-O \quad R \\ O \quad Si \\ -Si-O \quad R \\ O \end{array}$$

$$X=Cl, \; CH_3O \; or \; C_2H_5O$$

$$\begin{array}{l} O \\ -Si-O \quad X \\ O \quad Si \\ -Si-O \quad R \\ O \end{array}$$

(hydrolysis yields X = OH)

Scheme 11.1 — Common reactions employed to functionalize silica.

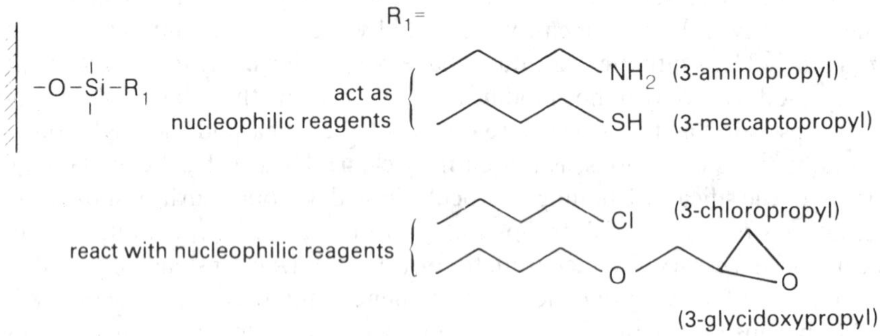

$$R_1 =$$

$$-O-Si-R_1$$

act as
nucleophilic reagents

$$\qquad NH_2 \quad \text{(3-aminopropyl)}$$
$$\qquad SH \quad \text{(3-mercaptopropyl)}$$

react with nucleophilic reagents

$$\qquad Cl \quad \text{(3-chloropropyl)}$$
$$\qquad O \qquad O$$

(3-glycidoxypropyl)

Scheme 11.2 — Useful reactive end-groups for ligand coupling.

with bare silica and building up the bonded CSP in a number of consecutive steps, or to do the opposite, i.e. synthesize a chiral selector containing a suitable spacer and silylating functionality and link this entity to the bare silica in the last step? There is no simple answer to this question and there are certainly advantages as well as disadvantages associated with both routes. It is well known, for example, that aminopropyl-silica can easily be prepared with a high surface coverage of functional groups, but that all these groups do not react in a subsequent step leading to immobilization of the chiral selector. This might give rise to problems of undesirable interactions with the analyte.

The strategy applied is often dictated by what seems to be sound and realistic from a synthetic point of view. In the following, some useful synthetic approaches, outlined in Scheme 11.3, will be discussed in some detail.

A. The "synthesis prior to immobilization" strategy:

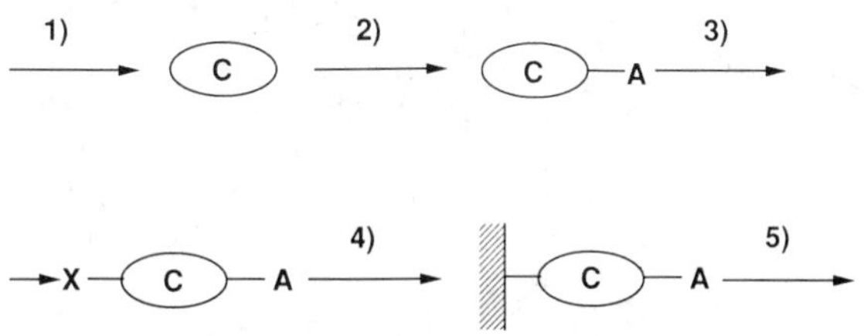

(1) preparation of the chiral starting material via synthetic and optical resolution procedures
(2),(3) chemical modifications of the chiral building block (X is a *reactive* functional group)
(4) immobilization to the silica (unmodified or modified)
(5) eventual removal of residual reactive groups by end-capping

Examples of steps (2)–(4) (see also section 7.2.3; Schemes 7.11–7.13);

(a) Homobifunctional chiral synthon:

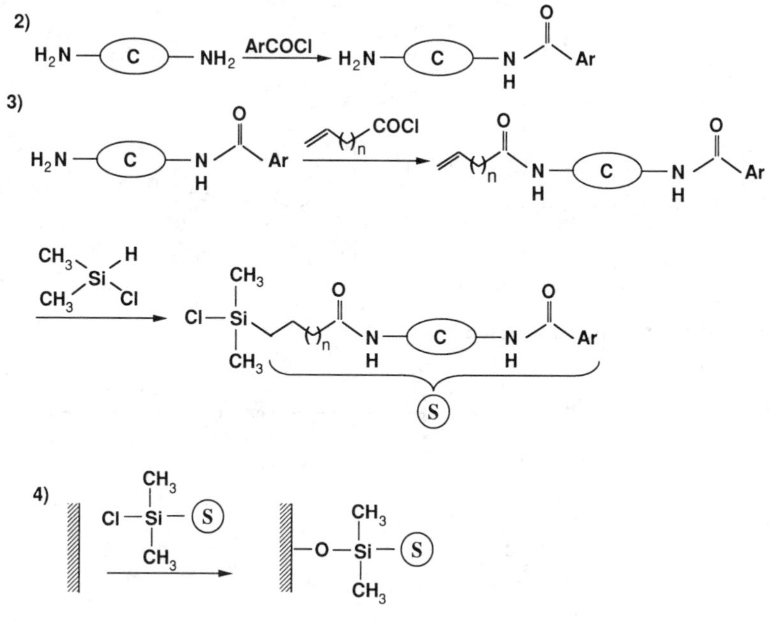

Scheme 11.3 — Some synthetic strategies used to obtain silica-bonded CSPs.
Continued page 264

(b) Heterotrifunctional chiral synthon:

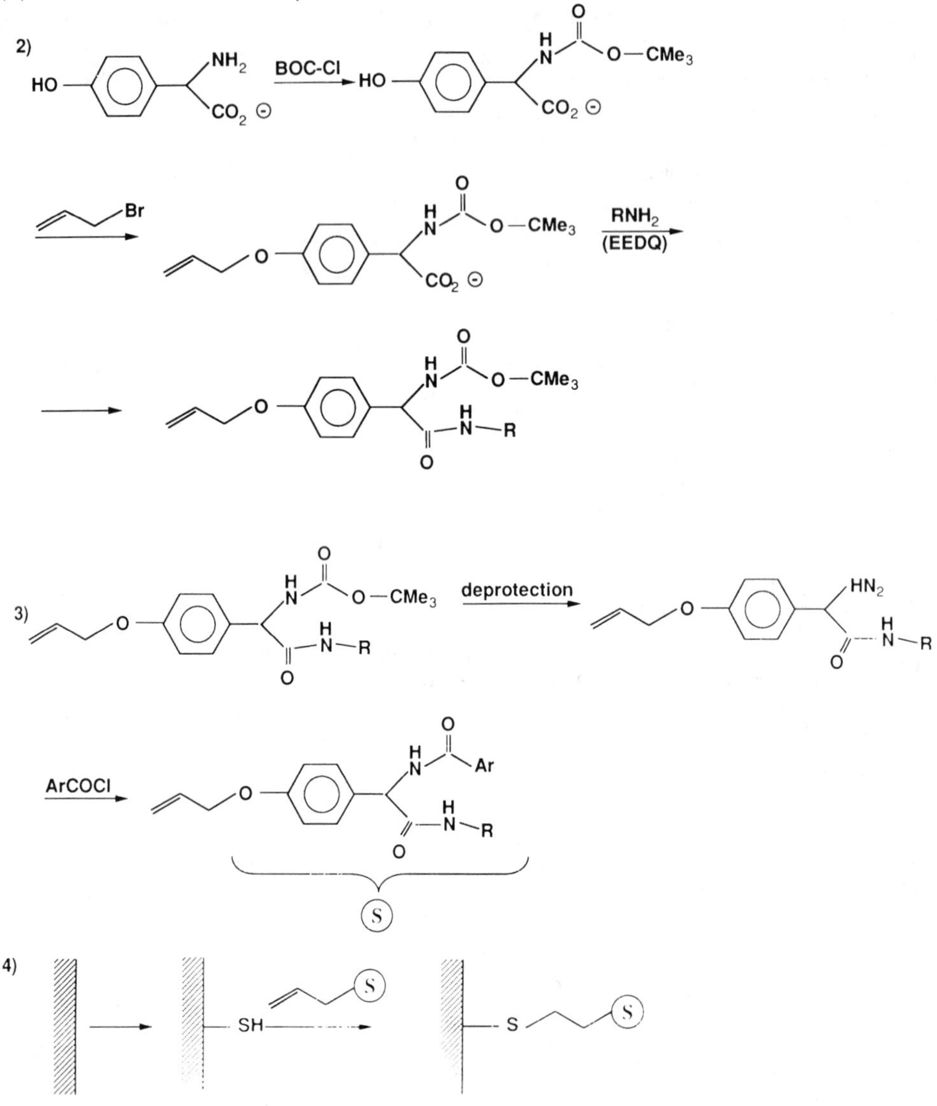

B. The solid-phase-synthesis strategy:

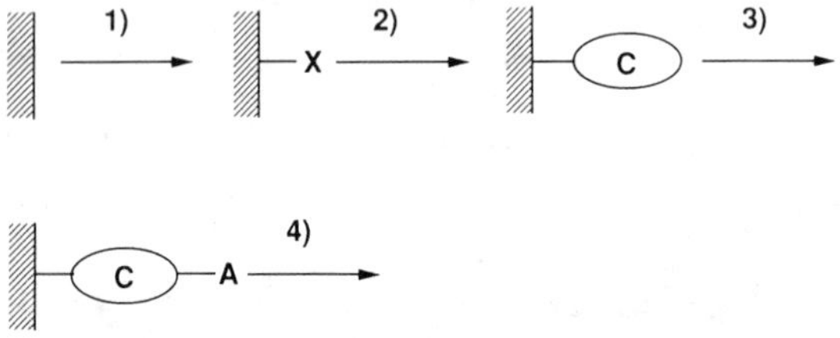

Scheme 11.3 (*Cont.*).

(1) introduction of a terminal reactive functional group
(2) coupling of the chiral structure element
(3) modification of the chiral structure by the introduction of suitable interacting functions (A)
(4) eventual removal of residual reactive groups by end-capping

Example: Use of an optically active diamine as the chiral part of the structure:

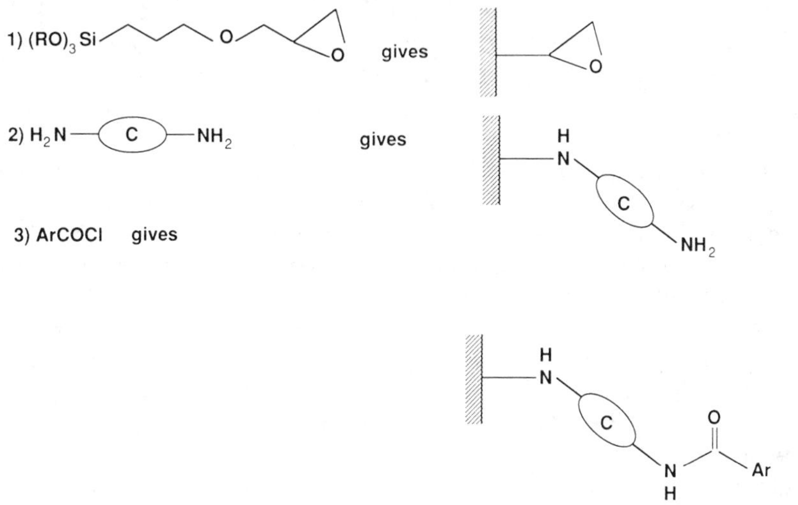

1) $(RO)_3$ Si ~~~~ O ~~~ △ O   gives

2) H_2N — C — NH_2 gives

3) ArCOCl   gives

Scheme 11.3 (*cont.*).

(a) *Addition reactions utilizing allyl-derivatized chiral selectors.* The main principle here is to make use of an anti-Markownikow addition of a terminal thiol group on the silica to an allyl group on the selector. Such reactions are radical-initiated and lead to very stable covalent bonds. This type of reaction was first applied to selectors of natural origin, *viz.* the *Cinchona* alkaloids (quinine, quinidine, cinchonidine), having an allylic function in their structure. Since allylic functions are easily introduced into many organic structures, this principle has also recently been used for other selectors, notably a series of L-tyrosine derivatives (see Scheme 11.3). The phenolic OH group in tyrosine is readily converted into an allyl ether function via reaction with allyl bromide. By this strategy the carboxyl group (which is used for bonding to silica in the Pirkle-type phases) is freely available for modification and design of new CSPs.

(b) *Hydrosilylation reactions.* In the presence of a catalyst (chloroplatinic acid) triethoxysilane adds to a sterically unhindered terminal double bond, yielding a compound that can be directly bound to bare silica. Since many chiral selectors can be modified to contain an alkyl chain with a terminal double bond, this reaction is highly attractive. In this way the complete organic structure is assembled before attachment to the silica, thus obviating the need for prior organic modification of the support.

So far this strategy has been used mainly for the introduction of relatively long hydrocarbon spacers (C_{11}) between the silica surface and the chiral selector. However, there is some general applicability of this method as indicated in Scheme

11.3. An interesting application of the technique is for the introduction of chiral selectors into polysiloxanes which can then be used for the coating of capillary columns (glass or fused silica) for LC and SFC [17]. It has been assumed that the coating partly involves covalent bonding of the polysiloxane to the silica surface due to the presence of Si–H bonds still remaining after the hydrosilylation reaction. This technique also permits a regulation of the film thickness in the capillary [17].

(c) *Chemical modification of the chiral structural element after binding to the support.* A representative example of a complete solid-phase synthesis strategy is found in the method devised by Gasparrini *et al.* [18] which is based on the use of optically active *trans*-1,2-diaminocyclohexane as the chiral structural element of the CSP. This diamine is readily available in both optically active forms ((−)-(R,R)- and (+)-(S,S)-) by large-scale classical resolution with tartaric acid [19]. As found in Scheme 11.3, the silica is first converted into a derivative having a terminal group (in this case an epoxide) which can react with a nucleophilic reagent. Then the chiral part (here the optically active 1,2-diaminocyclohexane) is immobilized through nucleophilic attack and ring-opening. The second amino group is free, however, and can be used for a variety of subsequent modifications. An obvious step is acylation of this amino group for the introduction of a π-acidic or π-basic substituent according to Pirkle's concept. So far the 3,5-(dinitrobenzoyl)-derivative has been the most investigated.

Below are given some useful detailed synthetic procedures which illustrate the general strategies adopted for preparation of chiral stationary phases immobilized onto silica.

11.3.1 Preparation of 3-glycidoxypropyl-silica

(a) *Acid pretreatment and drying* [20]. The silica is activated by refluxing in concentrated hydrochloric acid for 4 hr. It is then washed with distilled water until the washings are neutral and is finally dried in a vacuum desiccator.

(b) *Silylation* [21]. To porous silica (3 g), dried at 200°C overnight and then slurried in 100 ml of sodium-dried toluene, are added 80 μl of triethylamine and 4 ml of 3-glycidoxypropyltrimethoxysilane. The mixture is refluxed under nitrogen for 4 hr in a flask provided with a PTFE stirrer. The product is filtered off on a glass filter-funnel and washed with toluene, acetone and diethyl ether (100 ml of each) and finally sucked dry. The degree of silylation can be determined by hydrolysis of the epoxy groups to diol functions, with dilute sulphuric acid (pH 2, 90°C, 1 hr) followed by periodate oxidation. It will amount to ca. 320 μmole/g of silica.

11.3.2 Large–scale preparation of (R)-N-(3,5-dinitrobenzoyl)phenylglycine covalently silica-bonded sorbent [22]

(a) *Synthesis of 3-aminopropyl-silica.* A slurry of 4 kg of silica (58 μm) in 8 litres of benzene, contained in a 22-litre round-bottomed flask, is dried by azeotropic distillation of water (Dean and Stark trap). When there is no more collection of water, triethoxy-3-aminopropylsilane (400 g) is added from a tap-funnel over a period of 20 min and azeotropic distillation continued to remove liberated ethanol and water. The mixture is refluxed overnight and allowed to cool. The supernatant liquid is removed (by suction through a glass filter tube) and the remaining solid washed by repeated suspension in benzene and removal of the supernatant liquid by suction.

(b) *Preparation of the CSP*. Solid (R)-N-(3,5-dinitrobenzoyl)phenylglycine (346 g) and 1-ethoxycarbonyl-2-ethoxy-1,2-dihydroquinoline (EEDQ) (300 g), are added to the aminopropyl-silica gel (ca. 4 kg). Enough methylene chloride (ca. 6 litres) is added to suspend the solids when the mixture is stirred. The reaction mixture darkens in colour as the reagents dissolve. After 2 hr of stirring, the supernatant liquid is removed by suction and the CSP washed with six 2-litre portions of methanol. As the EEDQ can cause partial racemization, it is essential to remove the base by careful washing and to use short reaction times.

11.3.3   Preparation of (S)-(−)-α-N-(2-naphthyl)leucine [23]

A mixture of L-leucine (10 g, 76 mmole), β-naphthol (11 g, 76 mmole) and anhydrous sodium sulphite (9.6 g, 76 mmole) is placed in a pressure vessel equipped for magnetic stirring. Saturated sodium bisulphite solution (60 ml) is added and the bottle sealed and heated slowly to 115°C, with stirring. After 4 days the bottle is cooled, emptied into a 1-litre beaker and washed alternately with 5% sodium bicarbonate solution and acetone. The combined washings and reaction mixture are diluted to 500 ml with water and the pH adjusted to 8.5–9.0 (with sodium carbonate). The mixture is extracted with two 50-ml portions of methylene chloride to remove the unreacted β-naphthol. The organic phase is washed with two 20-ml portions of 5% sodium bicarbonate solution and the washings are combined with the aqueous layer. The pH of the aqueous phase is adjusted to 3–3.5 with $6M$ hydrochloric acid (evolution of CO_2 and SO_2) and the white precipitate collected by filtration. The filtrate is extracted repeatedly with ethyl acetate and the extracts are evaporated. The residue is added to the first product and the mixture is recrystallized from 95% ethanol to give 5.1 g (21.3 mmole, 25% yield) of the substituted leucine, m.p. 125°C, $[\alpha]_D^{20} = -168.7$ (c. 1.0, THF).

11.3.4   Hydrosilylation of (R)-N-(10-undecenoyl)-α-(6,7-dimethyl-1-naphthyl) isobutylamine [24]

To a 50-ml round-bottomed flask equipped with a reflux condenser and magnetic stirrer, are added 1.97 g of (R)-N-(10-undecenoyl)-α-(6,7-dimethyl-1-naphthyl)isobutylamine and 10 ml of triethoxysilane, under nitrogen. The mixture is warmed to 40°C and 0.8 ml of chloroplatinic acid solution (71.5 mg in 20 ml of 2-propanol) is added. The mixture is warmed to 90°C and stirred for 1 hr at this temperature. The excess of triethoxysilane is removed under vacuum and the residue is rapidly chromatographed on silica (dichloromethane/ethyl acetate, 10:1) to yield the hydrosilylated product (1.48 g) as a white solid (m.p. 52–54°C) in 53% yield.

11.3.5   Preparation of silica-bonded (S)-1-(α-naphthyl)ethylamine [25]

To a slurry of 25 g of aminopropyl-silica (LiChrosorb NH_2, 10 μm, Merck) in 30 ml of water, 2.5 g of succinic anhydride (2.5 g) are added with swirling. Sodium hydroxide ($2M$) is used to maintain the pH at 4.0 and the slurry is stirred at room temperature for 5 hr. The product silica is collected by filtration and washed with water, methanol and diethyl ether.

To 2.5 g of this product in 30 ml of THF, at 0°C, 1,1'-carbonyldiimidazole (3.0 g) is added and the mixture is stirred gently for 3 hr, after which (S)-1-(α-naphthyl)-ethylamine (3 g) is added and stirring is continued at room temperature for a further 5 hr. The silica sorbent is collected and washed exhaustively with THF, methanol and diethyl ether.

By the method described above, the chiral selector is linked to the aminopropyl-silica terminal through amide bonds to the succinic acid spacer. By a somewhat more complicated technique [26], this segment can be replaced by a terephthaloyl group, i.e. $-CH_2CH_2-$ is exchanged for $p-C_6H_4-$.

11.3.6   Preparation of silica-bonded polyacrylamide and polymethacrylamide [26]

A suspension of diol-silica (Lichrosorb Diol, 5 μm, 2.50 g) in 20 ml of anhydrous dioxan is prepared under nitrogen and stirred at room temperature for 30 min with a solution of methacrylic anhydride (1.50 g in 10 ml of anhydrous dioxan) and a solution of di-isopropylethylamine (0.80 g in 30 ml of anhydrous dioxan). After 24 hr at room temperature, the modified silica is isolated on a sintered-glass filter (G4), washed with 100 ml of dioxan and dried under high vacuum. To a suspension of 2.5 g of this product in 10 ml of anhydrous toluene is added, under nitrogen and with stirring, a solution of (S)-N-acryloylphenylalanine ethyl ester (7.5 g in 10 ml of toluene) and azobis(isobutyronitrile) (22 mg in 10 ml of toluene). After 15 min at 80°C the reaction is stopped by addition of 4-*tert*-butyl-1,2-dihydroxybenzene (200 mg in 10 ml of toluene). The silica adsorbent is collected with a sintered-glass filter and washed successively with toluene, dioxan, hexane and 2-propanol. After drying under high vacuum the product has the composition C 21.4%, H 3.1%, N 1.2%.

The same technique can also be used for immobilization of chiral polymers based on methacrylamide derivatives, provided the reaction time of the polymerization process is increased to ca. 30 min.

It is recommended that these chiral sorbents are packed in steel analytical columns, as slurries in 2-propanol (30 ml) at 450–500 bar pressure. For chromatography, hexane/dioxan mixtures are preferred as mobile phases.

11.3.7   Preparation of 3-aminopropylsilica-bonded *N*-(S)-1-(α-naphthyl)ethyl-aminocarbonyl-L-valine (a representative of Oi's urea phases) (27)

To a solution of 3.7 g of L-valine in 17 ml of 2*M* sodium hydroxide are added 8.9 g of (S)-1-(α-naphthyl)ethylisocyanate and 5 ml of THF, and this mixture is stirred at room temperature for 6 hr. The reaction mixture is washed with ethyl acetate and acidified with 6*M* hydrochloric acid. The white, crystalline material is then extracted with ethyl acetate. Removal of the solvent under reduced pressure affords crude *N*-(S)-1-(α-naphthyl)ethyl-aminocarbonyl-L-valine. The product is recrystallized from a mixture of ethyl acetate and hexane. M.p. 176–177°C (dec.).

To a solution of 1.6 g of this material in 35 ml of a mixture of THF and dioxan are then added 2.5 g of 3-aminopropylsilica (LiChrosorb-NH$_2$, 5-μm (Merck)) and 1.36 g of EEDQ with stirring at 0°C for 1 hr, and the mixture is then left overnight at room temperature with stirring. The modified silica is then collected by centrifugation and washed exhaustively with THF, methanol, chloroform and diethyl ether, and finally dried under reduced pressure. On the basis of C and N analyses, the degree of substitution may be calculated and typically amounts to 0.45 mmoles of *N*-(S)-1-(α-naphthyl)ethylaminocarbonyl-L-valine per gram of silica are found.

The diastereomeric CSP can also easily be prepared by the above procedure if the (S)-enantiomer of 1-(α-naphthyl)ethylisocyanate is replaced by the R-isomer. In this case the crystalline product obtained in the first step has a higher m.p., 189–90°C (dec.).

11.3.8   Preparation of a silica-bonded glycyl-L-proline CSP [28]

The silica (Develosil 100-5 (Nomura Chemical, Aichi, Japan)) is dried at ~120°C for 4 hr and then 20 g is suspended in 100 ml of dry toluene. To this suspension is then added 12 ml of 3-glycidoxypropyltrimethoxysilane and the mixture heated for 20 hr with continuous removal of the methanol formed. Isolation of the modified silica on a glass filter (G-4 porosity), washing with toluene and drying under vacuum gives 23.6 g. Suspension of 3.7 g of this product in 20 ml of dry methanol is then followed by addition of 512 mg of the sodium salt of glycyl-L-proline and stirring of the mixture at room temperature under nitrogen for 7 days. The final product is isolated by filtration and washed exhaustively with methanol. Elemental analyses gave: C 7.38, H 1.43, N 0.57%. For the 3-glycidoxypropyl-silica: C 7.28, H 1.47, N<0.1%.

11.3.9   Procedure for immobilization of proteins to aldehyde-silica [29]

Diol-silica is the starting support material in this preparation. It can either be bought from various suppliers or prepared from 3-glycidoxypropylsilica (see 11.3.1) by acid catalysed epoxide ring opening [30]. In a typical preparation, the epoxide-silica (*ca.* 10 g) is slurried in 150 ml of 0.1*M* sulphuric acid and the mixture stirred at 80°C for 2 hr. The modified silica is collected on a glass filter and washed with water until the filtrate is almost neutral. Then the product is washed with methanol followed by diethyl ether and finally dried at *ca.* 40°C. The degree of substitution amounts to 0.4–0.6 mmoles per g of silica.

A suspension of 5 g of diol-silica in 30 ml of water is treated with 0.35 g of periodic acid. After ultrasonication for 1 min followed by shaking for 2–3 hr at room temperature, the aldehyde–silica formed is collected on a glass filter and washed with water. To the wet aldehyde–silica is then added 35 ml of a phosphate buffer, pH 7.0, containing 0.75 g of protein (original procedure given for cellobiohydrolase-I (CBH-I)) and 0.13 mg of sodium cyanoborohydride. After 1 min of ultrasonication, the slurry is shaken for two days, then filtered, and the modified silica is washed with phosphate buffer (0.1*M*, pH 7.0). The amount of protein bound can be determined by UV absorption spectroscopy (absorbance difference obtained). The degree of substitution of CBH-I amounts to between 50 and 75 mg/g of silica.

11.3.10   Procedure for coupling of proteins to 3-aminopropyl-silica by use of *N,N'*-disuccinimidyl carbonate (DSC)

This procedure, first described by Miwa *et al.* [31] for the immobilization of ovomucoid onto silica, is exemplified below by the synthesis of an avidin-based chiral sorbent [32].

3-Aminopropyl-silica (LiChrosorb NH$_2$, 2 g) and N,N'-disuccinimidyl carbonate (3 g) are reacted overnight in 60 ml of the coupling buffer (0.1M sodium bicarbonate, pH 6.8) at room temperature. To ensure efficient mixing of the slurry during the reaction, a rotary evaporator without aspiration may be used. The activated silica gel is then washed with 50 ml of the coupling buffer and suspended in 30 ml of the same buffer. Then a solution of avidin (2 g of avidin in 30 ml of coupling buffer) is added gradually over a period of 30 min with shaking. The mixture is stirred for 2 hr and the protein-functionalized silica gel is thoroughly washed with coupling buffer. The coupling buffer is also used for slurry-packing of the material into the analytical column (4.6×150 mm).

A method for the isolation of avidin on a preparative scale from powdered egg white (20 kg) has also been described [32].

11.3.11   Techniques for coupling of tyrosine-derived CSPs to silica [33,34]

(a) *Synthesis of 3-mercaptopropyl-silica.* This nucleophilic silica material is prepared as described by Rosini *et al.* [35] or by Pirkle [36]. A 100-ml round-bottomed flask, fitted with a water separator (Dean and Stark trap), is charged with 4.80 g of HPLC-grade silica and 40 ml of toluene. After heating under reflux for *ca.* 20 hr, the solvent is removed under reduced pressure and replaced by 15 ml of toluene, 10 ml of pyridine and 15 ml of (3-mercaptopropyl)trimethoxysilane. The mixture is then heated to 100°C for *ca.* 30 hr. After cooling, the silica is collected on a glass filter and washed sequentially with acetone, diethyl ether, hexane and diethyl ether (*ca.* 50 ml of each organic solvent) and finally dried in a vacuum. The degree of substitution of this functionalized silica as determined by elemental analysis is usually of the order 0.9 mmoles/g.

(b) *Preparation of (S)-tyrosine-O-(2-propen-1-yl) butylamide* [33]. BOC-L-tyrosine (39.4 g, 0.14 mol) is dissolved in dry DMF (400 ml) and the solution cooled to 5°C. Then a sodium hydride suspension (60% in mineral oil, 14.56 g, 0.36 mole) is added in small portions under nitrogen. Stirring is maintained for 2 hr at 10°C and allyl bromide (18.6 g, 0.156 mole) is added dropwise. Stirring is continued at room temperature for 48 hr, ice water (1500 ml) is then added and stirring maintained for a further 1 hr. The aqueous solution is washed twice with benzene (250-ml portions) and then acidified with hydrochloric acid (6 M) in the cold (<10°C). The crude product is extracted into ethyl acetate (3×500 ml) and the solution washed with water and brine. Water is then removed by filtration through a hydrophobic filter. Concentration under vacuum yields an orange-coloured oil (37.3 g, 83%).

A part of this oil (16.6 g, 51.7 mmoles) is mixed with butylamine (3.77 g, 5.1 ml, 51.7 mmoles) and EEDQ (15.3 g, 62 mmole) in THF (250 ml) and magnetically stirred under nitrogen at room temperature for 48 hr. Concentration under vacuum yields a yellow solid that is recrystallized from di-isopropyl ether:2-propanol (30:1) to give 13.6 g (70%) of a white product of m.p. 117–8°C and $[\alpha]_{365} = +14$ (c. 2.0, THF).

Deprotection of the amino group is then carried out by dissolving the product (44.2 g, 117 mmole) in a saturated dioxan solution of anhydrous hydrogen chloride (400 ml). After magnetic stirring for 2 hr at room temperature (reaction may be monitored by TLC (CH$_2$Cl$_2$:MeOH, 95:5)), the solution is concentrated by vacuum evaporation, diluted with water (1 l) and washed with dichloromethane (200 ml). After alkalinization (with 2 M NaOH), the product is extracted twice with dichloromethane (300 ml portions). Washing of the combined extracts with water and brine, followed by evaporation of the solvent and recrystallization from diisopropyl ether (100 ml) gives a white solid (27.9 g, 86%) of m.p. 50–2°C and $[\alpha]_D = -47.7$ (c. 2.0, THF).

N-Aroylation of the above product is finally carried out by adding small portions of 3,5-dinitrobenzoyl chloride (26.5 g, 115 mmole) and propylene oxide (20 g, 23.3 ml, 345 mmole) simultaneously to a solution of the product (31.8 g, 115 mmole) in THF (200 ml). After stirring overnight under nitrogen, the solvent is removed in a vacuum. Trituration of the solid with acetonitrile:diisopropyl ether (2:8) and filtration gives a white product (49.5 g, 89%) of m.p. 176–7°C. Further purification by column chromatography and recrystallization gives m.p. 177–8°C and $[\alpha]_D = -30.5$ (c. 2.0, THF). This CSP material is 100% optically pure.

(c) *Attachment of the CSP to the mercaptopropyl-silica* [33]. A mixture of the mercaptopropyl-silica (7 g), the above CSP material (1.3 g, 2.8 mmoles) and AIBN (0.092 g, 0.56 mmole) in chloroform (*ca.* 70 ml) is refluxed under nitrogen and with mechanical stirring for 40 hr. After cooling, the modified silica is successively washed with chloroform, methanol, acetonitrile, methanol and ether. After drying, the pink-coloured sorbent shows a degree of substitution of 0.195 mmole/g, based on nitrogen analyses.

11.3.12   Immobilization of *N*-methylquininium iodide onto silica [37]
3-Mercaptopropyl-silica is prepared as described in Section 11.3.11 (*a*). The rest of the procedure is analogous to that in Section 11.3.11 (*c*). To a solution of 3.3 g (7.02 mmoles) of *N*-methylquininium iodide in 50 ml of DMF is added 115 mg (0.7 mmole) of AIBN and 5.28 g of the 3-mercaptopropyl-silica. The resulting slurry is maintained at *ca.* 90°C for 24 hr with stirring. After the solid material has been isolated by filtration or centrifugation, it is washed thoroughly with DMF to remove the excess of the alkaloid (this may be followed spectroscopically) and then with methanol, ether and pentane. An alkaloid content of the modified silica of *ca.* 11% (by weight) is obtained.

11.3.13   Preparation of silica-bonded derivatives of (R,R)- or
(S,S)-1,2-diphenylethane-1,2-diamine [38]
(a) *Synthesis of (S,S)-1,2-diphenylethane-1,2-diamine [(S,S)-DPEDA]* [39]. Benzil and cyclohexanone (1 equivalent of each) are reacted with ammonium acetate–acetic acid at 120°C for 1 hr, resulting in a high yield (97%) of the corresponding cyclic bis-imine of m.p. 105–6°C. This compound is then reduced stereospecifically with lithium (4 equivalents) in THF–liquid ammonia (4:5; 0.3 M in bis-imine) at −78°C for 2 hr with addition of ethanol (2 equivalents) in four portions to give the corresponding *trans*-imidazoline in 95% yield. A dichloromethane solution of this product is then treated successively with 2 M hydrochloric acid and aqueous base, which yields, after removal of the solvent, the racemic DPEDA in 97% yield. M.p. 81–2°C. The overall yield from benzil is 89%.

Resolution of DPEDA is readily achieved with the use of tartaric acid [40] and affords both the (R,R)- and (S,S)-forms of high optical purity (>99%).

(b) *Monoacylation and binding to silica.* (S,S)-DPEDA is reacted with 3,5-dinitrobenzoyl chloride (1.1 molar equivalents) in dichloromethane for 30 min at 25°C. Separation of the mono from the diacylated product is performed by Soxhlet extraction of the base with carbon tetrachloride. The monoamide is obtained in *ca.* 60% yield and has a m.p. of 163–4°C; its specific rotation is $[\alpha]_{546} = -45.6$ (c. 1, MeOH). A second acylation, with 10-undecenoyl chloride (1.1 molar equivalents) in a dichloromethane/aqueous potassium carbonate (1.5 molar equivalents) two-phase system for 30 min at 25°C, gives a product which is recrystallized from ethanol. The yield is 80%, m.p. 214–6°C, $[\alpha]_{546} = +61.5$ (c. 0.5, CH_2Cl_2). This product (2 g) is then hydrosilylated with dimethylchlorosilane (5 ml in 20 ml of dichloromethane) and hexachloroplatinic acid (2 mg in 0.1 ml of 2 propanol) added as a catalyst; the reaction time is 16 hr at 25°C. The reaction may be followed by observation of the disappearance of the olefinic protons in the NMR. Reaction of this product, without purification, with silica (2.1 g of Nucleosil Si 100, 5μm) in pyridine (50 ml) for 48 hr at 25°C gives the chiral sorbent. This is finally subjected to end-capping by reaction with *N*-(trimethylsilyl)imidazole (2 ml) in dry toluene (50 ml) for 5 min at 90°C. The degree of substitution amounts to *ca.* 0.3 mmole/g (180 mg/g), based on elemental analyses. The material is slurry-packed into stainless-steel columns in chloroform/dioxan.

11.3.14 *In situ* end-capping of a Pirkle chiral sorbent: silica-bonded (S)-undecanyl *N*-(2-naphthyl)alaninate [41]

A column (4.6×250 mm) packed with (S)-11-(triethoxysilyl)-1-undecanyl-*N*-(2-naphthyl)alaninate-reacted silica (degree of substitution *ca.* 0.15 mmole/g) is attached to a high/-pressure pump and first flushed with 100 ml of dichloromethane. Then 50 ml of the same solvent containing 0.81 g (5 mmoles) of hexamethyldisilazane (or 1.09 g (10 mmoles) of trimethylchlorosilane and 1.01 g (10 mmoles) of triethylamine) is pumped through the column at a rate of *ca.* 0.5 ml/min at room temperature. The column is then washed again with 100 ml of dichloromethane followed by 100 ml of methanol.

11.3.15 Coating of TLC plates with a chiral selector for ligand exchange [42]

A glass plate with alkylsilica gel (RP 18 TLC) is dipped into a 0.25% copper(II) acetate solution (in methanol:water 1:9 v/v) and dried. The plate is then immersed into a 0.8% solution of the chiral selector in methanol for 1 min. The plate is then air-dried.

Although the preferred chiral selector has been (2S,4R,2'RS)-*N*-(2'-hydroxydodecyl)-4-hydroxyproline, which is relatively easy to prepare [42], any of the long-chain *N*-alkyl-L-4-hydroxyprolines used by Davankov *et al.* for reversed-phase columns [43] (see Section 7.3.1) are suitable.

Suitable eluents for amino acid resolutions are methanol:water mixtures (1:1 v/v) containing *ca.* 25–70% of acetonitrile [42].

BIBLIOGRAPHY

K. K. Unger, *Porous Silica*, Elsevier, Amsterdam, 1979.

P. Hodge and D. C. Sherrington (eds.), *Polymer-supported Reactions in Organic Synthesis*, Wiley, New York, 1980.

R. W. Souter, *Chromatographic Separations of Stereoisomers*, CRC Press, Boca Raton, 1985.

REFERENCES

[1] G. Hesse and R. Hagel, *Liebigs Ann. Chem.*, 1976, 996.
[2] G. Hesse and R. Hagel, *Chromatographia*, 1976, **9**, 62.
[3] A. Mannschreck, H. Koller and R. Wernicke, *Kontakte (Darmstadt)*, 1985, No. 1, 40.
[4] Y. Okamoto, M. Kawashima, K. Yamamoto and K. Hatada, *Chem. Lett.*, 1984, 739.
[5] K.-H. Rimböck, M. A. Cuyegkeng and A. Mannschreck, *Chromatographia*, 1986, **21**, 223.
[6] Y. Okamoto, M. Kawashima and K. Hatada, *J. Chromatog.*, 1986, **363**, 173.
[7] W. A. König, S. Lutz, P. Mischnick-Lübbecke, B. Brassat, E. von der Bey and G. Wenz, *Starch/Stärke*, 1988, **40**, 472.
[8] I. Ciucanu and F. Kerek, *Carbohydr. Res.*, 1984, **131**, 209.
[9] A. L. German and E. C. Horning, *J. Chromatog. Sci.*, 1973, **11**, 76.
[10] J. Bouche and M. Verzele, *J. Gas Chromatog.*, 1968, **6**, 501.
[11] G. Blaschke and A.-D. Schwanghart, *Chem. Ber.*, 1976, **109**, 1967.
[12] A.-D. Schwanghart, W. Backmann and G. Blaschke, *Chem. Ber.*, 1977, **110**, 778.
[13] N. A. Adrova and L. K. Prokhorova, *Vysokomol. Soedin.*, 1961, **3**, 1509 (*Chem. Abstr.*, 1962, **56**, 10384e).
[14] Y. Okamoto, K. Suzuki, K. Ohta, K. Hatada and H. Yuki, *J. Am. Chem. Soc.*, 1979, **101**, 4769.
[15] Y. Okamoto, K. Suzuki and H. Yuki, *J. Polym. Sci. Polym. Chem. Ed.*, 1980, **18**, 3043.
[16] H. Yuki, Y. Okamoto and I. Okamoto, *J. Am. Chem. Soc.*, 1980, **102**, 6356.
[17] F.-J. Ruffing, J. A. Lux, W. Roeder and G. Schomburg, *Chromatographia*, 1988, **26**, 19.
[18] F. Gasparrini, D. Misiti, C. Villani, F. La Torre and M. Sinibaldi, *J. Chromatog.*, 1988, **457**, 235.
[19] F. Galsbøl, P. Steenbøl and B. Søndergaard Sørensen, *Acta Chem. Scand.*, 1972, **26**, 3605.
[20] J. S. Fritz and J. N. King, *Anal. Chem.*, 1976, **48**, 570.
[21] M. Glad, S. Ohlson, L. Hansson, M.-O. Månsson and K. Mosbach, *J. Chromatog.*, 1980, **200**, 254.
[22] W. H. Pirkle, D. W. House and J. W. Finn, *J. Chromatog.*, 1980, **192**, 143.
[23] W. H. Pirkle and T. C. Pochapsky, *J. Org. Chem.*, 1986, **51**, 102.
[24] W. H. Pirkle and M. H. Huyn, *J. Org. Chem.*, 1984, **49**, 3043.
[25] N. Oi, M. Nagase and T. Doi, *J. Chromatog.*, 1983, **257**, 111.
[26] G. Blaschke, W. Bröker and W. Fraenkel, *Angew. Chem.*, 1986, **98**, 808.
[27] N. Oi and H. Kitahara, *J. Liq. Chromatog.*, 1986, **9**, 511.
[28] M. Ohwa, M. Akiyoshi and S. Mitamura, *J. Chromatog.*, 1990, **521**, 122.
[29] I. Marle, P. Erlandsson, L. Hansson, R. Isaksson, C. Pettersson and G. Pettersson, *J. Chromatog.*, in the press.
[30] J. N. Kinkel, *Thesis*, Univ. Mainz, 1983.
[31] T. Miwa, M. Ichikawa, M. Tsuno, T. Hattori, T. Miyakawa, M. Kayano and Y. Miyake, *Chem. Pharm. Bull.*, 1987, **35**, 682.
[32] T. Miwa, T. Miyakawa and Y. Miyake, *J. Chromatog.*, 1988, **457**, 227.
[33] A. Tambuté, A. Bégos, M. Lienne, P. Macaudière, M. Caude and R. Rosset, *New J. Chem.*, 1989, **13**, 625.
[34] L. Siret, A. Tambuté, M. Caude and R. Rosset, *J. Chromatog.*, 1991, **540**, 129.
[35] C. Rosini, C. Bertucci, D. Pini, P. Altemura and P. Salvadori, *Tetrahedron Lett.*, 1985, **26**, 3361.
[36] W.H. Pirkle and J. A. Burke, III, *Chirality*, 1989, **1**, 57.
[37] C. Rosini, C. Bertucci, D. Pini, P. Altemura and P. Salvadori, *Chromatographia*, 1987, **24**, 671.
[38] G. Uray and W. Lindner, *Chromatographia*, 1990, **30**, 323.
[39] E. J. Corey, R. Imwinkelried, S. Pikul and Y. B. Xiang, *J. Am. Chem. Soc.*, 1989, **111**, 5493.
[40] O. F. Williams and J. C. Bailar, *J. Am. Chem. Soc.*, 1959, **81**, 4464.
[41] W. H. Pirkle and R. S. Readnour, *Chromatographia*, 1991, **31**, 129.
[42] K. Günther, *J. Chromatog.*, 1988, **448**, 11.
[43] V. A. Davankov, A. S. Bochkov, A. A. Kurganov, P. Roumeliotis and K. K. Unger, *Chromatographia*, 1980, **13**, 677.

Appendix
Main manufacturers of material for chiral gas and liquid chromatography

Company	Material (type of stationary phase)
Alltech Associates Inc., 2051 Waukegan Road, Deerfield, Ill. 60015, USA	RSL-007 capillary GC columns (polysiloxane bonded chiral amide phase
Applied Science Laboratories, Inc., P.O. Box 440, State College, Pa. 16801, U.S.A.	Chirasil-Val capillary GC columns Resolvosil HPLC columns (bovine serum albumin)
Astec (Advanced Separation Technologies Inc.), 37 Leslie Court, P.O. Box 297, Whippany, N.J. 07981, U.S.A.	Cyclobond HPLC columns (β-cyclodextrin)
J. T. Baker Research Products, 222 Red School Lane, Phillipsburg, N.J. 08865, U.S.A.	Bakerbond HPLC columns (DNB–phenylglycine and –leucine)
ChromTech AB, Norsborg, Sweden	Chiral-AGP HPLC columns
Daicel Chemical Industries, 8-1, Kasumigaseki 3-chome, Chiyoda-ku, Tokyo 100, Japan	Chiralcel HPLC columns (OA: cellulose triacetate; OB; cellulose tribenzoate; OC: cellulose trisphenylcarbamate; OE: cellulose tribenzyl ether; OK: cellulose tricinnamate Chiralpak HPLC columns [OT(+): poly(triphenylmethyl methacrylate); OP(+): poly(2-pyridyldiphenylmethyl methacrylate); WH: proline–copper complex; WM: amino-acid copper complex] Lipodex capillary GC columns
Macherey-Nagel GmbH, P.O. Box 307, D-5160 Düren, FRG	Resolvosil HPLC columns Chiral-1 HPLC columns Chiralplate TLC plates [Cu(II)–proline complex on RP silica gel] CEL-AC-40 XF (microcrystalline triacetylcellulose)

Company	Material (type of stationary phase)
E. Merck, Frankfurter Strasse 250, D-6100 Darmstadt, FRG	Microcrystalline triacetylcellulose
Perstorp Biochem Ideon, S-22370 Lund, Sweden	Conbrio-Tac microcystalline triacetylcellulose LC columns (including semipreparative versions)
Pharmacia LKB Biotechnology, Box 308, S-1626 Bromma, Sweden	Enantiopac HPLC columns
Regis Chemical Co., 8260 Austin Avenue, Morton Grove, Ill. 60053, U.S.A.	DNB–phenylglycine and –leucine
S.E.D.E.R.E., Vitry-sur-Seine Cedex, France	ChyRoSine-A and AD LC columns
Serva Feinbiochemica, Postfach 105260, D-6900 Heidelberg 1, FRG	HPLC columns: Chiral BDex (β-cyclodextrin) Chiral DNBPG-C (DNB-phenylglycine); Chiral DNBLL-C, Chiral DNBDL-C [DNB-leucine (D- and L-)]; Chiral HyproCu (hydroxproline–copper); Chiral Procu (proline–copper); Chiral ValCu (valine–copper)
Shinwa Kako, Kyoto, Japan	Ultron ES-OVM LC columns
Societé Française Cheromato Colonne, Paris, France	Chiral Protein 2 (HSA-silica) LC columns)
Sumitomo Chemical Co., Sumitomo Bld., 5–15 Kitahama, Higashi-ku, Osaka 541, Japan	Sumipax-CC GC chiral capillary columns: OA-200 {N,N'-[2,4-(6-ethoxy-1,3,5-triazinediyl)]-bis-(L-valyl-L-valine isopropyl ester)}; OA-300A {N,N'-[2,4-(6-ethoxy-1,3,5-triazinediyl)]-bis-(L-valyl-L-valyl-L-valine isopropyl ester)}; OA-500 [N-lauroyl-L-proline (S)-1-(1-napththyl)ethylamide]; OA-510 [(R,R)-*trans*-chrysanthemoyl (S)-1-(1-naphthyl)ethylamide]; OA-520 [O-lauroyl-(S)-mandelic acid (S)-1-(1-naphthyl)ethylamide]
	Sumipax chiral HPLC columns: OA-1000, -1100 [N-acyl derviatives of (S)-(1-naphthyl)ethylamine]; OA-2000 [N-(3,5-dinitrobenzoyl)-D-phenylglycine]; OA-2100 [(S)-2(4-chlorophenyl)isovaleryl D-phenylglycine amide]; OA-2200 [(R,R)-*trans*-chrysanthemoyl D-phenylglycine amide]; OA-3000 (N-*tert*-butylaminocarbonyl-L-valine); OA-3100 (N-3,5-dinitrophenylaminocarbonyl-L-valine); OA-4000, -4100 [(S)- and (R)-1-(1-naphthyl)ethylaminocarbonyl L-valine]; OA-4200, -4300 [(S)- and (R)-1-(1-naphthyl)ethylaminocarbonyl D-phenylglycine]; OA-4400, -4500 [(S)- and (R)-1-(1-naphthyl)ethylaminocarbonyl L-proline]
Supelco Inc., Supelco Park, Bellefonte, Pa. 16823, U.S.A.	SP-300 (lauroyl-L-valine-*tert*-butylamide) chiral GC phase; also available on 100/120 mesh Supelcoport) Supelcosil LC-(R)-Urea (1-phenylethylurea)

Index